Lecture Notes in Computer Science 12524

More information about this subseries at http://www.springer.com/series/7408

Loek Cleophas · Mieke Massink (Eds.)

Software Engineering and Formal Methods

SEFM 2020 Collocated Workshops

ASYDE, CIFMA, and CoSim-CPS
Amsterdam, The Netherlands, September 14–15, 2020
Revised Selected Papers

Springer

Editors
Loek Cleophas 🆔
Eindhoven University of Technology
Eindhoven, The Netherlands

Mieke Massink 🆔
CNR-ISTI
Pisa, Italy

ISSN 0302-9743   ISSN 1611-3349   (electronic)
Lecture Notes in Computer Science
ISBN 978-3-030-67219-5   ISBN 978-3-030-67220-1   (eBook)
https://doi.org/10.1007/978-3-030-67220-1

LNCS Sublibrary: SL2 – Programming and Software Engineering

This Springer imprint is published by the registered company Springer Nature Switzerland AG
The registered company address is: Gewerbestrasse 11, 6330 Cham, Switzerland

Preface

This volume contains the selected and revised versions of papers that have been presented at three international workshops co-located with the 18th edition of the International Conference on Software Engineering and Formal Methods (SEFM 2020). Because of the COVID-19 pandemic, both the conference and the workshops were held virtually. They were hosted by the Centre for Mathematics and Informatics (CWI) in Amsterdam, The Netherlands, and took place on September 14–17, 2020.

The SEFM 2020 international conference offered a virtual interactive platform for leading researchers and practitioners from academia, industry, and government to advance the state of the art in formal methods, to facilitate their uptake in the software industry, and to encourage their integration within practical software engineering methods and tools.

The work presented at the three international workshops focused on emerging areas of software engineering, software technologies, model-driven engineering, and formal methods and on interdisciplinary topics on the border of software and cognition. In particular, the contributions that are collected in this volume have been selected from the presentations at the following workshops:

- **ASYDE 2020** – Second International Workshop on Automated and verifiable Software sYstem DEvelopment. Organised by: Marco Autili, University of L'Aquila, Italy; Federico Ciccozzi, Mälardalen University, Västerås, Sweden; Francesco Gallo, University of L'Aquila, Italy; Marjan Sirjani, Mälardalen University, Västerås, Sweden. The ASYDE workshop series is the result of a follow-up action, thanks to the work of the Steering Committee members, bringing together and consolidating the following previous events: OrChor'14, SCFI'15, SCART'15, VeryComp'16. The review procedure consisted of a single round of peer review, single blind, with 3 reviews per submission.
- **CIFMA 2020** – Second International Workshop on Cognition: Interdisciplinary Foundations, Models and Applications. Organised by: Pierluigi Graziani, University of Urbino, Italy; Pedro Quaresma, University of Coimbra, Portugal. The review procedure consisted of two rounds of peer review, single blind, with 4 reviews per submission.
- **CoSim-CPS 2020** – Fourth International Workshop on Formal Co-Simulation of Cyber-Physical Systems. Organised by: Cinzia Bernardeschi, University of Pisa, Italy; Cláudio Gomes, Aarhus University, Denmark; Paolo Masci, National Institute of Aerospace (NIA), USA; and Peter Gorm Larsen, Aarhus University, Denmark. The review procedure consisted of a single round of peer review, single blind, with 3 reviews per submission.

We are grateful to all organisers of the workshops at SEFM 2020 for their selection of interesting topics and presentations, despite the difficult circumstances due to the pandemic. We would also like to thank the members of the respective Programme

Committees and reviewers for their thorough and careful reviews, for organising the programme for each workshop, and for making the compilation of this volume possible.

We thank all authors of contributions and all attendees of the workshops, as well as the Keynote Speakers for adapting their excellent presentations and discussions to the new virtual setting. Special thanks go the hosting institution CWI in Amsterdam, its organising team, and, in particular, the general chairs of the SEFM 2020 conference, Frank de Boer and Antonio Cerone, for their incredible work to make this event possible under the continuously changing circumstances and uncertainties created by the pandemic.

November 2020   Loek Cleophas
                                                                 Mieke Massink

Contents

CoSIM-CPS 2020

ASYDE 2020

Organization

ASYDE 2020 – Workshop Chairs

Marco Autili	University of L'Aquila, Italy
Federico Ciccozzi	Mälardalen University, Sweden
Francesco Gallo	University of L'Aquila, Italy
Marjan Sirjani	Mälardalen University, Sweden

ASYDE 2020 – Programme Committee

Luciano Baresi	Politecnico di Milano, Italy
Steffen Becker	University of Stuttgart, Germany
Domenico Bianculli	University of Luxembourg, Luxembourg
Antonio Brogi	University of Pisa, Italy
Giovanni Denaro	University of Milano-Bicocca, Italy
Antinisca Di Marco	University of L'Aquila, Italy
Amleto Di Salle	University of L'Aquila, Italy
Ehsan Khamespanah	University of Tehran, Iran
Marina Mongiello	Polytechnic University of Bari, Italy
Cristina Seceleanu	Mälardalen University, Sweden
Meng Sun	Peking University, China
Apostolos Zarras	University of Ioannina, Greece

ASYDE 2020 – Steering Committee

Farhad Arbab	CWI, The Netherlands
Marco Autili	University of L'Aquila, Italy
Federico Ciccozzi	Mälardalen University, Sweden
Dimitra Giannakopoulou	NASA, USA
Pascal Poizat	Sorbonne University, France
Massimo Tivoli	University of L'Aquila, Italy

Model Translation from Papyrus-RT into the NUXMV Model Checker

Sneha Sahu[1]([✉]) [iD], Ruth Schorr[1], Inmaculada Medina-Bulo[2] [iD], and Matthias Wagner[1] [iD]

[1] Frankfurt University of Applied Sciences, 60318 Frankfurt am Main, Germany
{sneha.sahu,rschorr,mfwagner}@fb2.fra-uas.de
[2] Departamento de Ingeniería Informática, Universidad de Cádiz, Puerto Real, Spain
inmaculada.medina@uca.es

Abstract. Papyrus-RT is an eclipse based modelling tool for embedded systems that makes use of the Model-Driven Engineering approach to generate executable C++ code from UML-RT models. The UML-RT state diagrams are very similar to Finite State Machines used in the NUXMV model checker (an extension of the NUSMV symbolic model checker). In this paper we present an approach for automated verification of the UML-RT models by exporting them mechanically into equivalent NUXMV models with positive results.

Keywords: Model translation · UML-RT · State transition diagrams · Model-checking · Finite State Machines (FSM) · Symbolic Model Verification (SMV) · NUXMV

1 Introduction

Papyrus-RT [2] is an industrial-grade modelling tool based on the concept of Model Driven Engineering (MDE) [11]. It makes use of the UML-RT [23] graphical representations for describing model behaviour. Interaction between components, referred to as *capsules* in Papyrus-RT, takes place through connected ports and trigger controls, the semantics of which are described with the help of state-transition diagrams. The designed model serves as the key artifact for generating an executable C/C++ project [22,24]. Each system model consists of one 'Top Capsule', which is used to define the overall system connections. The sub-components further have their own hierarchical system of elements and corresponding state-diagram to describe its semantics. Papyrus-RT offers inbuilt protocols for logging and timer functionalities and also allows users to define custom protocols for their individual system design. These protocols are basically a set of rules in terms of incoming and outgoing messages to be followed by ports for communication.

The correctness of the semantics of the executable code, generated from the Papyrus-RT model, relies completely upon the designer, who can choose to go through the simulation logs. However, the possibility to introduce automated

© Springer Nature Switzerland AG 2021
L. Cleophas and M. Massink (Eds.): SEFM 2020 Workshops, LNCS 12524, pp. 3–20, 2021.
https://doi.org/10.1007/978-3-030-67220-1_1

verification techniques to a model-driven development tool would be an achievement in the domain of model based development, as it can ensure the correct system behaviour against a set of formal properties defined out of the system requirements. With this as the motivation, this paper discusses an approach for possible automation.

Papyrus-RT state-diagrams resemble very closely to finite-state machines (FSMs), thereby making the NUXMV Model Checker [3,6] a perfect candidate for the desired translation. NUXMV is an extension of the NuSMV Symbolic Model Checker [7,8] which allows verification of infinite-state systems based on algorithms from Satisfiability Modulo Theory (SMT) [1].

In the following sections, we start with briefly introducing the selected case studies, followed by a detailed explanation of our approach and its outcome. Then some of the existing works in this direction are discussed, followed by conclusions and future work.

2   Case Studies

For the study of this research work, we selected three sample models, covering multi-threading, multiplicity, hierarchical state-diagrams and asynchronous communications with or without payload.

The first case study is the classical 'ping-pong' model consisting of two units communicating over one connection. The details of the model design can be studied from the Papyrus-RT tutorial [25].

The second case study, is the classical 'Rock-Paper-Scissor' game, played between two players and judged by one 'Referee'. The players are created as two instances of the same capsule definition in the UML-RT model and there is no direct communication between them. The referee controls all communications and also decides about the winner based on certain rules provided as C++ code snippets in the UML-RT state diagram. The model for this case study was taken from the resources for Papyrus-RT provided in the MODLES'17 conference [12]. The system translation becomes moderately complex in this case because of the introduction of capsule and port multiplicity, payload along connections, hierarchical state diagrams and the use of logical C++ codes for computations in addition to the normal send and logging code blocks. Also, the asynchronous mode of communication is clearly visible with this example.

The third case study, the Insulin Pump System, is a real world example and also a safety critical medical device. The model was designed referring to Ian Sommervielle's overview and requirement specifications [26]. Here, we also introduced Papyrus-RT's multi-threading functionality along with all the other previously mentioned features.

Figure 1 shows an overview of the modified version of the Insulin Pump, consisting of three main units − a power-supply unit referred to as the *battery*; a *reservoir* for storing insulin; and a *bgController* for dosage computation. The tasks of the sensor (to read current blood glucose (BG) level), needle assembly (to take blood sample as well as inject insulin) and the pump (to pass insulin

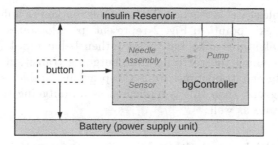

Fig. 1. Simplified insulin pump system

from reservoir to needle) have been integrated within the 'bgController'. A *button* in the main system maintains communication between all components and helps decide about the overall system status – 'Running' or 'Stopped'. There is also a communication channel between the 'bgController' and the 'reservoir' in order to exchange information about insulin injection.

3   Translation Approach

The Papyrus-RT system model is basically a set of capsules, communicating in real-time. Each capsule comprises of a hierarchy of sub-components, namely - attributes, capsules, ports, connections, and state-diagrams.

Various capsules within a system communicate based on message triggers on the incoming ports. By default every capsule is assigned to one default controller, which maintains a message queue of all outgoing messages from its capsules. These messages get further directed to the target capsule in FIFO order. Messages are handled sequentially within each controller and for parallel processing multi-threading is possible using separate controllers [22, 24].

The translation is achieved in two stages—in first stage an intermediate FSM [18] is created corresponding to each capsule's state diagram and in the next stage an SMV model is designed out of the FSMs and UML-RT model elements. Any non-trivial C++ action code in the state machine is handled as an abstraction of what is being done.

3.1   UML-RT State-Diagrams to Finite-State-Machines

Each state machine in the UML-RT model is converted into a finite-state machine $M = (Q, Q_0, \Sigma, \delta, F)$, where Q is the set of all states, Q_0 is the set of initial states, Σ is the set of state variables or attributes, δ represents the transition function and F is the set of final states.

Dealing with 'pseudo States'. UML-RT models can have two kinds of states – 'stable' states and 'pseudo' states. A transition in UML-RT state diagrams stops only at a 'stable state', while in case of a 'pseudo state' it continues to run until the next 'stable state' is reached. The symbols marked as 'initial', 'decision', 'entry point' and 'exit point' in Fig. 2 represent 'pseudo states'. For the FSM creation, we treat all states alike, but preserve their behavioural differences with the help of guards along the corresponding outgoing transitions. Also, in case there are any outgoing transitions from the previous 'stable' state, having a different set of triggers, then this is replicated as an outgoing transition of the current 'pseudo' state as well.

Handling of Multi-layered State Diagram. Papyrus-RT allows multiple layers of state-diagrams, thereby creating a hierarchical state-diagram, which is removed by merging states into the parent state-machine.

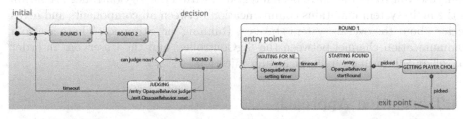

(a) State Diagram for a capsule   (b) State Diagram for sub-state *Round1*

Fig. 2. Sample state diagram from Papyrus-RT

Transitions. Every transition in Papyrus-RT is associated with a set of 'Trigger', 'Effect' and 'Guard' as can be seen in Fig. 3. Triggers act like checks that initiate any transition and hence are added as guard to the transition in the FSM. An 'Effect' is a set of instructions to be performed during the transition and is therefore noted as actions for the transition. Any non-port related code-snippets is noted as abstraction in the FSM.

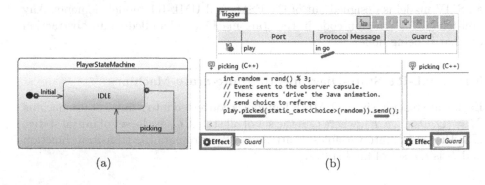

Fig. 3. State Diagram along with corresponding Transition Trigger and Effect.

Message Synchronization. In Papyrus-RT state-diagram, it is possible not to have a visible time gap between received and sent messages, even though in reality there may be some delay. This becomes significant in case the following transition actions are not just internal, but also communicate with other capsules of the system. Hence, in the intermediate FSM model it must be ensured that a receive and a send operation is never performed around the same state—

- a new waiting state is created to split the sync messages into two transitions 'ta' and 'tb'.
- guards of the original transitions stay with 'ta' while a new guard to check for the turn of current capsule is added to 'tb'.
- the set of actions of the actual transition is moved to 'tb'.

For example, in Fig. 3, the transition 'picking' is triggered based on incoming message 'go' and at the same time also sends out message 'picked' with parameter. The two actions here are split by introducing a new state 'wait_1' as shown in figure Fig. 4 and the corresponding outgoing transitions would be triggered when controller is either free or gives control to this capsule.

Figure 4 shows how an intermediate FSM would appear for the 'Player-StateMachine' in Fig. 3.

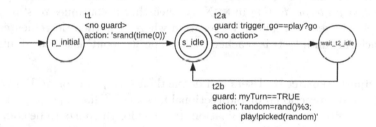

Fig. 4. FSM corresponding to the State Diagram from Fig. 3.

RTState *entry/exit* Behaviours. Papyrus-RT also allows to define certain behavioural logic within each 'RT state' in the form of state-entry and state-exit behaviours. After introduction of the 'wait_*' states, the actions from the *entry behaviour* are added to the incoming state transition, while those of the *exit behaviour* are included as part of the outgoing transition. So, if there were any *exit behaviour* for the 'Idle' 'RTstate' from Fig. 3a, then those actions would be a part of the transition 't2a' in Fig. 4 and not 't2b'.

Thus, each capsule in the model is mathematically expressed as a finite-state machine $M = (Q, Q_0, \Sigma, \delta, F)$, where—

- $Q = Q_s \cup Q_p \cup Q_w$ is a finite set of states, where Q_s is a set of stable states, Q_p is a set of pseudo states and Q_w is the set of new wait states introduced for separating incoming and outgoing messages.
- $Q_0 \subseteq Q$ is a set of non-empty initial states (the pseudo state called 'initial').

– Σ is the set of attributes including ports and timer.
– $\delta : Q_i \times \Sigma \to Q_j$ is a deterministic transition function. Each transition is associated with a set of guards and actions depending on their UML-RT counterparts.
– $F \subseteq Q$ is the set of final or accepted states.

3.2 Translation into SMV Model

SMV (Symbolic Model Verifier) system was the first model-checking tool, developed by CMU (Carnegie Mellon University, Pittsburgh). It was used for verification of finite state systems against specifications in temporal logic based on Binary Decision Diagrams (BDDs) [10,21].

The set of FSMs together with the UML-RT model, is then mapped into a NUXMV model as follows—

Timing Protocol. Papyrus-RT supports an inbuilt timer in the form of a 'Timing Protocol'. The only practically usable information we could obtain from the eclipse forum and published articles on Papyrus-RT was its usage around setting a timeout value and using that as a trigger for transitions [13,22]. The logic for the 'Timing' protocol was implemented by introducing a corresponding Module called *protocol_ timing* in NUXMV, such that it continues to stay at '−1' unless explicitly set to a higher value, from where it is then decremented by 1 unit at each run until '−1' is reached again. The 'timeout' is set at '0' unit.

User Defined Protocols. Users can define their own protocols in Papyrus-RT with multiple unidirectional or bidirectional messages. The Capsule ports make use of these protocols for communication. In order for the ports to be connected, it is mandatory that they follow the same protocol and one of them is conjugated.

For each user defined protocol in Papyrus-RT, a new MODULE 'protocol_ *' is introduced in NUXMV following these guidelines—

– Module must have three associated parameters, one each for *conjugation property*, *message name* and *message payload*. The payload is restricted to at most one parameter per message and only integers. It could also be possible to use enumerations or booleans, but this has not yet been explored.
– The *message name*, prefixed with its direction, is translated into a DEFINE expression that uses the *conjugation property* and the *message name* as shown in Listing 1.1. When conjugation is true, an incoming message becomes outgoing (line #10) and vice-versa (lines #3 and #8). There should be no behavioural change for a bidirectional message with change in conjugation and is therefore always TRUE when the relevant message is passed (line #4).
– The *message payload*, prefixed with the carrying *message name*, is also translated into a DEFINE expression that holds the parameter value. This should be '0' when the corresponding message is FALSE (line #9 in Listing 1.1).

Figure 5 shows a couple of user defined protocols in Papyrus-RT and Listing 1.1 is their corresponding translations following the above rules.

Listing 1.1. Modules in NUXMV for user defined protocols.

```
1   MODULE protocol_commands(conjugated, msg, param)
2     DEFINE
3       out_start := (conjugated = FALSE & msg = start);
4       inout_stop := (msg = stop);
5
6   MODULE protocol_insulin(conjugated, msg, param)
7     DEFINE
8       out_ask := (conjugated = FALSE & msg = ask);
9       out_ask_dose := (out_ask ? param : 0);
10      in_inject := (conjugated = TRUE & msg = inject);
```

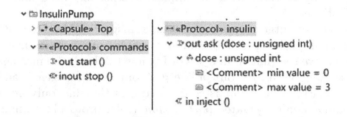

Fig. 5. Sample of user defined protocols in Papyrus-RT.

Capsules. Each Module in the system represents a FSM, having a set of initial states, next states and transitions. The 'Top Capsule' is translated into the 'Module Main', while every other capsule is translated into an equivalent Module of its own in NUXMV.

For each 'CapsulePart', which is an instance of another Capsule in the system, a VARiable of the corresponding Module type is created in the parent Module. The inputs to these VARiable instances is based on the 'Connectors' as in the layout of the parent Capsule.

- An input parameter 'myTurn' is introduced for every Capsule except the 'Top Capsule'. See lines #6 and #11 of Listing 1.2.
- For every incoming connection on an external port, one input parameter 'in_*' is added. This means that for the 'power' capsule in Fig. 6, the corresponding module declaration should have 3 input parameters, one for each of the three incoming connections as can be seen in line #6 of Listing 1.2.
- For each internal or external port, except 'log', a new 'port_*' VARiable of its corresponding protocol type (Module) is added.
  Refer lines #8, #13 and #15 in Listing 1.2.

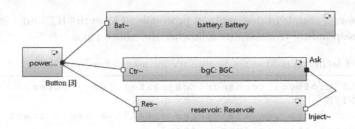

Fig. 6. Layout of a sample 'Top Capsule' showing capsulePart.

- In case a port uses the 'timing protocol', then this is supported in NUXMV with two additional VARiables – *timerResetFlag* of boolean type INITialized as 'FALSE', and *timerResetValue* of bounded integer type INITialized to '−1', as shown in lines #16 and #17 of Listing 1.2. This approach for timer handling is only relevant with the implementation discussed under "Timing Protocol" above.
- For every type of outgoing port messages, a 'msg_*' VARiable of enum type is declared, composed of a 'null' and all possible outgoing message options. For example, we see that in line #9 of Listing 1.2, both the message options are include along with *null*, but in line #14, only *stop* message is added. This is because in Fig. 6 port *Bat* is conjugated and thus has only one outgoing message unlike non-conjugated port *Button*. (refer protocol 'commands' from Fig. 5.) The message VARiables are INITialized with 'null'.
- For every parameter associated with any outgoing port message, a new 'param_*' VARiable of integer type The parameter range could be provided in the Papyrus-RT model as comments as shown in Fig. 5. The parameter VARiables are INITialized with 0.
- A *state* VARiable of enum type is introduced and consists of a set of all 'RTstates' along with the other states as explained earlier in Sect. 3.1 – stable states, pseudo states, newly created wait states and the merged-states to remove hierarchical state structure.

Listing 1.2. Modules in NUXMV corresponding to Capsules from Papyrus-RT Model.

```
1  MODULE main
2    VAR
3      power : capsule_Power(TURN_P, battery.port_Bat,
           reservoir.port_Res, bgC.port_Ctr);
4      battery : capsule_Battery(TURN_B, power.port_Button);
5    ...
6  MODULE capsule_Power(myTurn, in_Bat, in_Res, in_Ctr)
7    VAR
8      port_Button : protocol_commands(FALSE, msg_Button, 0)
           ;
9      msg_Button : {null, start, stop};
10   ...
```

```
11  MODULE capsule_Battery(myTurn, in_Button)
12    VAR
13      port_Bat : protocol_commands(TRUE, msg_Bat, 0);
14      msg_Bat : {null, stop};
15      port_Timer: protocol_timing(timerResetValue,
              timerResetFlag);
16      timerResetValue: -1..100;
17      timerResetFlag : boolean;
18      bLvl : 0..100;
19      state : {sInit, sMid_1, sFinal};
20    ...
```

Capsule Attributes. Every attribute in the Papyrus-RT model the domain must be clearly defined. Our research is limited to the use of only integers, booleans and enum—

- For boolean, the domain is always a set of 'null', 'true' and 'false'.
- For integers, the domain-range can be defined by a pair of read-only attributes marking the max and min values.
- For enumerations, the set of elements as defined.

In the NUXMV translation, a VARiable is declared for each writable attribute and a DEFINE expression 'const_*' is added for each of the accompanying read only attributes, set to their default values. The VARiable itself is added to INITialize section with its default value.

Controller Message Queue. For each controller 'x' used in the Papyrus-RT system, a VARiable 'msgPtr_x' of integer type is introduced in the 'Main Module' which acts as a pointer to the message Queue Head. DEFINE variables 'TURN_C*', one for every CapsulePart, is introduced and used as input parameters to the capsules. This serves the role of 'Connectors'. Another DEFINE variable 'FREE_x' is introduced for each controller. These definitions together with corresponding NEXT assignments ensure that only one 'capsulePart' per controller as a valid turn at a time, thus maintaining a sequential processing of the sync messages within each controller.

Line #5 onwards in Listing 1.3 shows the NUXMV code related to controller for a system consisting of two capsules and only the default controller. The count of capsules is set as the upper bound for the controller's message pointer.

Listing 1.3. Controller logic in NUXMV for message queue handling.

```
1    MODULE main
2      VAR
3        c1 : capsule_1(TURN_C1, c1.outPort);
4        c2 : capsule_2(TURN_C2, c2.outPort);
5        msgPtr_x: 0..2;
6      INIT msgPtr_x = 0
```

```
 7        ASSIGN
 8          next(msgPtr_x) := case
 9              c1.WAITING & FREE_x : 1;
10              c2.WAITING & FREE_x : 2;
11              TRUE : 0;
12          esac;
13        DEFINE
14          FREE_x := (msgPtr_x = 0);
15          TURN_C1 := (msgPtr_x = 1);
16          TURN_C2 := (msgPtr_x = 2);
```

The DEFINE variables are fed as input to the corresponding capsule module and helps to determine if it is their turn for the next transition. On the individual capsule side, i.e. in the corresponding Modules, DEFINE expression for 'WAITING' (relevant for lines #9 and #10 from Listing 1.3) is set to become TRUE whenever the capsules land in one of the 'wait_*' states. In case there are no such states then it is set to FALSE.

RTStateMachine Transitions. For every transition in the intermediate FSM, a DEFINE expression is declared as conjunction of the starting state and the corresponding guards. Additional DEFINE expressions are also included for guards in the form of triggers.

An additional transition 't_none' is also defined, which becomes true when none of the actual FSM transitions are valid. This is required in NUXMV to prevent the 'Empty FSM' warning in absence of valid next transitions in the state machine. As an effect of this transition, all VARiables except those being used as input to other modules, preserve their present values. This implies that under the TRANSition section for 't_none' the NEXT of outgoing messages and parameters as well as variables associated with the 'timer' and any other such fields are assigned back to their default INITial values. The same rules are also followed when dealing with transitions that are silent about the change of values for such VARiables, i.e. reset to initial default values.

Actions of a Transition. At this point all functional code blocks of Papyrus-RT – Transition Effect, RTState entry and RTState exit, have been moved into the action of some transition in the intermediate FSM. These are translated into NEXT assignments for each local variable either in the TRANSition Section or ASSINGment section. TRANSition section was preferred for the local 'state' variable as well as any integer type variable undergoing mathematical computations. In case, any new variable is introduced as a part of any transition action, then it is expressed in the DEFINE section as 'newVar_t*_*', to be modified accordingly when the corresponding start state is valid. However, it is better to avoid such introductions.

In the translated model, it is important that there exists exactly one NEXT definition for every local variable and any avoid ambiguity. For example, the actions for the transitions in the FSM from Fig. 4 could be translated as shown in Listing 1.4.

Listing 1.4. Translation of transition actions in NUXMV.

```
1     TRANS t_none -> next(state) = state;
2     TRANS t1 -> next(state) = s_idle;
3     TRANS t2 -> next(state) = wait_1;
4     TRANS t3 -> next(state) = s_idle;
5     ASSIGN
6     next(msgPlay):= case
7       t_none: null;
8       t1 | t2: msgPlay;
9       t3: picked;    -- send picked(random)
10     esac;
11     next(paramPlay):= case
12       t_none: 0;
13       t1 | t2: paramPlay;
14       t3: {1,2,3};   -- rand()%3
15     esac;
```

While modelling the NEXT assignments for code snippets from Papyrus-RT – *Effects, Entry* and *Exit*, it is important be careful about when to use the present value of a variable and when to use its next value. If the value of any variable is changed within a transition's action and reused at another point post the changes in the same set of actions, then the new value must be referred in the NUXMV model. For example, in Fig. 7, the highlighted value are updated and reused within the same transition action. In the corresponding NUXMV translation—

- 'this→firstPlayerChoice' is an attribute to the capsule and hence already a VARiable exists. So, its NEXT assigment can be simply updated with 'choice' which is received with an incoming message.
- 'result' is a new variable, and hence a new DEFINE expression is created for it. But since it is using the recently updated value of 'this → firstPlayerChoice', the expression in NUXMV would have to use 'next(this → firstPlayerChoice)' and not the direct value.
- Again it can be seen that the value for already existing attribute 'this → firstPlayerScore' is changed based on the updated value of 'result'. However, since 'result' was a new addition to the transition and a DEFINE expression was created, its value will be computed in run-time with every reference. So here, although it is modified recently, using its direct value will still give the desired outcome.

```
picked (C++)
    int player = msg->sapIndex();
    log.show("Player %d played: ", player);
    static const char* choices[] = {"Rock", "Paper", "Scissors"};
    log.show("%s", choices[choice]);
    if (player == 1) {
        this->firstPlayerChoice = choice;
    }
    else if (player == 2){
        this->secondPlayerChoice = choice;
    }
    log.show("\n");
    int result = (3 + this->firstPlayerChoice - this->secondPlayerChoice) % 3;
    if(result == 1) {        this->firstPlayerScore++;    }
    else if(result == 2) {        this->secondPlayerScore++;    }
    this->round++;
```

Fig. 7. Code Snippet from Papyrus-RT showing multiple computations.

Exactly One Active TRANSition. During the translation of the Papyrus-RT model into NUXMV, we have included an additional 't_none' transition into the model to ensure that the system does not run into an 'Empty FSM' state, when none of the actual FSM transitions are valid (the less than one case).

Similarly, with the usage of the TRANS sections, it is required that not more than one TRANSition becomes valid at any point. Otherwise the NUXMV compiler cannot handle the ambiguity. Hence, an INVARiant is defined to check the property that of all the TRANSitions in the Module, exactly one is valid at any instance.

3.3 Model Restrictions

During the formulation of the above approach, certain challenges were identified with modelling methods in Papyrus-RT that could hinder with a correct translation into NUXMV. In order to avoid any unnecessary complications a set of restrictions have been adopted for Papyrus-RT model design—

1. the behavioural logic, in the form of C++ code snippets along Transition *Effects* and State *Entry/Exit*, should be kept short and simple. Better if possible with only the port related instructions and direct attribute assignments. Basic mathematical computations in if-else format is also a workable option.
2. preference to using UML-RT design elements over code-snippets.
3. messages within user-defined Protocols to be limited to carry at max one parameter of integer type only with clear information on the possible value range. The value '0' usually should imply an empty payload.
4. any incoming payload should not be directly used, but instead read into a designated local attribute, which in turn can be used for computations.
5. a single attribute should be updated only ones along the complete transition set between two stable states.
6. every integer type attribute to be accompanied with a pair of read-only integer attributes to indicate the domain maximum and minimum values.

4 Evaluation

The system behaviour for the models in Papyrus-RT was checked through simulation of specific test scenarios. The corresponding trace from translations in NUXMV were then compared to ensure behavioural similarities with their Papyrus-RT versions and then later the NUXMV versions were verified using a set of formalised temporal properties.

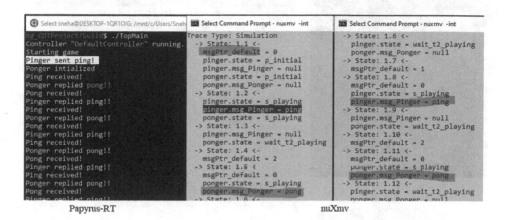

Fig. 8. Trace comparison - Ping Pong model

The Ping Pong Example was a very straight forward model without any complication and with only one executable message in the controller queue at any given instance. The behavioral order in the simulation of the translated NUXMV version matched with that of the Papyrus-RT model, as can be seen from Fig. 8. In both cases system initiates with a 'ping' message, followed by a 'pong' and then alternating one after the other. This behaviour was then also verified using temporal specifications satisfying the properties – "never directly should a 'ping' be followed by another 'ping' and a 'pong' by another 'pong' message in the overall system" and "a 'ping' should always eventually be responded by a 'pong' and vice-versa".

For the Rock Paper Scissors Model as well, we were able to get a NUXMV translation with similar structure and simulation. Here, we did also make use of behaviour abstraction on some parts of the code snippet. For instance, the combined use of '$srand(time(0))$' and '$intx = rand()\%3$' in C++, basically assigns a randomly chosen value between 1–3 to the variable 'x'. This had to be replaced with a 'set' assignment in NUXMV.

Figure 9 shows a simulation from Papyrus-RT as well as the relevant sections from an equivalent NUXMV translation. In the particular execution, the two rounds from Papyrus-RT simulation were computed in 24 NUXMV states or runs. States 3.10–3.12 correspond to picking of choices by the players in round 1, while states 3.20–3.22 correspond to that of round 2, with the final judgement after round 2 in State 3.24.

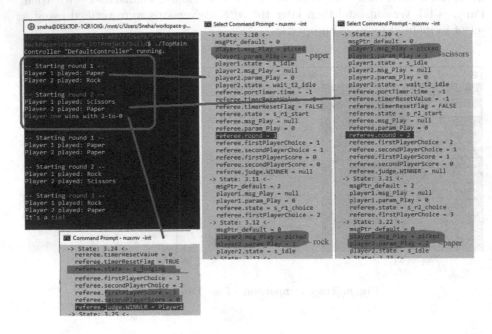

Fig. 9. Simulations for Rock Paper Scissors model

The behavioural similarities of the simulation traces were compared for 3 main test scenarios,—Test 1, a 'tie' case; Test 2, winner is found in two out of three rounds, as shown in Fig. 9; Test 3, all three rounds are played.

As for the behavioural verification of the model, a set of temporal specifications corresponding to the below properties were used

- Each 'player' always has to wait for a signal from the 'referee' to be able to play its round.
- A 'player's' choice can never be outside of the given domain of choices.
- Whenever the score of a 'player' is incremented by the 'referee', it must always comply with the game rules, i.e. 'Rock wins over Scissor', 'Scissor wins over Paper' and 'Paper wins over Rock'.

In the Insulin Pump Case also a similar approach for trace comparison and property verification was adopted. Most of the system properties verified were related to correct dose computation, dose injection and maintenance of a synchronised communication between the sub-parts. In addition, it was also verified that "a running system is always able to maintain a normal blood glucose level eventually". Tests for trace comparison included the scenarios – Test 1, system stops due to battery failure; Test 2, insufficient level of insulin in reservoir causes system failure; Test 3, simulation is able to bring down high blood glucose within the normal range.

Fig. 10. Simulation states for Insulin Pump model

Although it was possible to achieve equivalence with respect to the model structure and semantics of the Papyrus-RT version, slight behavioural variations between the non-communicating units 'battery' and 'bgc' was observed in case of NUXMV. For instance, let us consider the section of the simulation traces in Fig. 10, which shows the start and stop situation from the two versions. In both simulations the system stops functioning because of limited insulin level in reservoir, i.e. at 19 units and manages to bring down blood glucose to 125 units. However, the 'battery' level at this point drops to 16 in the Papyrus-RT version, while in NUXMV, the drop is further down to 11. In other words, the 'bgC' unit in Papyrus-RT version seems to function at a faster pace compared to its counterpart in NUXMV. On a deeper analysis of the simulation trace, it was found that this mismatch in behaviour was due to the variation in implementation of the timing protocol. The one inbuilt in Papyrus-RT queues up the timer reset requests and hence is designed to process each one of them, whereas the one in our approach works by overwriting. This implies that the translation model will behave differently whenever multiple timer reset actions lead into the same state of the FSM. This could be handled by either implementing a different timer logic, or by restricting the model design to not use another timer reset unless the first one is exited.

5   Related Work

In [19] and [20], the researchers present a translation for a subset of the UML state diagrams, while [5] and [27] focus on those of UML-RT. Except [20], where the researchers have developed a self-contained model checking tool for their approach, the other three are targeted towards using PROMELA (SPIN) [14,15] as the language for model checking with implementations based on IBM's RSARTE tool [16]. In our approach, we focus on creating a SMV model that can be used with the NUXMV model checker. As far as we could make out from [5], UML-RT constructs such as hierarchical states, signal payloads, multiplicity of model elements, pseudostates and guard conditions are not considered in their translation, unlike ours. In case of Zrowska and Dingel's work [27], they also use an intermediate Functional FSM somewhat similar to that in our approach, which they say separates state machine behaviour from its actions and therefore can support different action languages. However, when describing the implementation only a subset of C++ comprising of basic variable operations, if statements and while loops is included.

A work that supports formal verification of its models using the NUXMV model checker is found in the model-based development tool called AutoFocus3 (AF3) [17]. The tool uses NUXMV together with another command-line verification tool called OCRA [9]. Modelling in the tool is based on the FOCUS model of computation [4] with a global clock and a synchronous execution of all components. This is different from the UML-RT models of Papyrus-RT, where each capsule has its own local timer and inter-capsule communications are asynchronous, controlled by global message queue systems.

6   Conclusion and Future Work

In this paper, we have talked about a detailed approach on translation of UML-RT models of Papyrus-RT along with its implications on three example models, ranging from very simple to moderate complexities. After multiple attempts with Papyrus-RT modelling and the corresponding translations into NUXMV, we were able to come up with a refined translation methodology that is applicable in coordination with the modelling requisites listed in Sect. 3.3. Based on our experience with the above discussed case studies and a few others like the Conveyer Belt System and the Traffic Light system for a 4-way crossing, we can say that UML-RT translation into equivalent NUXMV models for well designed using our approach. As for multi-threaded systems, the present approach for timer handling could be an issue and needs enhancement. Another aspects that can be looked up is the handling of enumerations as payload for communication.

Backward translation of the NUXMV counter examples to be usable for rectification of design flaws and automation of the translation approach are two of the long term future aspects of this research work.

References

1. Barrett, C., Tinelli, C.: Satisfiability modulo theories. In: Handbook of Model Checking, pp. 305–343. Springer, Cham (2018). https://doi.org/10.1007/978-3-319-10575-8_11
2. Beaton, W.: Eclipse Papyrus for Real Time (Papyrus-RT). projects.eclipse.org, July 2017. https://www.eclipse.org/papyrus-rt/. Accessed 31 Aug 2020
3. Bozzano, M., et al.: nuXmv 1.1. 1 User Manual. FBK-Via Sommarive 18, 38055 (2016).   https://es-static.fbk.eu/tools/nuxmv/index.php?n=Documentation. Home. Accessed 31 Aug 2020
4. Broy, M., Stølen, K.: Specification and Development of Interactive Systems: Focus on Streams, Interfaces, and Refinement. Springer, New York (2012). https://doi.org/10.1007/978-1-4613-0091-5
5. Carlsson, M.G., Johansson, L.G.: Formal verification of UML-RT capsules using model checking. Master's thesis, Chalmers University of Technology (2009)
6. Cavada, R., et al.: The NUXMV symbolic model checker. In: Biere, A., Bloem, R. (eds.) CAV 2014. LNCS, vol. 8559, pp. 334–342. Springer, Cham (2014). https://doi.org/10.1007/978-3-319-08867-9_22
7. Cavada, R., Cimatti, A., Keighren, G., Olivetti, E., Pistore, M., Roveri, M.: NuSMV 2.6 Tutorial. FBK-irst-Via Sommarive 18, 38055 (2010)
8. Cimatti, A., et al.: NuSMV 2: an OpenSource tool for symbolic model checking. In: Brinksma, E., Larsen, K.G. (eds.) CAV 2002. LNCS, vol. 2404, pp. 359–364. Springer, Heidelberg (2002). https://doi.org/10.1007/3-540-45657-0_29
9. Cimatti, A., Dorigatti, M., Tonetta, S.: OCRA: Othello Contracts Refinement Analysis Versions 1, 3 (2015)
10. Clarke, E., McMillan, K., Campos, S., Hartonas-Garmhausen, V.: Symbolic model checking. In: Alur, R., Henzinger, T.A. (eds.) CAV 1996. LNCS, vol. 1102, pp. 419–422. Springer, Heidelberg (1996). https://doi.org/10.1007/3-540-61474-5_93
11. Favre, J.M.: Towards a basic theory to model model driven engineering. In: 3rd Workshop in Software Model Engineering, WiSME, pp. 262–271. Citeseer (2004)
12. Hili, N., Posse, E., Dingel, U., Beaulieu, A.: Supporting Material For EclipseCon'17 Unconference - Modeling & Analysis In Software Engineering. School of Computing, Queen's University (2017). https://flux.cs.queensu.ca/mase/papyrus-rt-resources/supporting-material-for-eclipsecon17-unconference/. Accessed May 2020
13. Hili, N., Posse, E., Dingel, J.: Calur: an action language for UML-RT. In: 9th European Congress on Embedded Real Time Software and Systems (ERTS 2018), Toulouse, France, January 2018. https://hal.archives-ouvertes.fr/hal-01739675
14. Holzmann, G.J.: The model checker SPIN. IEEE Trans. Softw. Eng. 23(5), 279–295 (1997)
15. Holzmann, G.J.: The SPIN Model Checker: Primer and Reference Manual, vol. 1003. Addison-Wesley, Reading (2004)
16. IBM Knowledge Center: IBM Rational Software Architect RealTime Edition. https://www.ibm.com/support/knowledgecenter/SS5JSH_9.5.0/com.ibm.xtools.rsarte.legal.doc/helpindex_rsarte.html. Accessed July 2020

17. Kanav, S., Aravantinos, V.: Modular transformation from AF3 to nuXmv. In: MODELS (Satellite Events), pp. 300–306 (2017)
18. Koshy, T.: Finite-State-Machines. In: Discrete Mathematics with Applications, pp. 771–802. Elsevier (2004)
19. Latella, D., Majzik, I., Massink, M.: Automatic verification of a behavioural subset of UML statechart diagrams using the SPIN model-checker. Formal Aspects Comput. **11**(6), 637–664 (1999)
20. Liu, S., et al.: A formal semantics for complete UML state machines with communications. In: Johnsen, E.B., Petre, L. (eds.) IFM 2013. LNCS, vol. 7940, pp. 331–346. Springer, Heidelberg (2013). https://doi.org/10.1007/978-3-642-38613-8_23
21. McMillan, K.L.: The SMV system. In: Symbolic Model Checking, pp. 61–85. Springer, Boston (1993). https://doi.org/10.1007/978-1-4615-3190-6_4
22. Posse, E.: PapyrusRT: modelling and code generation. In: Workshop on Open Source for Model Driven Engineering (OSS4MDE 2015) (2015)
23. Posse, E., Dingel, J.: An executable formal semantics for UML-RT. Softw. Syst. Model. **15**(1), 179–217 (2016)
24. Posse, E., Rivet, C.: Papyrus-RT: high-level view of the general transformation architecture. Survey of Requirements Management Standards - Eclipsepedia, February 2017. http://wiki.eclipse.org/Papyrus-RT/Developer/Design/0.8/Codegen_High_Level_Overview. Accessed 24 Apr 2019
25. Rivet, C., Posse, E., Toolan, D.: Getting Started with Papyrus for Real Time v1.0. Survey of Requirements Management Standards - Eclipsepedia, September 2017. https://wiki.eclipse.org/Papyrus-RT/User/User_Guide/Getting_Started. Accessed 24 Apr 2019
26. Sommerville, I.: An Insulin Pump Control System. Software Engineering 10th Edition, December 2014
27. Zurowska, K., Dingel, J.: Symbolic execution of UML-RT state machines. In: Proceedings of the 27th Annual ACM Symposium on Applied Computing, SAC 2012, pp. 1292–1299. Association for Computing Machinery, New York (2012). https://doi.org/10.1145/2245276.2231981

Modeling and Verification of Temporal Constraints for Web Service Composition

Maya Souilah Benabdelhafid[1,2]([envelope]), Houda Boubaker[2], and Mahmoud Boufaida[2]

[1] Ecole Supérieure de Comptabilité et Finances, 5 Chemin Forestier, SMK, Constantine, Algeria
mbenabdelhafid@escf-constantine.dz
[2] LIRE Laboratory, Abdelhamid Mehri Constantine 2 University, Constantine, Algeria

Abstract. This paper aims to verify temporal constraints for Web service composition. The expected deployment of such verification when composing services strongly depends on the development of an adequate solution that guarantees a high level of service quality to the system users. Given the importance of e-commerce solutions for Algerian citizens that are favorable to it due to the current confinement situation during the Covid-19 pandemic, we develop a Web service composition that studies the speed distribution of Every Consumer Goods in Algeria by using the Timed Colored Petri Nets formalism. Once the temporal constraints are identified and the formal model is developed, we analyze the performance by creating a monitor on which multiple simulations are performed by using the software CPN Tools allowing the collection of several time data, which are evaluated thereafter using the Java Framework.

Keywords: Service composition · Temporal constraints · Timed Colored Petri Nets · Formal verification · Monitoring · CPN tools · Covid-19

1 Introduction

As a core technique of Service Oriented Architecture (SOA) [1], service composition is a powerful approach to enhance the flexibility of the system as a whole in a way that new components and functionality pieces, that can be easily integrated. It combines a series of services, and at run-time, for each service, components are integrated and invoked. Temporal constraints [2] are seen as an important quality criterion in Business to Business Web service compositions, since lead time is a key performance indicator reflecting the competitiveness of enterprises [3]. One of the major benefits of applying formal approaches is the possibility of verifying whether service compositions meet specific requirements and properties [4]. Particularly, this mechanism has been identified as a prospective area where the modeling power of Petri Nets (PN) can be used in the dynamic representation and online monitoring of activities and resources, which have been proven very helpful.

PN, which represent a formal conceptual tool to specify concurrent systems, are used to formally model and verify services, making them easy to specify and model message exchange between services, service composition, and other aspects. Colored

L. Cleophas and M. Massink (Eds.): SEFM 2020 Workshops, LNCS 12524, pp. 21–29, 2021.
https://doi.org/10.1007/978-3-030-67220-1_2

PNs (CPN) offer many advantages to compose services because they are able to model a system where many activities take place concurrently and asynchronously. Jensen and Kristensen [5] introduced Timed CPN (TCPN) to verify discretely timed systems. Let us notice that TCPN is the same language of CPN with the addition of a stamp t statement that terminates successfully t time units after it has started. The powerful simulation software CPN Tools [6] can directly collect and analyze experiment data from the CPN based model for performance analysis. Therefore, TCPN and CPN Tools are widely used in the performance analysis of a large and complex service composition. For instance, Franceschetti and Eder [7] proposed a technique for the providers of composed services with temporal parameters to compute the broadest ranges for the temporal input parameters such that all service invocations where the parameters are within these ranges are dynamically controllable. Also, in [8], the authors proposed a TCPN based approach for modeling and evaluating the business processes from temporal viewpoints. They first model the business processes without time constraints, including various business and legal regulations along with resource based constraints. They added the temporal modules to them with minimal modification of the original models. However, the authors have given a brief description on how the evaluation should be done whereas in [9], we have already proposed a detailed approach in which CPN are used to model the behavior of the system. Our previous efforts were concentrated to show the passage from CPN to TCPN when modeling Web service composition.

In order to handle temporal constraints, we propose in this paper a temporal constraints-aware service composition by using TCPN formalism, which is supported by the software CPN Tools. First, an e-commerce application is discussed where we highlight the importance of temporal constraints and e-commerce applications for Algerian citizens that suffer in this moment from COVID 19 pandemic. Then, we make use of TCPN for modeling these constraints that will be after verified. The verification process is based on the monitor creation on which several simulations are performed allowing collecting time data that will be transmitted to Java framework in order to be analyzed.

The remainder of this paper is organized as follows. Section 2 illustrates our motivation through a real scenario. Section 3 introduces the temporal constraint concept by making use of the scenario and describes the temporal constraints modeling and verification approach. Section 4 defines the first step of the approach based on the TCPN modeling. Section 5 talks about the created monitor and Sect. 6 reveals the third step which is about the results visualization. Finally, Sect. 7 concludes this research work and discusses future directions.

2 Motivation Scenario

Despite the widespread use of e-commerce, shopping Everyday Consumer Goods (ECG) [10] online remains so far limited in Algeria. COVID-19 confinement in Algeria shows that households are still committed to traditional forms of retail, for which going out home remains necessary. Substantial efforts are being made by the commerce ministry in order to control delivery supply chains, both in terms of time span and correct pricing. The online sale has just started in many parts of the country, and various items are ordered, mostly via social networks and delivered on time. These actions are helping to enforce the confinement but still insufficient and poorly organized. Consequently, the Algerian Government needs to raise the general awareness of e-commerce. In this context, the present paper examines an e-commerce scenario that considers temporal constraints. The proposed application allows a client to send two messages: the order to the Customer Web Service (CWS) and the payment order to the Bank Web Service (BWS). Then, the BWS returns to the client a payment verification, which is a security code number for verifying the client identity within 15 to 30 s. The client must respond by sending a payment confirmation that will be checked by the BWS within 15 to 18 s. If the verification is unsuccessful, then the BWS must address an error confirmation notification to the client. Otherwise, it must check the balance within 30 to 60 s. If the payment transaction fails, then the BWS must return an error payment notification. Otherwise, it contacts the CWS by sending a payment notification in within 24 to 30 s, which indicates that the payment operation is completed successfully. Eventually, the CWS sends an electronic bill to the client. After 27 to 32 min, the CWS must transmit the order to the Warehouse Web Service (WWS) in order to check its availability. The WWS responds by Exist Notification message, which means that the order is available or by Not Exist Notification otherwise. The availability check operation takes within 2 to 3 min. If the order is available, then the CWS sends a delivery order to the Delivery Web Service (DWS), so as to deliver the goods to the client within 2 to 4 days. When the DWS terminates, it confirms its performing by sending the delivery confirmation to the CWS. In the other case (the order is not available), the CWS must transmit a supply order to the Provider Web Service (PWS) for providing the missing items and sending them to distribution centers within 2 to 3.30 h. When the WWS receives the supply, it confirms the supplying operation by sending a supply notification to the CWS. Once in a while, the order can have shipping damaged. Consequently, the client reports the CWS by addressing an item in which he explains the reason for return. After that, he receives a mailing label which is an electronic sticker that must be printed by the client and stuck on the outside of the order package. The CWS must send a return order to the DWS for returning the damaged order to the distribution center and receives a confirmation. The return process takes within 5 to 10 days of the receipt of shipment by the client. Finally, a refund order with the same items that the original order will be created, so the CWS must restart the process another time.

3 Temporal Constraints Modeling and Verification

For the above ECG example, we define the following temporal constraints by assuming that a minute as a basic time unit (Table) 1.

- **TC1:** The BWS must send the payment verification to the client within (15–30 s). This is a local temporal constraint, which defines the estimated execution of the activity "send the payment verification" of the BWS.
- **TC2:** The BWS must verify the confirmation within (15–18 s). This local temporal constraint defines the execution time of the activity "verify confirmation".
- **TC3:** The BWS must verify the payment within (30–60 s). It is a local constraint, which specifies the operating time of the activity "verify payment".

Table 1. Some temporal constraints.

Temporal constraint	Web service	Type	Time Unit (TU)
TC1	BWS	Local	0.25–0.5 TU
TC2	BWS	Local	0.25–0.3 TU
TC3	BWS	Local	0.5–1.0 TU
TC4	BWS	Local	0.4–0.5 TU
TC5	CWS	Local	27–32 TU
TC6	CWS/WWS	Global	2–3 TU
TC7	CWS/PWS/WWS	Global	120–210 TU
TC8	CWS/DWS	Global	2880–5760 TU
TC9	CWS/DWS	Global	7200–14400 TU

- **TC4:** The BWS sends a payment notification to the CWS that must be per- formed within (24–30 s). This local temporal constraint defines the execution time of the activity "send payment notification to CWS".
- **TC5:** The CWS must transmit the order to the WWS after 27 to 32 min of sending the bill to the client. This is also a local temporal constraint but it is used to specify the expected delay between two activities which are "send the bill to client" and "send the order to the WWS".
- **TC6:** The availability check operation takes within 2–3 min. It determines the delay between sending the order to the WWS and the receipt of the Exist or Not Exist Notification. Here, we are referring to a global temporal constraint, which represents the time required for the CWS to invoke the operations of the WWS.
- **TC7:** The supply process, which takes within (2–3.30 h) between sending the supply order to PWS and the receipt of supply notification. It is also a global temporal constraint, which represents the time required for the CWS to invoke the operations related to both of PWS and WWS.
- **TC8:** The delivery process determinates within (2–4 days) forming a delay between sending the order to the DWS and the receipt of the delivery confirmation.
- **TC9:** The return process of the damaged order specifies (5–10 days) as the delay between the sending of the item to the CWS and the receiving the confirmation.

For handling the temporal violations, we propose a process that includes three steps:

1 **TCPN-based Modeling:** The formal modeling of each Web service is performed by making use of the TCPN semantics based on the color set and the time step concepts. The set local temporal constraints are first modeled. Then, the TCPN composition model is performed and the global temporal constraints are next considered. Multiple simulations are performed frequently to check whether the formal model behaves as expected.

2 **CPN Tools-based Monitoring:** We make use of the "User defined" component in the software CPN Tools. This monitor is associated to all TCPN transitions forming our ECG application.

3 **Java Framework-based Evaluation:** We run a simulation at the CPN Tools. Meanwhile, Java receives the transmitted data through the monitor in order to visualize a clear simulation. It uses the transferred model of the transitions in order to compute the execution time of each temporal constraint during the simulation and comparing them with those added during the modeling. Finally, a Java library is used for generating a graph, which shows the execution time for the transitions associated to temporal constraints over steps. If some constraints are not verified, then we should return towards the model for correct it.

4 TCPN-Based Modeling

First, several color sets are defined in order to construct the TCPN of each Web Service; $\Sigma = \{DATA, BOOL, REAL, STK, T1, T2, T3, T4, T5, T6, T7, T8, T9\}$, where:

- DATA: is the color set that represents the exchanged messages;
- BOOL: is the color set which represents the declaration of Booleans;
- REAL: is the color set which represents a declaration of real values;
- STK: is the color set which represents the percentage decrease of the stock;
- The real color set T1 (respectively T2, T3, T4, T5, T6, T7, T8, T9) representing a duration between [0.25, 0.5] TU (respectively [0.25, 0.3] TU, [0.5, 1.0] TU, [0.4, 0.5] TU, [27.0, 32.0] TU, [2.0, 3.0] TU, [120.0, 210.0] TU, [2880.0, 5760.0] TU, [7200.0, 14400.0] TU).

We also define the set of functions which are related the local temporal constraints that must be assigned to the individual TCPN transitions:

- The function *fun delayPayVer() = T1.ran()*, which is assigned to the transition S_Verification_t_C of the BWS;
- The function *fun delayConf() = T2.ran()*, which is assigned to the transition Verify_Confirmation of the BWS;
- The function *fun delayVerPayment() = T3.ran()*, which is assigned to the transition Verify_Payment of the BWS;
- The function *fun delayPayNot() = T4.ran()*, which is assigned to the transition S_PayNotification_T_CWS of the BWS;
- We add a local constraint to the output arc of the transition "S_Bill_T_C" of the CWS by assigning the function *fun delaySendOrder() = T5. ran()*, which means that the order will not be confirmed until after 27 to 32 min.

Next, we join the TCPN of each Web Service to obtain a formal model dedicated to the composition. In addition, we define the functions which are related to global temporal constraints. They are dedicated to the latter that are set between two activities:

- The function *fun delayExt() = T6.ran()* is assigned to the transition S_ExistNotification_T_CWS and the transition S_NotExistNotification_T_CWS. It determines the time between the sending of the Order to the WWS and the receipt of the Exist Notification or the Not Exist Notification;
- The function *fun delaySupp() = T7.ran()* is assigned to the transition S_SuppNotification_T_CWS. It determines the time between the sending of the Supply Order to the PWW and the receipt of the Supply Notification;
- The function *fun delayOrder() = T8.ran()* is assigned to the transition S_Order_T_C. It determines how long it takes for the DWS to deliver the order to the customer;
- The function *fun delayReturn() = T9.ran()* is assigned to the transition S_RetConfirmation_T_CWS. It is a temporal constraint related to the return of a damaged order.

As a result, we obtain a TCPN model of the ECG application, which considers the nine set temporal constraints (see Fig. 1). As it is shown, the initial marking consists primary on three tokens: Order message in the place Order_received, Payment Order message in the place Pay_Order_received and true Boolean in the place damaged. They are associated to timestamp (see @ given in green boxes) equal to 0.0.

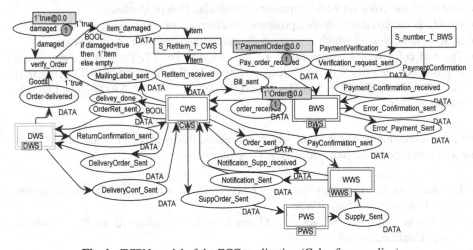

Fig. 1. TCPN model of the ECG application (Color figure online)

5   CPN Tools-Based Monitoring

Once the system is modeled, we use the monitoring palette of CPN Tools 4.0 in order to create a monitor "User defined". As we can see in Fig. 2, we define firstly the function "init ()", which establishes the connection communication between CPN Tools and the Java Framework.

Then, we make use of two well-known functions: Observer and Predicate. The Observer function returns the Web Service, the transition name, the temporal values

```
▼Java monitor
    Type: User defined
  ►Nodes ordered by pages
  ▼Init
      fun init () =
        if !connected = true
        then (ConnManagementLayer.closeConnection("Conn 1");
                  connected := false)
      else ()
  ►Predicate
  ►Observer
  ▼Action
      fun action (s1,s2,s3,s4) =
        (if not(!connected)
        then (ConnManagementLayer.acceptConnection("Conn 1",9000);
                  connected:=true)
        else ();
        send_to_Java(s1,s2,s3,s4))
  ▼Stop
      fun stop () =
        if !connected = true
        then (ConnManagementLayer.closeConnection("Conn 1");
                  connected := false)
        else ()
```

Fig. 2. User defined monitor of the ECG application

and the step number when the transition is fired. It uses Time and Step functions. Concerning the Predicate function, each time it returns true, the data is extracted by the Observer function and transmitted to the defined Action function, which checks the connection to the application for transferring the collected data.

Finally, the Stop function is called when the simulation is finished. Once it is executed, the connection with the Java framework is closed. We note that the Action function invokes Send_to_Java function. This latter uses another function called ConnManagementLayer.send, which is available in the Comms/CPN functions. It allows the sending of the data (the Web service it belongs to, the transition name, the time model and the step number when the transition is fired) that are related to each transition to Java.

6   Java Framework-Based Evaluation

This evaluation consists of receiving the transmitted data through the created monitor. It contains three essential parts as follow:

Simulations. Once the collected data are received by Java, they are used for visualizing TCPN model simulations at Java application in the run-time. When a transition is fired on the CPN Tools level, it will be transmitted in real time and displayed.

Time Computing. Here, the execution times of transitions that are associated to local or global temporal constraints are computed during the simulation. They are then compared to the set constraints. Let us compare the set temporal constraints and the computed time data. Table 2 reveals that the returned data are in the interval of the time that is associated to the set temporal constraints (see columns 2 and 3).

Temporal Constraints Visualization: At this level, we use the "JFreeChart" library for creating a line graph, which depicts the execution time of each transition associated to local or global temporal constraints over steps. Figure 3 represents the obtained graph

Table 2. Comparison of the set temporal constraints with the computed ones.

Temporal constraint	Time Unit (TU)	Computed time using Java
TC1	0.25–0.5 TU	0.397006354829 TU
TC2	0.25–0.3 TU	0.297161121642 TU
TC3	0.5–1.0 TU	0.869714341759 TU
TC4	0.4–0.5 TU	0.45778840620 TU
TC5	27–32 TU	29.48714989867 TU
TC6	2–3 TU	2.3845970725 TU/2.1641991 TU
TC7	120–210 TU	203.0698617334 TU
TC8	2880–5760 TU	5285.316646771TU/4880.0243730 TU
TC9	7200–14400 TU	13392.2057475 TU

containing three line charts. The red line chart shows the execution time of the transitions during the simulation whereas the blue (respectively the green) line represents the execution time in the case when the transition is associated to the set temporal constraints but with minimum (respectively maximum) values. For instance, we notice that the TC8, which is related to the delivery process is defined by between 40 and 44 that takes around 5000 TU during the simulation. This value is between the minimum value 2880 TU and the maximum value 5760 TU, which are defined by the blue and green lines chart respectively. For the other constraints, it is shown plainly that the red curve of simulation is always between the blue and the green curves. That is to say that the temporal constraints are verified in this step another time. Hence, the Java framework based evaluation allows analyzing successfully the temporal accuracy and reliability of the Web service composition.

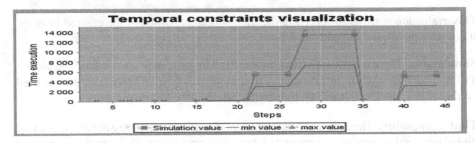

Fig. 3. Temporal constraints visualization (Color figure online)

7   Conclusion

The importance of an adequate management of temporal aspects of process aware information systems is beyond dispute [11]. This paper proposes a development process able to model and verify temporal constraints for Web service composition by using TCPN formalism, CPN Tools software and Java Framework. In doing so, and by an illustrative example, this work reveals that TCPN models based composition can be transferred into Java framework able to allow the developer to verify several temporal constraints. In future, the performance challenge will be discussed for the same case study but for other criteria such as security.

References

1. Wu, Z.: Service Computing: Concept, Method and Technology. Academic Press, Cambridge (2014)
2. Deng, S., Huang, L., Wu, H., Wu, Z.: Constraints-driven service composition in mobile cloud computing. In: 2016 IEEE Inter Conference on Web Services, pp. 228–235. IEEE (2016)
3. Song, W., Ma, X., Cheung, S.C., Dou, W., Lu, J.: A public-view approach to timed properties verification for B2B web service compositions. In: 2009 IEEE International Conference on Services Computing, pp. 427–434. IEEE (2009)
4. Zhu, M., Li, J., Fan, G., Zhao, K.: Modeling and verification of response time of QoS-aware web service composition by timed CSP. Proc. Comput. Sci. **141**, 48–55 (2018)
5. Jensen, K., Kristensen, L.M.: Coloured Petri Nets: Modelling and Validation of Concurrent Systems. Springer, Heidelberg (2009). https://doi.org/10.1007/b95112
6. Jensen, K., Kristensen, L.M., Wells, L.: Coloured Petri Nets and CPN Tools for modelling and validation of concurrent systems. Int. J. Softw. Tools Technol. Transfer **9**(3–4), 213–254 (2007)
7. Franceschetti, M., Eder, J.: Checking temporal service level agreements for web service compositions with temporal parameters. In: 2019 IEEE International Conference on Web Services (ICWS), pp. 443–445. IEEE (2019)
8. Shinkawa, Y., Shiraki, R.: Temporal evaluation of business processes using timed colored Petri Nets. In: FedCSIS Position Papers, pp. 161–168 (2017)
9. Benabdelhafid, M.S., Bérard, B., Boufaida, M.: Analysing timed compatibility of web service choreography. Inter J. Crit. Comput.-Based Syst. **7**(3), 253–278 (2017)
10. Pernot, D.: Internet shopping for Everyday Consumer Goods: an examination of the purchasing and travel practices of click and pickup outlet customers. Res. Transp. Econ., 100817 (2020)
11. Eder, J.: Computing history-dependent schedules for processes with temporal constraints. In: Dang, T.K., Küng, J., Takizawa, M., Bui, S.H. (eds.) FDSE 2019. LNCS, vol. 11814, pp. 145–164. Springer, Cham (2019). https://doi.org/10.1007/978-3-030-35653-8_11

Modeling Attack-Defense Trees' Countermeasures Using Continuous Time Markov Chains

Karim Lounis[1](✉) and Samir Ouchani[2]

[1] QRST, School of Computing, Queen's University, Kingston, ON, Canada
lounis@cs.queensu.ca
[2] LINEACT, École d'Ingénieur CESI, Aix-en-Provence, France
souchani@cesi.fr

Abstract. ADTrees (Attack-Defense Trees) are graphical security modeling tools used to logically represent attack scenarios along with their corresponding countermeasures in a user-friendly way. Many researchers nowadays use ADTrees to represent attack scenarios and perform quantitative as well as qualitative security assessment. Among all different existing quantitative security assessment techniques, CTMCs (Continuous Time Markov Chains) have been attractively adopted for ADTrees. ADTrees are usually transformed into CTMCs, where traditional stochastic quantitative analysis approaches can be applied. For that end, the correct transformation of an ADTree to a CTMC requires that each individual element of an ADTree should have its correct and complete representation in the corresponding CTMC. In this paper, we mainly focus on modeling countermeasures in ADTrees using CTMCs. The existing CTMC-model does not provide a precise and complete modeling capability, in particular, when cascaded-countermeasures are used. Cascaded-countermeasures occur when an attacker and a defender in a given ADTree recursively counter each other more than one time in a given branch of the tree. We propose the notion of tokenized-CTMC to construct a new CTMC-model that can precisely model and represent countermeasures in ADTrees. This new CTMC-model allows to handle cascaded-countermeasure scenarios in a more comprehensive way.

Keywords: Attack-defense trees · CTMCs · Security graphical models · Stochastic models · Quantitative security assessment

1 Introduction

ADTrees (Attack-Defense Trees) [1] are defined as a graphical methodology used to represent attack scenarios by systematically representing the different actions

QRST (Queen's Reliable Software Technology) Laboratory.
LINEACT (Laboratoire d'Innovation Numérique pour les Entreprises et les Apprentissages au service de la Compétitivité des Territoires).

L. Cleophas and M. Massink (Eds.): SEFM 2020 Workshops, LNCS 12524, pp. 30–42, 2021.
https://doi.org/10.1007/978-3-030-67220-1_3

that an attacker may undertake to realize a security breach, and the different actions that a defender may apply to mitigate the attacker's actions. ADTrees can be interpreted using different types of semantics, generally classified into stochastic (e.g., continuous time Markov chains (CTMCs) [2] and stochastic Petri-nets (SPNs) [3]) and non stochastic (e.g., propositional, multiset, equational, and De Morgan [1]). They have proven to be simple, easy to use, and yet powerful in their modeling capability. Also, they can be automatically generated [4] and applied for security assessment using a publicly available tool, called ADTool [5].

Figure 1 illustrates a simple ADTree that can be followed to generate a Bluecutting attack[1]. In an ADTree, attacks are shown by red circles ($\bigcirc$), whereas defenses are depicted in green squares ($\square$). Attack and defense refinements are depicted by solid lines ($\bigcirc\!-\!\bigcirc$ and $\square\!-\!\square$), whereas attack and defense mitigations (counterattacks) are represented by dashed lines ($\bigcirc\cdots\square$ and $\square\cdots\bigcirc$). The root node in an attack-defense tree represents the final goal of the attacker. The intermediary nodes show subgoals. Finally, leaf nodes represent basic "atomic" attacks or defenses. Attack as well as defense nodes can be refined further into more detailed attack or defense actions. The refinement can be either conjunctive (i.e., all actions must occur), disjunctive (i.e., one out

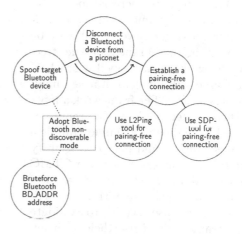

Fig. 1. ADTree for disconnecting a legitimate Bluetooth device from a Bluetooth piconet, a.k.a., Bluecutting attack. (Color figure online)

of many actions occur), sequentially conjunctive (i.e., all actions must occur following an order), or parallel disjunctive (i.e., a combination of optional actions occur). For example, in the ADTree of Fig. 1, the attack node "Disconnect a Bluetooth device from a piconet", i.e., the root node, is refined using the sequential conjunctive refinement (represented by an arrow linking the refinements). Such a refinement enforces that the attack action "Establishing a pairing-free connection" can be performed if and only if the attack action "Spoof target Bluetooth device" is conducted and completed. The attack node "Establishing a pairing-free connection" is further disjunctively refined into two other attack actions. This refinement allows the attacker to either perform the attack node

---

[1] Bluecutting attack is a denial of service attack on Bluetooth wireless technology. In this attack the attacker creates a new connection (pairing-free connection) with a remote device to force the latter to disconnect from another device [6–8].

"Use L2Ping tool for pairing-free connection" or the attack node "Use SDPtool for pairing-free connection".

Among the modeling capabilities of ADTrees that make them favorable over other existing graphical models, the ability to model cascaded-countermeasures. Cascaded-countermeasures occur when an attacker and a defender in a given ADTree recursively counter each other more than one time in a given branch of the tree. For example, in the ADTree of Fig. 1, the branch composed of attack nodes "Spoof target Bluetooth device" and "Bruteforce Bluetooth BD_ADDR address", and the defense node "Adopt Bluetooth non-discoverable mode" constitutes an example of a cascaded-countermeasure (viz., Fig. 2).

We note that there is no restriction on the number of cascaded-countermeasures' levels that can be created in one branch of the tree. In [2], we have developed a stochastic framework for quantitative analysis of ADTrees. The framework adopts ADTree methodology to represent attack scenarios in a simple graphical representation, and performs security quantitative assessment using CTMC analytical approach. We have performed a security quantitative analysis on a case study to validate the framework. Nevertheless, when cascading-countermeasures appear in a given attack scenario, the CTMC-based model used in the framework does not precisely and completely represent the cascaded-countermeasures scenario as we will see in the next sections. Such a limitation leads to an incorrect and incomplete security assessment. Thus, in this paper, we enrich the old CTMC-model with new items to represent countermeasures,

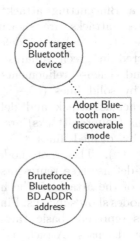

Fig. 2. ADTree tree for spoofing a legitimate Bluetooth device. The initial attack "Spoof target Bluetooth device" is mitigated with a countermeasure "Adopt Bluetooth non-discoverable mode", which is in turn countered by the attacker with another attack "Bruteforce Bluetooth BD_ADDR address".

in general, and cascading-countermeasures scenarios, in particular, in a more precise and complete way using the notion of tokenized-CTMC.

The remainder of the paper is organized as follows: Sect. 2 briefly provides a background of ADTrees, CTMCs, and introduces the notion of tokenized-CTMC. In Sect. 3 we define a new CTMC-model for countermeasures in terms of tokenized-CTMC. We conclude the paper in Sect. 4.

2 ADTrees and CTMCs

2.1 ADTrees

ADTrees (Attack-Defense Trees), viz., Fig. 1, is a graphical methodology used to systematically represent security scenarios gathering both attacker actions and defender actions in one graphical tree-based layout. Recall that in ADTrees, attacks are shown by red circles ($\bigcirc$), whereas defenses are depicted in green squares ($\square$). Attack and defense refinements are depicted by solid lines ($\bigcirc-\bigcirc$ and $\square-\square$), whereas attack and defense mitigations (counterattacks) are represented by dashed lines ($\bigcirc \ldots \square$ and $\square \ldots \bigcirc$). In the case of a cascaded-countermeasure, an attacker could counterattack a countermeasure. The counterattack is attached to the countermeasure with dashed lines ($\bigcirc \ldots \square \ldots \bigcirc$). The root node in an attack-defense tree represents the final goal of the attacker. The intermediary nodes show subgoals. Finally, leaf nodes represent basic "atomic" attacks and defenses. The attacker and the defender are commonly referred to as the proponent/opponent depending on the scenario. Also, due to the possibility for an attacker to counterattack a countermeasure and for a defender to mitigate the counterattack, the attacker and the defender are sometimes called players [1].

Each action performed by a player can be refined into more detailed actions, called refinements. These refinements can be of four types: (1) Conjunctive, representing necessary actions to be performed. (2) Disjunctive, which represents exclusive optional actions to be performed. (3) Sequential conjunctive, representing necessary actions to be performed in a particular order. (4) Parallel disjunctive, which represents combination of optional actions [3]. This latter refinement can be interpreted as a generalized form of Refinement 2.

Formally, ADTrees are defined as closed-terms, called ADTerms, over a signature $\Sigma = (\mathbb{S}, \mathbb{F})$, where $\mathbb{S} = \{p, o\}$ represents the set of player types, and $\mathbb{F} = \{(\wedge_k^{s \in \mathbb{S}})_{k \in \mathbb{N}}, (\vee_k^{s \in \mathbb{S}})_{k \in \mathbb{N}}, (\overrightarrow{\wedge}_k^{s \in \mathbb{S}})_{k \in \mathbb{N}}, (\widetilde{\vee}_k^{s \in \mathbb{S}})_{k \in \mathbb{N}}, c^{s \in \mathbb{S}}\} \bigcup \mathbb{B}$ is a set of function symbols union the set of basic actions (i.e., $\mathbb{B}$) for both players. The unranked functions $(\wedge_k^s)_{k \in \mathbb{N}}, (\vee_k^s)_{k \in \mathbb{N}}, (\overrightarrow{\wedge}_k^s)_{k \in \mathbb{N}}$, and $(\widetilde{\vee}_k^s)_{k \in \mathbb{N}}$, where $s \in \mathbb{S}$, represent the conjunctive refinement ($\wedge$), the disjunctive refinement ($\vee$), the sequential conjunctive refinement ($\overrightarrow{\wedge}$), and the parallel disjunctive refinement ($\widetilde{\vee}$) for the proponent and the opponent, respectively. The binary functions $c^{s \in \mathbb{S}}$ connect an action of a given type $s \in \mathbb{S}$ with an action of the opposite type $\bar{s} \in \mathbb{S}$. Conventionally $\bar{p} = o$ and $\bar{o} = p$. Therefore, we can write any ADTree using the following recursive definition:

$$t ::= b^s \mid \vee^s (t, \ldots, t) \mid \wedge^s(t, \ldots, t) \mid \overrightarrow{\wedge}^s(t, \ldots, t) \mid \widetilde{\vee}(t, \ldots, t) \mid c^s(t, t)$$

If we consider the ADTree of Fig. 2, then by denoting the atomic actions "Spoof target Bluetooth device" by b_0^p, "Bruteforce Bluetooth BD_ADDR address" by b_1^p, and the countermeasure "Adopt Bluetooth non-discoverable mode" by b_0^o, the corresponding ADTerm of the ADTree will be expressed as follows:

$$t = c^p(b_0^p, c^o(b_0^o, b_1^p))$$

2.2 Enumerated Continuous Time Markov Chains

Markov chains are stochastic processes used to model systems' behavior where probabilistic events may occur. They are called Markovian since the predictions are based only on the current state of the system, and not on any previous state. A Markov process that transits from one state to another via an exponential rate is called a CTMC (Continuous Time Markov Chain). In [2], we have introduced the notion of enumerated-CTMC, which is a standard CTMC with an explicit distinction between the different states of a given CTMC (i.e., the initial, intermediate, and final states). We have used the enumerated-CTMC to develop a stochastic framework that transforms ADTrees into CTMCs for security quantitative analysis. Furthermore, in [3], Lounis et al., extended the framework to adopt Petri-Nets and to perform qualitative security analysis of ADTrees.

Definition 1. An *enumerated continuous time Markov chain* M is a tuple (S, S_0, S_*, G), where:

 - S is a finite set of states,
 - $S_0 \subset S$ is a finite set of initial states,
 - $S_* \subset S$ is a finite set of final states,
 - $G: S \times S \to \mathbb{R}$ is the infinitesimal generator matrix which gives the rate of transition between two given states s and s'.

We note that there exists a set of intermediate states that we denote by $S_m \subset S$, where $S = S_0 \cup S_m \cup S_*$ and $S_0 \cap S_m \cap S_* = \emptyset$.

2.3 Countermeasures in ADTrees

For a better clarification, we present the concept of countermeasure in ADTree methodology. We explain how this term can be interchanged with attacks and defenses, in particular, when cascaded-countermeasures are used.

Definition 2. A countermeasure can be seen as an action $b^s \in \mathbb{B}$ undertaken by a player $s \in \mathbb{S}$ to prevent another action $b^{\bar{s}} \in \mathbb{B}$ performed by the opposite player $\bar{s} \in \mathbb{S}$ from occurring.

Example 1. If we consider the ADTree in Fig. 2, then the action "Adopt Bluetooth non-discoverable mode" is a countermeasure for the attacker's action "Spoof a target Bluetooth device". At the same time, the action "Bruteforce Bluetooth BD_ADDR address" is respectively a countermeasure (in the sense of a counterattack) for the action "Adopt Bluetooth non-discoverable mode", which is performed by the defender. Therefore, a countermeasure is any action performed by a given player (proponent/opponent) against his opposite player (opponent/proponent) in an ADTree.

As there are many types of defensive countermeasures (see the classification in [3]), for the sake of simplicity, in this paper, we only consider the delayed-type countermeasure introduced in [2]. In fact, this type of countermeasures is actually the type that leads to an incorrect representation of the cascaded-countermeasures scenario when the old CTMC-model is adopted for ADTrees, i.e., the enumerated CTMC-model [2].

2.4   CTMC Model for Countermeasures

In [2], we developed a CTMC-based model for ADTrees. In that model, we represented each action, whether it was an attack or a defense, by a single-transition CTMC (viz., CTMC in Row 1 and 2 of Fig. 3), and represented the (delayed-type) countermeasure function with a birth–death process (viz., CTMC in Row 3 of Fig. 3). With such representation, a delayed-type countermeasure action mitigates the attack action once the attack is executed. For instance, assuming that an attacker has cracked a password and is currently connected through telnet[2] to a server (attack), then if the administrator modifies the password (countermeasure), the attacker will be disconnected from the server. This forces the attacker to restart the attack from scratch, which includes cracking the new password to reconnect to the server through telnet.

To clarify the modeling limitation in the old CTMC-model, let us first consider the upper part ADTree of Fig. 2 (i.e., without the attack node "Bruteforce Bluetooth BD_ADDR address"). The resulting ADTerms and CTMCs representation are illustrated in the first three rows of Fig. 3. We can see that the adopted model is straightforward, not hard to understand, and easy to evaluate using CTMC analytical approach. In fact, the interpretation of the birth-death process in this case is: the attacker starts performing the attack (birth), once done, the countermeasure can be executed (death), canceling the success of the attack. Of course, the attacker would move forward over the CTMC if the countermeasure is not executed. Nevertheless, if we try to model the whole ADTree of Fig. 2 using the same CTMC-model (viz., Row 4 of Fig. 3), we lose track of certain executed actions. As can be seen in the last row of Fig. 3, the two attack actions "Spoof target Bluetooth device" and "Bruteforce Bluetooth BD_ADDR address" are both represented by a single-transition. Also, their execution rate (i.e., exponential rate in CTMCs) has been aggregated. This consequently prevents us from keeping complete track of attack scenarios that occur, in particular, during the security assessment process. If that single attacker's transition is executed, we cannot determine exactly which atomic action was executed, i.e., was it the b_0^p or b_1^p. This would unfortunately result in a less precise quantitative analysis. Hence, we need to enrich the CTMC-model with more information to be able to handle such type of scenarios and have a more precise modeling capability and complete security assessment.

---

[2] Telnet is an application-layer protocol that allows remote access to computer systems over a network. The telnet service runs on the communication port 23.

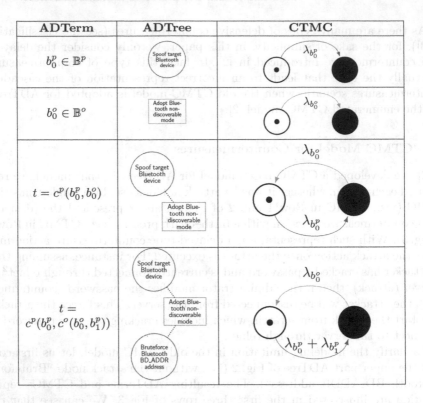

Fig. 3. Formal (Column 1), graphical (Column 2), and CTMC-representation (Column 3) for a basic attack (Row 1) and countermeasure (Row 2), countermeasure application (Row 3), and cascaded-countermeasure (Row 4) in the old CTMC-model, where ⊙ represents the initial state, and ● represents the final states. Also, the red transition ⌒ represents an attack transition in a CTMC, whereas the green transition ⌒ represents a countermeasure transition in a CTMC. (Color figure online)

2.5 Tokenized Continuous Time Markov Chain

To precisely handle the modeling limitation discussed in the previous subsection, we extend the old CTMC-model [2] with a new notion of colored indexed-tokens. Throughout the rest of the paper, we arbitrarily and interchangeably refer to the attacker as the proponent and the defender as the opponent.

Definition 3. *Colored indexed-tokens* are indexed elements of a set $\mathbb{C} = \mathbb{C}^p \bigcup \mathbb{C}^o$ *that can take one of the two colors: Red (•) or Green (∘). Arbitrarily, we use the red color (•) to refer to the proponent and the green color (∘) for the opponent.*

Definition 4. *Let* $\mathbb{B}$ *be the set of basic actions and let* $\mathbb{C} = \{•_0, \ldots, •_n, ∘_0, \ldots, ∘_m\}$ *be the set of colored indexed-tokens, for* $n, m \in \mathbb{N}$. *Then, an action-coloring is a function* $\sigma : \mathbb{B} \to \mathbb{C}$, *which associates for each basic action* $b \in \mathbb{B}$ *a singleton of a colored indexed-token, i.e.,* $\{•\}$ *or* $\{∘\}$.

Example 2. If we consider the ADTree $t = c^p(b_0^p, c^o(b_0^o, b_1^p))$, where the attacker is the proponent and the defender is the opponent, we can write: $\sigma(b_0^p) = \{\bullet_0\}, \sigma(b_1^p) = \{\bullet_1\},$ and $\sigma(b_0^o) = \{\circ_0\}$.

Finally, we associate to each state in a given enumerated-CTMC an ordered set of colored indexed-tokens to indicate which action/actions has/have been successfully achieved at that state. This will determine whether the proponent or the opponent is the vanquisher at a specific state of the system. Note that these tokens are stored within an ordered set in order to keep track of the order of achievement of the actions, which is necessary to determine the vanquisher.

Definition 5. Let S be *the set of states* of a given enumerated-CTMC, and let $\mathbb{C}$ be a *set of colored indexed-tokens*. Then, a state-coloring is a function $\tau\colon S \to \mathcal{P}(\mathbb{C})$, which associates for each state $s \in S$ an ordered set of colored indexed-tokens from $\mathcal{P}(\mathbb{C})$.

Example 3. If we consider the first row in Fig. 3, which represents the attack action $b_0^p \in \mathbb{B}^p$, then the colored indexed-token associated to the states of its CTMC is $\tau(S_0^{b_0^p}) = \{\} = \emptyset$ and $\tau(S_*^{b_0^p}) = \sigma(b_0^p) = \{\bullet_0\}$.

An enumerated-CTMC associated with a states-coloring function is called a tokenized-CTMC, or T-CTMC for short.

Definition 6. A *tokenized continuous time Markov chain* M' is a tuple (M, τ), where M is an *enumerated-CTMC* and τ is a *states-coloring function*.

3   Tokenized-CTMC for Countermeasure Modeling

To have a more precise and complete model to represent countermeasures, we slightly modify the existing CTMC-model for the countermeasure in such a way so that we allow the countermeasure and the countered action to evolve in parallel. This means that at the initial state, both the countered action and the countermeasure can be executed, which is not the case in the birth-death process representation, where the countermeasure could only be executed when the countered action is performed. The new CTMC-model for countermeasure application is depicted in Row 1 of Fig. 4.

Using the tokenized-CTMC model, we formally express the countermeasure function using a new unranked function $r^s\colon \mathbb{M}' \times \mathbb{M}' \to \mathbb{M}'$ (note that, here we use a different symbol r^s for the countermeasure so that it does not get confused with the classical countermeasure function c^s). This new function takes two tokenized-CTMCs $(M^s, M^{\bar{s}}) \in \mathbb{M}' \times \mathbb{M}'$ as inputs, one representing the proponent/opponent action and the other one representing the opponent/proponent action, where $M^s \in \mathbb{M}'$ is a tokenized-CTMC. Then, it links them in such a way so that the attack and the countermeasure can evolve in parallel (both actions can occur at the same time) and at the same time takes care of the order of

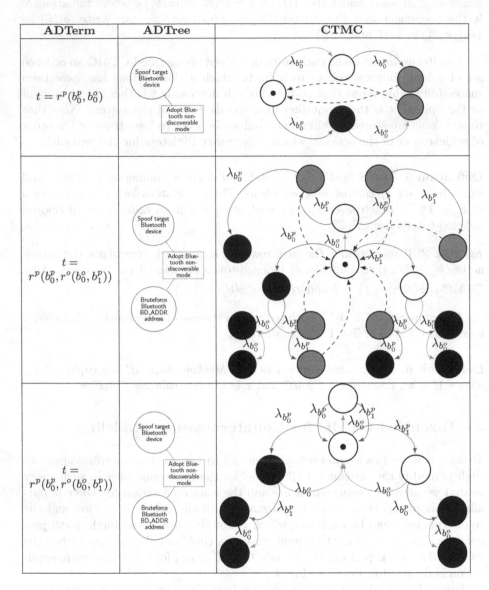

Fig. 4. Formal (Column 1), graphical (Column 2), and CTMC-representation (Column 3), for the new countermeasure function (Row 1), the cascaded-countermeasure function application (Row 2), and the optimized cascaded-countermeasure function application (Row 3), where ⊙ represents the initial state, ● represents the final states, O represents the intermediary states, and ⬤ represents the dump states. Also, the red transition ⌢ represents an attack transition in a CTMC, whereas the green transition ⌢ represents a countermeasure transition in a CTMC. (Color figure online)

execution (keeping track using the colored indexed-tokens). Therefore, the constructed tokenized-CTMC after applying function $\kappa = r^s(M^s, M^{\bar{s}})$, is defined as $M = (S^\kappa, S_0^\kappa, S_*^\kappa, G^\kappa, \tau^\kappa)$ where the set of states S^κ is generated by:

$$
S_0^{M^s} \times S_0^{M^{\bar{s}}} \; \bigcup \; S_m^{M^s} \times S_0^{M^{\bar{s}}} \; \bigcup \; S_0^{M^s} \times S_m^{M^{\bar{s}}} \; \bigcup \; S_*^{M^s} \times S_0^{M^{\bar{s}}}
$$
$$
\bigcup
$$
$$
S_0^{M^s} \times S_*^{M^{\bar{s}}} \; \bigcup \; S_m^{M^s} \mathbin{\square} S_m^{M^{\bar{s}}} \; \bigcup \; S_*^{M^s} \mathbin{\square} S_m^{M^{\bar{s}}} \; \bigcup \; S_*^{M^s} \mathbin{\square} S_*^{M^{\bar{s}}}
$$

and $\square$ is a kind of sequential Cartesian product operator, such that:

$$
\{a\} \mathbin{\square} \{(b,c)\} = \{(a,b,c),(b,a,c),(b,c,a)\}
$$

The generator matrix G^κ is given as follows, where $S_\emptyset$ is a particular set of states, called dump states, which we will explain later on, and Δ is the asymmetric difference between two sets:

$$
G^\kappa(s_i, s_j) = \begin{cases} -\sum_{i \neq j} G^\kappa(s_i, s_j) & \text{if } i = j \\ -1 & \text{if } (s_i, s_j) \in S_\emptyset^\kappa \times S_\emptyset^\kappa \\ 0 & \text{if } i \neq j \text{ and } |s_i \Delta s_j| > 2 \\ G^{idf}(S^{idf} \cap \{s_i\}, S^{idf} \cap \{s_j\}) & \text{Otherwise} \end{cases}
$$

Overall, this formulation consists of determining which transition goes from state s_i to state s_j. Since every transition t_i^{idf} belongs to only one CTMC M_{idf}, the execution of t_i^{idf} will only affect the states of M_{idf}. Thus, in general, there will be only one transition (one rate). Notwithstanding, the rate $G^{idf}(S^{idf} \cap \{s_i\}, S^{idf} \cap \{s_j\}) \neq 0$ of the identified transition t_i^{idf} exists only if $|\tau(s_j) \setminus \tau(s_i)| = 1$, i.e., the destination state s_j contains all the colored indexed-tokens of s_i in addition to one new colored indexed-token representing the transition itself.

Definition 7. A *colored-indexed token evaluator* is a function $\xi \colon \mathcal{P}(\mathbb{C}) \to \mathcal{P}(\mathbb{C})$, which selects a subset of colored indexed-tokens from a larger set of colored indexed-tokens w.r.t. the following rules, where $c_{i \in \mathbb{N}}, c_{j \in \mathbb{N}}, c_{k \in \mathbb{N}} \in \mathbb{C}$ are *colored indexed-tokens*, and $c_{i \in \mathbb{N}}$ *the selected ones*:

- $\nexists j, k \in \mathbb{N} \mid r^s(\sigma^{-1}(c_i), \sigma^{-1}(c_j)) \wedge r^s(\sigma^{-1}(c_j), \sigma^{-1}(c_k))$.
- $\exists j, k \in \mathbb{N}, j > k > i \mid r^s(\sigma^{-1}(c_i), \sigma^{-1}(c_j)) \wedge r^s(\sigma^{-1}(c_j), \sigma^{-1}(c_k))$.
- $\exists j, k \in \mathbb{N}, k > i > j \mid r^s(\sigma^{-1}(c_i), \sigma^{-1}(c_j)) \wedge r^s(\sigma^{-1}(c_j), \sigma^{-1}(c_k))$.
- $\exists j, k \in \mathbb{N}, i > j, j > k \mid r^s(\sigma^{-1}(c_i), \sigma^{-1}(c_j)) \wedge r^s(\sigma^{-1}(c_j), \sigma^{-1}(c_k))$.

Example 4. If we consider the ADTree $t = r^p(b_0^p, r^o(b_0^o, b_1^p))$, from Row 2 of Fig. 4, where the attacker is the proponent and the defender is the opponent, then for a given set of colored indexed-tokens $\{\circ_0, \bullet_1, \bullet_0\}$ associated to a given state $s' \in S$ and by applying Rule 3 and Rule 4, we obtain $\xi(\{\circ_0, \bullet_1, \bullet_0\}) = \{\bullet_1, \bullet_0\}$. Using Rule 3, we obtain $\{\bullet_1\}$, and using Rule 4, we obtain $\{\bullet_0\}$.

Therefore, to determine the initial state, the intermediate states, and the final states, we use the colored indexed-tokens associated to each state along with the colored indexed-token evaluator.

- $S_0^\kappa = \{s' \in S^\kappa \mid \xi(\tau(s')) = \emptyset\}$.
- $S_*^\kappa = \{s' \in S_\kappa \mid \forall c \in \xi(\tau(s')), c \in \mathbb{C}^s\}$.
- $S_m^\kappa = \{s' \in S^\kappa \mid s' \notin S_0^\kappa \bigcup S_*^\kappa\}$.

The colored indexed-tokens of a given set $s \in S^\kappa$ are generated by the union of the colored-indexed tokens of the sets involved in the generation of the set s. For example, if we consider the ADTree $t = r^p(b_0^p, b_0^o)$, where the attacker is the proponent and the defender is the opponent, we have:

- $\tau(S_0^\kappa) = \tau(S_0^{M^s}) \bigcup \tau(S_0^{M^{\bar{s}}}) = \{\} \bigcup \{\} = \emptyset$.
- $\tau(S_*^\kappa) = \tau(S_*^{M^s}) \bigcup \tau(S_*^{M^{\bar{s}}}) = \{\bullet_0\} \bigcup \{\} = \{\bullet_0\}$.
- $\tau(S_2^\kappa) = \tau(S_*^{M^s}) \bigcup \tau(S_*^{M^{\bar{s}}}) = \{\} \bigcup \{\circ_0\} = \{\circ_0\}$.
- $\tau(S_3^\kappa) = \tau(S_*^{M^s}) \bigcup \tau(S_*^{M^{\bar{s}}}) = \{\bullet_0\} \bigcup \{\circ_0\} = \{\bullet_0, \circ_0\}$.
- $\tau(S_4^\kappa) = \tau(S_*^{M^{\bar{s}}}) \bigcup \tau(S_*^{M^s}) = \{\circ_0\} \bigcup \{\bullet_0\} = \{\circ_0, \bullet_0\}$.

The set S^κ is composed of the initial state, which contains both the initial state of the proponent and the opponent. The final state however, consists of states where the main player (proponent/opponent) is the vanquisher. These states are identified using the colored indexed-tokens of the main player. The remaining states are the intermediate states where neither the proponent nor the opponent is the vanquisher. Therefore, if we consider the ADTree $t = r^p(b_0^p, b_0^o)$, where the attacker is the proponent and the defender is the opponent, then we can observe that: if the proponent (attacker), executes its action $b_0^p \in \mathbb{B}^p$, it tries to spoof a target Bluetooth device by for example scanning for its Bluetooth device address (i.e., BD_ADDR). Then, if it does not find the Bluetooth device address before the countermeasure (defense: turn device into non-discoverable mode) is executed, the attacker will not be able to find the device address. Similarly, if the countermeasure (defense) is executed before the attack, then the execution of the attack will certainly fail (as the target Bluetooth device is in non-discoverable mode).

Although the proponent has planned to counter the countermeasure as in Row 2 of Fig. 4 (i.e., $t = r^p(b_0^p, r^o(b_0^o, b_1^p))$), the execution of the second attack (b_1^p) may not succeed as the target device may use a different BD_ADDR address when set on non-discoverable mode. Therefore, the previously used BD_ADDR address is useless. We represent the stages where generally the assets (e.g., discovered BD_ADDR address) become useless as gray states, which we call dump states. They are elements of the dump set, which is denoted as $S_\emptyset$. All dump states are linked to the initial state using an immediate transition (viz., dashed arcs in Row 2 of Fig. 4). These states are merged afterwards to obtain the final CTMC (viz., Row 3 of Fig. 4). We define the set of dump state as follows, where $\kappa = r^s(M^s, M^{\bar{s}})$, and $\Lambda = [r^p(\tau^{-1}(c), \tau^{-1}(c')) \vee r^o(\tau^{-1}(c'), \tau^{-1}(c))]$:

$$S_\emptyset^\kappa = \{s' \in S^\kappa \mid \exists(c, c') \in \mathbb{C}^p \times \mathbb{C}^o, c, c' \in \tau(s') \wedge \Lambda\}$$

Finally, at the end of the composition, the states linked with a rate ϵ will be merged together, and their outgoing chains will be deleted, as shown in Row 3, Column 3, of Fig. 4.

4   Conclusion

Attack-Defense Trees (ADTrees) are graphical security modeling tools that are largely adopted by the security research community [7,8]. They allow to represent attack scenarios in a logical, user-friendly, and easy to understand graphical layout. Also, they allow to perform qualitative as well as quantitative security assessment. For quantitative security assessment, CTMCs (Continuous Time Markov Chains) are generally adopted, where an ADTree is transformed into a CTMC so that classical analytical analysis can be performed. To that end, a CTMC-based framework has been developed for ADTrees in [2]. Nevertheless, certain ADTrees, in particular, those containing cascaded-countermeasures, introduce a limitation in the framework. This limitation results in an incomplete CTMC-model and an incorrect security assessment of ADTrees.

In this paper, we have proposed a new CTMC-model to represent countermeasures in ADTrees. We have introduced the notion of tokenized-CTMC to precisely represent countermeasures and handle the order in which actions are occurring, allowing a correct and complete modeling and evaluation of ADTrees that contain cascaded-countermeasures.

We emphasize that the proposed solution is still subject to state explosion problem. In fact, in the old CTMC-model, modeling a coutermeasure function required two Markovian states, whereas in the proposed T-CTMC model, the modeling of a countermeasure requires four states. We plan to address this issue in a future work. Also, we plan to study the possibility of combining our framework with the frameworks presented in [9,10], for automatic vulnerability discovery and attacks generation.

References

1. Kordy, B., Mauw, S., Radomirović, S., Schweitzer, P.: Foundations of attack–defense trees. In: Degano, P., Etalle, S., Guttman, J. (eds.) FAST 2010. LNCS, vol. 6561, pp. 80–95. Springer, Heidelberg (2011). https://doi.org/10.1007/978-3-642-19751-2_6
2. Jhawar, R., Lounis, K., Mauw, S.: A stochastic framework for quantitative analysis of attack-defense trees. In: Barthe, G., Markatos, E., Samarati, P. (eds.) STM 2016. LNCS, vol. 9871, pp. 138–153. Springer, Cham (2016). https://doi.org/10.1007/978-3-319-46598-2_10
3. Lounis, K.: Stochastic-based semantics of attack-defense trees for security assessment. In: The proceedings of the 9th International Workshop on Practical Applications of Stochastic Modeling, vol. 337, pp. 135–154. Elsevier (2018)
4. Jhawar, R., Lounis, K., Mauw, S., Ramírez-Cruz, Y.: Semi-automatically augmenting attack trees using an annotated attack tree library. In: Katsikas, S.K., Alcaraz, C. (eds.) STM 2018. LNCS, vol. 11091, pp. 85–101. Springer, Cham (2018). https://doi.org/10.1007/978-3-030-01141-3_6

5. Gadyatskaya, O., Jhawar, R., Kordy, P., Lounis, K., Mauw, S., Trujillo-Rasua, R.: Attack trees for practical security assessment: ranking of attack scenarios with ADTool 2.0. In: Agha, G., Van Houdt, B. (eds.) QEST 2016. LNCS, vol. 9826, pp. 159–162. Springer, Cham (2016). https://doi.org/10.1007/978-3-319-43425-4_10

6. Lounis, K., Zulkernine, M.: Connection dumping vulnerability affecting bluetooth availability. In: Zemmari, A., Mosbah, M., Cuppens-Boulahia, N., Cuppens, F. (eds.) CRiSIS 2018. LNCS, vol. 11391, pp. 188–204. Springer, Cham (2019). https://doi.org/10.1007/978-3-030-12143-3_16

7. Lounis, K., Zulkernine, M.: Bluetooth low energy makes just works not work. In: The 3rd Cyber Security and Networking Conference, pp. 99–106 (2019)

8. Lounis, K., Zulkernine, M.: Attacks and defenses in short-range wireless technologies for IoT. IEEE Access J. **8**, 88892–88932 (2020)

9. Ouchani, S.: Ensuring the functional correctness of IoT through formal modeling and verification. In: Abdelwahed, E.H., Bellatreche, L., Golfarelli, M., Méry, D., Ordonez, C. (eds.) MEDI 2018. LNCS, vol. 11163, pp. 401–417. Springer, Cham (2018). https://doi.org/10.1007/978-3-030-00856-7_27

10. Ouchani, S., Lenzini, G.: Generating attacks in SysML activity diagrams by detecting attack surfaces. J. Ambient Intell. Human. Comput. **6**(3), 361–373 (2015). https://doi.org/10.1007/s12652-015-0269-8

Automated Validation of State-Based Client-Centric Isolation with TLA+

Tim Soethout[1,2]([envelope]) [ID], Tijs van der Storm[2,3], and Jurgen J. Vinju[2,4] [ID]

[1] ING Bank, Amsterdam, The Netherlands
tim.soethout@ing.com
[2] Centrum Wiskunde & Informatica, Amsterdam, The Netherlands
tim.soethout@cwi.nl
[3] University of Groningen, Groningen, The Netherlands
[4] Eindhoven University of Technology, Eindhoven, The Netherlands

Abstract. Clear consistency guarantees on data are paramount for the design and implementation of distributed systems. When implementing distributed applications, developers require approaches to verify the data consistency guarantees of an implementation choice. Crooks *et al.* define a state-based and client-centric model of database isolation. This paper formalizes this state-based model in TLA+, reproduces their examples and shows how to model check runtime traces and algorithms with this formalization. The formalized model in TLA+ enables semi-automatic model checking for different implementation alternatives for transactional operations and allows checking of conformance to isolation levels. We reproduce examples of the original paper and confirm the isolation guarantees of the combination of the well-known 2-phase locking and 2-phase commit algorithms. Using model checking this formalization can also help finding bugs in incorrect specifications. This improves feasibility of automated checking of isolation guarantees in synthesized synchronization implementations and it provides an environment for experimenting with new designs.

Keywords: Distributed systems · Model checking · Isolation guarantees

1 Introduction

Automatically generating correct and performant implementations from high-level specifications is an important challenge in computer science and software engineering. Ideally one makes high-level specifications, which completely describe the functional and relevant parts of an application, without having to bother with low-level implementation details at the same time. Implementation is left to specialized tools and approaches that benefit from automated model checking and other debugging tools.

A benefit of high-level specifications is that they enable more specialized and fine-tuned implementations than general purpose implementation strategies,

© Springer Nature Switzerland AG 2021
L. Cleophas and M. Massink (Eds.): SEFM 2020 Workshops, LNCS 12524, pp. 43–57, 2021.
https://doi.org/10.1007/978-3-030-67220-1_4

which in essence have to take into account all possible variations of operations users can define. High-level domain knowledge offers the potential to automatically generate and optimize code, e.g. removing locks and blocking for improved performance when it can derive that this is never necessary the specific situation.

Such optimizations often involve managing concurrency and parallelism on accessing data. These optimizations of course need to be correct w.r.t. the specification: data consistency needs to be guaranteed. Application logic defines the functional consistency and transaction isolation manages the consistency of concurrent operations. Historically, isolation concerns have been outsourced to database systems, using general purpose transactions and similar constructs. These databases generally support ACID transactions, with a variety of isolation guarantees [2, 7], where Serializability is the strongest guarantee.

In order to optimize the performance of specialized implementations, some parts of the general purpose transaction mechanism incorporated either in the application itself or in the database implementation. When developing these specialized implementations of higher-level specifications, we need to be sure that they guarantee the ACID properties, or, if not, to what extent. The seminal definition of isolation levels is given by Adya [1]. Adya uses transaction histories, where transactions have dependencies on each other based on accessing the same data. If a cycle can be found in the graph of these dependencies, an isolation anomaly is present. Crooks *et al.* [6] model a state-based and client-centric approach to isolation and prove that it is equivalent to Adya's formalization.

Various tools are available which try to find or visualize isolation anomalies [14, 16, 18]. Many rely on specific scripted error scenarios to show anomalies. The ELLE tool [14] can be used to validate of traces of implementations using Adya's formalization, but still required careful setup and tuning of a test setup. It infers the histories Adya requires from client-centric observed transactions. Crooks' formalization is defined from a client-centric perspective and is directly defined in terms of observed transactions. The state-based and client-centric isolation definitions of Crooks *et al.* are referenced as Crooks' Isolation (CI) throughout this paper.

This paper describes an approach using formal methods to (semi-) automatically validate the isolation level of observed transactions using CI. First, we give an introduction to CI and a formalization of it in TLA$^+$. Next we discuss how this formalization is used to validate the consistency guarantees of a transaction algorithm using two-phase commit (2PC) with two-phase locking (2PL), and use it to find a specification bug.

The formalization of CI and the TLA$^+$ model checker enable rapid checking of multiple isolation levels of different synchronization algorithms. This technique can be used to both validate observed transactions from run-time systems and of formalizations of algorithms.

The main contributions of this paper are:

1. Formalization of the core of CI in TLA$^+$ and updated definitions to allow incremental model checking (Sect. 3).
2. Reproduction of the claims and properties [6] using model checking (Sect. 4).

$$\begin{Bmatrix} A \mapsto 100 \\ B \mapsto 100 \end{Bmatrix} \xrightarrow{T_1} \begin{Bmatrix} A \mapsto 150 \\ B \mapsto 50 \end{Bmatrix} \xrightarrow{T_2} \begin{Bmatrix} A \mapsto 165 \\ B \mapsto 55 \end{Bmatrix}$$

with S_0 above the first set, S_1 above the second, S_2 above the third.

Fig. 1. Example execution with initial state S_0 for transactions $T_1 = \langle r(A, 100),$ $r(B, 100), w(A, 150), w(B, 50)\rangle$ and $T_2 = \langle r(A, 150), r(B, 50), w(A, 165), w(B, 55)\rangle$.

3. Formalization of 2PL/2PC in TLA$^+$ and validation of Serializability using model checking of the CI TLA$^+$ formalization (Sect. 5).
4. An example of finding isolation bugs in the algorithm specification of 2PL/2PC (Sect. 5.3).

Section 6 discusses results, limitations and future work based on this approach. We conclude in Sect. 7. All source code can be found on Zenodo [24].

2 Background: State-Based Client-Centric Consistency

Crooks *et al.* [6] define a state-based and client centric consistency model (CI) for reasoning about isolation levels. It defines predicates to state if a set of observed transactions occurs under a given isolation level. The main concepts of CI are transactions and executions. A transaction is a sequence of operations, consisting of reads and writes which includes observed keys and values: $r(k, v)/w(k, v)$. An execution represents a possible ordering of a set of transactions with the resulting intermediate database states. A state is a mapping from all database keys to a specific value. Within an execution each following state only differs in the values written by the intermediate transaction on the previous or parent state.

Figure 1 shows an example execution of two bank accounts A and B, which both have a balance of €100 in the initial state S_0. Transaction T_1 is money transfer: €50 is deposited from account A and withdrawn from account B, realized using two reads and two writes. Transaction T_2 is paying of interest: 10% of the balance is added to both accounts; this transaction also involves two reads and two writes. Note that from a starting state and an ordering of transactions the other states can be derived by applying the intermediate transaction's writes.

For a set of observed transactions T to satisfy an isolation level I, a commit test CT for I should hold for a possible execution e of T: $\exists e : \forall t \in T : CT_I(t, e)$. The commit test for serializability, for example, is that all reads in a transaction must be able to have read their value from the direct parent state. In our example all the values of T_1's and T_2's read operations are the same as their parent state's values for each corresponding key, e.g. T_1's $r(A, 100)$ can read from T_1's parent S_0's $A \mapsto 100$.

Another isolation level is Snapshot Isolation, where the commit test requires that all reads of a single transaction can be read from the same earlier, not necessarily parent, state, which represents the database snapshot.

3   Formalizing CI in TLA⁺

TLA⁺ [20] is a formal specification language for action-based modeling of programs and systems. PLUSCAL [19] is an abstraction on top of TLA⁺ for concurrent and distributed algorithms and compiles to TLA⁺. In practice TLA⁺ is used to model distributed algorithms and systems [5,9,11,21,22]. TLA⁺ models states and transitions. A specification defines an initial state and atomic steps to a next state. Complex state machines and their transitions can be represented this way. Multiple concurrently-running state machine define their local steps and the global next step non-deterministically picks one machine to progress each step. This captures all possible interleavings of these multiple machines.

CI is formalized as properties that hold on a TLA⁺ state. This enables querying the system if an initial database state together with an a set of observed transactions satisfies an isolation level, e.g., `Serializability(initalState,setOfTransactions)`. When using TLA⁺ to formally specify an algorithm, this isolation property is added as an invariant during model checking. TLA⁺'s model checker TLC can then check the isolation guarantees at every state in the algorithm's execution and produce a counter example if the invariant is violated.

To formalize CI, we assume the following TLA⁺ definitions:

```
State        = [Keys → Values]
Operation    = [op: {"read", "write"}, key: Keys, value: Values]
Transaction  = Seq(Operation)
ExecutionElem = [parentState: State, transaction: Transaction]
Execution    = Seq(ExecutionElem)
```

The system `State` is modeled as a mapping from keys to values. `Keys` and `Values` are left abstract on purpose here, since they differ per concrete model. In TLA⁺ sets and set membership are often used. `[Keys → Values]` represents the set of possible tuples of `Keys` and `Values`, we bind this to `State` to easily reference this later in the specification. `Operations` are a read or write of a value on a key and a `Transaction` is a sequence of these operations. An `Execution` is represented as a sequence of transactions with their parent state.

As intuitively sketched earlier CI checks if values could have been read from earlier states. The following definition of $\mathcal{RS}$ ("read states") captures this for an execution e and an operation $o = r(k,v)$:

$$\mathcal{RS}_e(o) = \left\{ s \in S_e \middle| \overbrace{\left(s \xrightarrow{*} s_p\right)}^{a} \wedge \left(\overbrace{(k,v) \in s}^{b1} \vee \overbrace{\left(\exists w(k,v) \in \Sigma_T : w(k,v) \xrightarrow{to} r(k,v)\right)}^{b2} \right) \right\}$$

Read states are a subset of the states in the execution S_e, which are: (a) up to and including the parent state s_p in the execution; (b1) have the same key and value as the operation $o = r(k,v)$; or (b2) there exists a write operation $w(k,v)$ with the same key and value earlier in the same transaction's operations (Σ_T).

Listing 1. TLA$^+$ ReadStates

```
 1 ReadStates(execution, operation, transaction) =
 2   LET Se = SeqToSet(executionStates(execution))
 3       sp = parentState(execution, transaction)
 4   IN { s ∈ Se: \* s ∈ S_e
 5       ∧ beforeOrEqualInExecution(execution, s, sp) \* a: s →* s_p
 6       ∧ ∨ s[operation.key] = operation.value \* b1: (k,v) ∈ s
 7            \* b2: ∃w(k,v) ∈ Σ_T
 8         ∨ ∃ write ∈ SeqToSet(transaction):
 9            ∧ write.op = "write" ∧ write.key = operation.key
10            ∧ write.value = operation.value
11            \* b2: w(k,v) →^to r(k,v)
12            ∧ earlierInTransaction(transaction, write, operation)
13         ∨ operation.op = "write"
14   }
```

The TLA$^+$ version of this definition is shown in Listing 1. These read states are defined for each operation given an execution. TLA$^+$'s syntax allows grouping of conjunctions ($\land$) and disjunctions ($\lor$) by vertical indentation. The function `executionStates` denote the sequence of states in an `execution`. `parentState` extracts the parent state of a `transaction` given an `execution`. `LET .. IN` has the standard semantics. The rest of `ReadStates` (Lines 4 to 5) follows the CI definition quite literally, except that the third alternative (Line 13) is not captured in the CI definition for $\mathcal{RS}$ above, but represents the "convention [that] write operations have read states too" [6] to include all states up until the parent state for writes.

A state is complete when all reads of a transaction could have read their values from it. It is the intersection of the states in which each operation of the transaction could read from. The following definition is extended to take into account transactions without operations to support the iterative construction of transactions, starting with the empty ones:

$$\text{COMPLETE}_{e,T}(s) \equiv s \in \left(\bigcap_{o \in \Sigma_T} \mathcal{RS}_e(o) \cap \left\{ s' \in S_e | s' \xrightarrow{*} s_p \right\} \right)$$

We omit the TLA$^+$ version (`Complete`) for the sake of brevity, but it closely follows the mathematical definition, just like `ReadStates` did compared to $\mathcal{RS}$.

A *commit test* $CT_I(T, e)$ determines if a set of transactions T is valid under an isolation level I and execution e. For a set of transactions to satisfy an isolation level, there needs to exist at least one possible ordering, for which the commit test holds for all transactions. Transactions describe the values that a client observes including the actual values read and written. The values observed by the client are compatible with an ordering of the transactions that satisfies the isolation level. This is why it is sufficient for a single possible execution ordering to satisfy the commit test. The specific commit test for an isolation level I abstracts over which reads are valid for I.

Listing 2. TLA$^+$ helper definitions for CI.

```
WriteSet(transaction) == \* W_T = {k|w(k,v) ∈ Σ_T}
  LET writes == { operation ∈ SeqToSet(transaction) : operation.op = "write" }
  IN  { operation.key : operation ∈ writes }

NoConf(execution, transaction, state) == \* NO-CONF_T(s) ≡ Δ(s,s_p) ∩ W_T = ∅
  LET Sp    = parentState(execution, transaction)
      delta = { key ∈ DOMAIN Sp : Sp[key] ≠ state[key] }
  IN  delta ∩ WriteSet(transaction) = {}

Preread(execution, transaction) == \* PREREAD_e(T) ≡ ∀o ∈ Σ_T : RS_e(o) ≠ ∅
  ∀ operation ∈ SeqToSet(transaction): ReadStates(execution, operation, transaction) ≠ {}

strictBefore(t1, t2, timestamps) == timestamps[t1].commit < timestamps[t2].start \* T_1 <_s T_2
beforeOrEqualInExecution(execution, state1, state2) ==  \* s_1 →* s_2
  LET states = executionStates(execution)
  IN  Index(states, state1) <= Index(states, state2)
```

Different isolation-level commit tests are shown in Table 1, both mathematically and in TLA$^+$. Note that the CI definitions and their TLA$^+$ counterparts are very similar. The definitions of `NoConf`, `Preread`, `strictBefore` and `beforeOrEqualInExecution` can be found in Listing 2.

4 CI Examples

The static examples of the CI-paper are reproduced using TLA$^+$'s model checker TLC and the `ASSUME` operator. The model checker checks if the assumed property is valid. Figure 2 shows a minimal example of transactions `ta` to `te`, which are checked for four different isolation levels given initial state `s0`. TLC checks the assumptions and all evaluate to `TRUE`. The source code [24] reproduces more checks on this example.

Bank Transfer Example. The bank transfer example introduced by Crooks *et al.*, shows the difference between Snapshot Isolation and Serializability. Alice and Bob simultaneously take money out of their joint current and savings accounts, both from the other account. The bank requires the sum of the balances of both accounts to stay positive.

The following execution contains the transactions
$T_{alice} = \langle r(S,30), r(C,30), w(C,-10) \rangle$ and $T_{bob} = \langle r(S,30), r(C,-10),$ *abort*$\rangle$. A serializable implementation requires T_{bob} to abort. T_{alice} reads both balances of C and S and withdraws €40 from C. T_{bob} reads the result and aborts because not enough balance is available for his withdraw of €40 from S:

$$
\begin{matrix} & S_1 & & S_2 & & S_3 \\ \left\{ \begin{matrix} C \mapsto 30 \\ S \mapsto 30 \end{matrix} \right\} & \xrightarrow{T_{alice}} & \left\{ \begin{matrix} C \mapsto -10 \\ S \mapsto 30 \end{matrix} \right\} & \xrightarrow{T_{bob}} & \left\{ \begin{matrix} C \mapsto -10 \\ S \mapsto 30 \end{matrix} \right\} \end{matrix}
$$

The TLA$^+$ code to check this is shown on the right of Fig. 2.

Table 1. Commit tests and corresponding TLA$^+$ definitions.

Isolation Level	Commit Test	TLA$^+$ definition
Serializability	COMPLETE$_{e,T}(s_p)$	Complete(e, T, parentState(e, T))
Snapshot Isolation	$\exists s \in S_e.$	
	COMPLETE$_{e,T}(s_p) \wedge$	$\exists$ s $\in$ toSet(states(e)):
	NO-CONF$_T(s)$	Complete(e, T, s) $\wedge$ NoConf(e, T, s)
Read Committed	PREREAD$_e(T)$	Preread(e ,T)
Read Uncommitted	True	**TRUE**
Strict Serializability	COMPLETE$_{e,T}(s_p) \wedge$	
	$\forall T' \in \mathcal{T}:$	LET Sp = parentState(e, t)
	$T' <_s T \Rightarrow s_{T'} \xrightarrow{*} s_T$	IN Complete(e, T, Sp) $\wedge$
		$\forall$ otherT $\in$ transactions(e):
		strictBefore(otherT, T, timestamps) $\Rightarrow$
		beforeOrEqualInExecution(
		e, parentState(e, otherT), Sp)

```
\* Initial State, all 0
s0 == [k ∈ {x,y,z} ↦ 0]
\* Helper functions for operations
r(k,v) ==
    [op ↦ "read",  key ↦ k, value ↦ v]
w(k,v) ==
    [op ↦ "write", key ↦ k, value ↦ v]

ta == << w(x,1) >>
tb == << r(y,1), r(z,0) >>
tc == << w(y,1) >>
td == << w(y,2), w(z,1) >>
te == << r(x,0), r(z,1) >>

trs == {ta, tb, tc, td, te}
ASSUME Serializability(s0, trs)
ASSUME SnapshotIsolation(s0, trs)
ASSUME ReadCommitted(s0, trs)
ASSUME ReadUncommitted(s0, trs)
```

```
\* Initial state of Current and
    ↪ Savings accounts.
bInit == (C :> 30) @@ (S :> 30)

talice ==
    << r(S,30), r(C, 30), w(C,-10) >>
tbob   ==
    << r(S,30), r(C,-10)
    (* w(S,-10) does not happen *) >>
bTrx == {talice, tbob}

ASSUME Serializability(bInit, bTrx)
ASSUME SnapshotIsolation(bInit, bTrx)
ASSUME ReadCommitted(bInit, bTrx)
ASSUME ReadUncommitted(bInit, bTrx)
```

Fig. 2. Running example (left) and serializable bank account example (right) from Crooks *et al.* [6] in TLA$^+$.

The same example is considered under Snapshot Isolation with transactions $T_{alice} = \langle r(S, 30), r(C, 30), w(C, -10)\rangle$ and $T_{bob} = \langle r(S, 30), r(C, 30), w(S, -10)\rangle$. Both T_{alice} and T_{bob} read from S_1 and find that there is enough total balance available. They both withdraw €40 from respectively C and S:

$$\left\{ \begin{array}{l} C \mapsto 30 \\ S \mapsto 30 \end{array} \right\} \xrightarrow{T_{alice}} \left\{ \begin{array}{l} C \mapsto -10 \\ S \mapsto 30 \end{array} \right\} \xrightarrow{T_{bob}} \left\{ \begin{array}{l} C \mapsto -10 \\ S \mapsto -10 \end{array} \right\}$$

S_1 S_2 S_3

Snapshot Isolation allows this because both T_{alice} and T_{bob} read from a valid snapshot or complete state and there is no conflict in their writes, because they write to different accounts. However, this violates the overall invariant that the sum of the balances should remain positive. This is the write skew isolation anomaly [1]. This can be checked by using a specification similar to the right-hand side of Fig. 2, with modified transactions, and assuming Serializability is **FALSE** .

5   Model Checking Algorithms Using CI

In contrast to the previous, static examples, where TLA$^+$'s state steps are not used, we now look at a TLA$^+$ specification of a transactional protocol (2PL/2PC) using states. At each step of the algorithm TLC checks if the isolation guarantees hold.

5.1   Formalizing 2PL/2PC

Two-Phase Commit (2PC) combined with Two-Phase Locking (2PL) forms a protocol used to implement ACID transactions. 2PC takes care of atomicity of a transaction and 2PL provides Serializable isolation. We extend the formalization of 2PC by Gray and Lamport [9] to support multiple parallel transactions via 2PL.

We model 2PL/2PC in the PLUSCAL algorithm language, which is compiled down to regular TLA$^+$, but provides a higher-level notation, closer to imperative programming languages. PLUSCAL describes multiple possibly different processes with atomic steps. During model checking, one of the processes takes a single step, which allows processes to be interleaved. The model checker makes sure all possible interleavings are explored.

The PLUSCAL encoding of 2PL/2PC consists of two types of processes: transaction managers and transaction resources. The actual number of processes is defined by model constants `transactions` and `resources`. Message passing is modeled by a monotonically growing set of messages. This means that messages are never lost, but a recipient process might handle them out of order or not at all.

Listing 3 shows the definition of the transaction manager. There is a `tm` process for each of the `transactions`. PLUSCAL processes do atomic steps, each represented by a label such as `INIT:`. A label can intuitively be viewed as a state in the process' state machine. All statements within a step are done as a single step.

Listing 3. PLUSCAL specification of 2PL/2PC manager

```
fair process tm ∈ transactions
begin
  INIT: sendMessage([id ↦ self, type ↦ "VoteRequest"]);
  WAIT: either \* receive commit votes
    await ∀ rm ∈ resources: [id ↦ self, type ↦ "VoteCommit", rm ↦ rm] ∈ msgs;
    sendMessage( [id ↦ self, type ↦ "GlobalCommit"]);
    goto COMMIT;
  or \* receive at least 1 abort votes
    await ∃ rm ∈ resources: [id ↦ self, type ↦ "VoteAbort", rm ↦ rm] ∈ msgs;
    sendMessage([id ↦ self, type ↦ "GlobalAbort"]);
    goto ABORT;
  or \* or timeout, solves deadlock when transactions lock each others resources
    sendMessage([id ↦ self, type ↦ "GlobalAbort"]);
    goto ABORT;
  end either;
  ABORT: goto Done; COMMIT: goto Done;
end process
```

A transaction manager first sends out the VOTEREQUEST message by adding a tuple with the transaction's identifier `self` and the message label `"VoteRequest"` to the `msgs` set. Then its next step is WAIT in which three alternatives (`either ... or`) can occur: 1) either it receives messages of type `"VoteCommit"` of each resource occurring in the set of messages, and sends GLOBALCOMMIT; 2) or one message of type `"VoteAbort"` and sends GLOBALABORT; 3) or it times out and aborts (to prevent deadlock). The `await` construct ensures that a step only happens if its precondition is fulfilled. TLC makes sure that all alternatives are explored. `goto`'s are added to explicitly label the steps for readability in the model checker's execution. `Done` is a special PLUSCAL label, which represents the process being completed.

The PLUSCAL specification of a transaction resource, shown in Listing 4, is slightly more involved. The resource process has local variables (Lines 1 to 6) to keep track of stopping, votes, commits, aborts and resource state. The state is used for CI as a symbolic state, represented by an integer.

When the resource is started (Lines 7 to 18), it `either` does nothing (`skip`) and decrements `maxTxs`, `or` receives a `"VoteRequest"` message. `with tId ∈ transactions\ voted` denotes choosing a transaction ID `tId` from the set of `transactions` minus the transactions already `voted` for. The resource can then either VOTECOMMIT or VOTEABORT. The `voted` local variable keeps track of the transactions it has already voted for and is updated to make sure to only vote once per transaction.

Next, it becomes READY (Lines 19 to 30) and waits on either GLOBALCOMMIT or GLOBALABORT, but only for transactions which it voted for, and has not committed yet. It keeps track of the `committed` and `aborted` transactions in order to not send duplicate messages and to later check the atomicity of the transactions. Each `while` iteration, decrements `maxTxs` to ensure termination.

Listing 4. 2PL/2PC resource in TLA$^+$

```
1 fair process tr ∈ resources
2 variables maxTxs = 5,      \* Limit number of transactions to limit search space
3             voted = {},      \* Transactions which this resource voted for
4             committed = {}, \* Committed transactions
5             aborted = {},    \* Aborted transactions
6             state = 0;       \* Counter to represent state changes for CC
7 begin TR_INIT:
8 while maxTxs >= 0 do
9    either skip; \* skip to not deadlock
10   or \* Wait on VoteRequest
11     with tId ∈ transactions \ voted do
12       await [id ↦ tId, type ↦ "VoteRequest"] ∈ msgs;
13       either \* If preconditions hold, VoteCommit, else VoteAbort
14         sendMessage([id ↦ tId, type ↦ "VoteCommit", rm ↦ self]);
15         voted := voted ∪ {tId};
16       or sendMessage([id ↦ tId, type ↦ "VoteAbort", rm ↦ self]);
17         voted := voted ∪ {tId}; aborted := aborted ∪ {tId}; goto STEP;
18       end either; end with;
19   READY: \* Wait on Commit/Abort
20     either with tId ∈ voted \ committed do  \* receive GlobalCommit
21         await [id ↦ tId, type ↦ "GlobalCommit"] ∈ msgs;
22         committed := committed ∪ {tId};
23         operations[tId] := operations[tId] o << r(self, state), w(self, state+1) >>;
24         state := state + 1;
25       end with;
26     or with tId ∈ voted \ aborted do \* receive GlobalAbort
27         await [id ↦ tId, type ↦ "GlobalAbort"] ∈ msgs;
28         aborted := aborted ∪ {tId};
29       end with; end either; end either;
30     STEP: maxTxs := maxTxs - 1;
31 end while; end process;
```

In order to model check CI it captures the read and written values in `operations` (Line 23) and updates its local `state`. Both reads and writes are added on commit and not on vote, because if reads are added on vote, it could be the case that the resource reads a later committed value when responding to the VOTEREQUEST later which will always be aborted anyway. This results in a violation of Serializability for the CI check, while it is technically never an observed value.

5.2 Model Checking 2PL/2PC

As sanity check for the formalization of 2PL/2PC, first atomicity and termination are checked:

```
Atomicity ≡
    ∀ id ∈ transactions: pc[id]=Done ⇒
        ∀ a1, a2 ∈ resources: ¬ id ∈ aborted[a1] ∧ id ∈ committed[a2]

AllTransactionsFinish ≡ ◇(∀ t ∈ transactions: pc[t] = Done)
```

For atomicity, when all transactions are completed (process counter `pc` is `Done`), for all pairs of resources it should not ($\neg$) be the case that a transaction is aborted by one resource, but committed by the another. So all should either committed or aborted the transaction. Property `AllTransactionsFinish` makes sure that eventually (<>) all transactions complete.

To model check the isolation guarantees an instance of the CI formalization is added, which gives access to the previously defined isolation level tests (see Sect. 4), given the initial state and the observed transactions.

```
ccTransactions = Range(operations)   \* Operations without transaction ids
InitialState = [k ∈ resources ↦ 0] \* Initial state is 0 for all resources


CC = INSTANCE ClientCentric WITH Keys ← resources, Values ← 0..10
Serializable      = CC!Serializability(InitialState, ccTransactions)
SnapshotIsolation = CC!SnapshotIsolation(InitialState, ccTransactions)
ReadCommitted     = CC!ReadCommitted(InitialState, ccTransactions)
ReadUncommitted   = CC!ReadUncommitted(InitialState, ccTransactions)
```

In this case all cases are valid when we run the TLC model checker for `transactions == {t1, t2}` and `resources == {r1, r2, r3}`.

The model checker then checks the isolation guarantees for each step of the algorithm. When the isolation test fails, it presents a counter example. Table 2 gives an intuition on the relative time durations of the TLC model checker on different numbers of transactions and resources. The model checker checks the four CI isolation levels (Serializability, Snapshot Isolation, Read Committed, Read Uncommitted) on each of the model's steps. It never invalidates the checks, so it traverses the entire state space.

Table 2. Run time durations of TLC on CI checks for different number of transactions and resources n of 2PL/2PC. Results on MacBook Pro (13-inch, 2016) with 3,3 GHz Intel Core i7 with 4 worker threads and allocated 8 GB RAM on AdoptOpenJDK 14.0.1+7, on TLC 2.15 without profiling and using symmetry sets for constants.

#tx	$n = 1$	$n = 2$	$n = 3$
1	7 s	9 s	19 s
2	8 s	21 s	5 m 55 s
3	11 s	1 m 53 s	3 h 21 m 54 s

5.3   2PL/2PC Bug Seeding

To additionally stress the formalization presented above, we have introduced a subtle, but realistic bug in the definition of transaction resource. When the resource is in the ready state and waiting on a GLOBALCOMMIT or GLOBAL-ABORT message from the transaction manager, the resource should only wait for these messages when it is the actual transaction it voted for. This is guaranteed by with tId ∈ voted \ committed in Listing 4 Line 20. The bug is to replace this with with tId ∈ transactions \ committed. This means tId can faultily represent a never-seen before transaction as well.

When this model is checked with two transactions and resources, all of the invariants hold and no problem is found. However, with three transactions and two resources the Serializability invariant is violated and a counter example with 20 steps is found within half a minute; this trace shown in Fig. 3. The example shows that due to this bug it is possible for a resource to side-step an in progress transaction, by responding to the GLOBALCOMMIT of a different transaction.

First t1 and t2 request to vote and r1 votes to commit for t1, then t2 aborts due to timeout with GLOBALABORT(T2). r1 then uses this abort to abort its waiting on t1. This is possible because with tId ∈ transactions \ committed allows r1. It receives the GLOBALABORT(T2), aborts and steps to receive the next transaction. The model checker requires some more steps to find non-serializable behavior, when the other transactions t1 and t3 commit and their effects are applied in different order on r1 and r2, hence the system is not serializable.

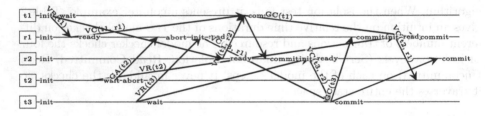

Fig. 3. Non-serializable trace found for bugged 2PL/2PC specification. Horizontal lines represent processes over time with state changes. Arrows represent messages sent and received. Message labels are abbreviations of 2PC messages: VOTEREQUEST, VOTECOMMIT, GLOBALABORT and GLOBALCOMMIT.

These kinds of bugs during specification can occur naturally, for example when specializing algorithms for specific applications with the goal of added efficiency [25]. Using CI in model checking helps us find bugs while designing new algorithms and also for validating claims of existing algorithms.

6   Discussion and Future Work

The formalization of CI in TLA⁺ is relatively straightforward. The definitions for the base abstractions, such as State and Execution, influence the whole formalization. Staying as close as possible to the mathematical model however, results in quite verbose output, since there are no labels on transactions. The definition on read states was improved to support incremental model checking, starting with empty transactions.

The main limitation of using model checking to find isolation violations is the state explosion when the numbers of processes grows. As seen in Table 2, running times grow rapidly and model checking becomes infeasible when more transactions are added. Since the model checker evaluates the isolation guarantees in every algorithm state we assume, however, that most isolation violations can be found in small examples. The small scope hypothesis [13] supports this saying that most bugs have small counterexamples. Nevertheless, we can not be entirely sure that anomalies that only occur in larger interactions and longer traces are found by the current approach, but it gives us confidence in the the the checked isolation level, while keeping it feasible.

There is a lot of research focusing on proving distributed consistency properties. Model checking tools, such as Uppaal [3], Spin [12], LTSMin [4], mCRL2 [10] and TLA$^+$ [20] are used to verify distributed systems and algorithms as well as real-world implementations and protocols [8,11,21,22].

There are also many approaches [2,17,23,28] that try to balance the trade-off between performance and data-consistency by choosing different isolation guarantees. Our work adds to this knowledge by providing a reusable framework to investigate and model check distributed consistency protocols.

To further evaluate the usefulness of our approach for real-life systems, it would be insightful to reproduce known isolation bugs in older versions of database implementations, such as found by Jepsen [14,15] and Bailis *et al.* [2]. In order to do this we could either create one or more clients that capture the observed transactions, or instrument the database to store this information for offline model checking.

Furthermore the scripts of the isolation anomalies of Hermitage [18] can be reproduced as TLC model checks to strengthen (our formalization of) CI. The TLA$^+$ Toolbox also features a theorem prover. The CI formalizations could be extended by proving certain properties, such as reproducing the proofs on equivalence with Adya's formalization and proving conformity to isolation levels for specific algorithms.

Generating performant and correct implementations from high-level specifications is an attractive goal in software engineering, as it would bring the benefits of (semi-)automatic verification to correct-by-construction implementation.

For instance, the Rebel domain-specific language has been used to specify realistic systems (for instance, in the financial domain), from which highly scalable implementations are generated using novel consistency algorithms [25–27]. It is however, a far from trivial endeavour to state and prove isolation guarantees of some of these algorithms. CI can be extended to support operations on a semantically higher level than reads and writes, such as the semantically richer operations used in Rebel. A TLA$^+$ formalization can then be used to allows for rapid prototyping of synchronization implementation alternatives for Rebel, while leveraging the higher-level semantics [29]. The checking of isolation guarantees can then be automated.

7   Conclusion

This paper formalizes Crooks' state-based client-centric isolation model (CI) in TLA$^+$ in order to check conformance to isolation levels using model checking. The running examples of Crooks *et al.* [6] are reproduced and validated in TLA$^+$. An example of a transaction implementation using two phase locking (2PL) and two phase commit (2PC) is formalized in TLA$^+$. The TLC model checker is used to automatically show conformance to the CI formalization. The CI formalization is also used to find a bug in the algorithm's formalization.

Formalizing CI in TLA$^+$ enables automatic validation of isolation guarantees of synchronization implementations by mapping their algorithms to read and write operations. It can be used both for checking isolation conformance of run-time traces of (distributed) systems and of formal specification of algorithms.

References

1. Adya, A.: Weak consistency: a generalized theory and optimistic implementations for distributed transactions. Ph.D. thesis, Massachusetts Institute of Technology, Department of Electrical Engineering and Computer Science (1999)
2. Bailis, P., Davidson, A., Fekete, A., Ghodsi, A., Hellerstein, J.M., Stoica, I.: Highly available transactions - virtues and limitations. Proc. VLDB Endow. **7**(3), 181–192 (2013). https://doi.org/10.14778/2732232.2732237
3. Bengtsson, J., Larsen, K., Larsson, F., Pettersson, P., Yi, W.: UPPAAL—a tool suite for automatic verification of real-time systems. In: Alur, R., Henzinger, T.A., Sontag, E.D. (eds.) HS 1995. LNCS, vol. 1066, pp. 232–243. Springer, Heidelberg (1996). https://doi.org/10.1007/BFb0020949
4. Blom, S., van de Pol, J., Weber, M.: LTSMIN: distributed and symbolic reachability. In: Touili, T., Cook, B., Jackson, P. (eds.) CAV 2010. LNCS, vol. 6174, pp. 354–359. Springer, Heidelberg (2010). https://doi.org/10.1007/978-3-642-14295-6_31
5. Brooker, M., Chen, T., Ping, F.: Millions of tiny databases. In: Bhagwan, R., Porter, G. (eds.) 17th USENIX Symposium on Networked Systems Design and Implementation, NSDI 2020, Santa Clara, CA, USA, 25–27 February 2020, pp. 463–478. USENIX Association (2020)
6. Crooks, N., Pu, Y., Alvisi, L., Clement, A.: Seeing is believing. In: Proceedings of the ACM Symposium on Principles of Distributed Computing, pp. 73–82. ACM, July 2017. https://doi.org/10.1145/3087801.3087802
7. Fekete, A., Liarokapis, D., O'Neil, E.J., O'Neil, P.E., Shasha, D.E.: Making snapshot isolation serializable. ACM Trans. Database Syst. **30**(2), 492–528 (2005). https://doi.org/10.1145/1071610.1071615
8. Gomes, V.B., Kleppmann, M., Mulligan, D.P., Beresford, A.R.: Verifying strong eventual consistency in distributed systems. Proc. ACM Program. Lang. **1**(OOPSLA), 1–28 (2017). https://doi.org/10.1145/3133933
9. Gray, J., Lamport, L.: Consensus on transaction commit. ACM Trans. Database Syst. **31**(1), 133–160 (2006). https://doi.org/10.1145/1132863.1132867
10. Groote, J.F., Mousavi, M.R.: Modeling and Analysis of Communicating Systems. MIT Press, Cambridge (2014)
11. Gustafson, J., Wang, G.: Hardening Kafka replication (2020). https://github.com/hachikuji/kafka-specification

12. Holzmann, G.J.: The SPIN Model Checker - Primer and Reference Manual. Addison-Wesley, Boston (2004)
13. Jackson, D.: Software Abstractions - Logic, Language, and Analysis. MIT Press, Cambridge (2006)
14. Kingsbury, K., Alvaro, P.: Elle: inferring isolation anomalies from experimental observations. CoRR abs/2003.10554 (2020)
15. Kinsbury, K.: Jepsen: distributed systems safety research (2020). http://jepsen.io/
16. Kinsbury, K.: Knossos (2020). https://github.com/jepsen-io/knossos
17. Kleppmann, M.: Designing Data-Intensive Applications: The Big Ideas behindReliable, Scalable, and Maintainable Systems. O'Reilly, Sebastopol (2016)
18. Kleppmann, M.: Hermitage: testing transaction isolation levels (2020). https://github.com/ept/hermitage
19. Lamport, L.: The PlusCal Algorithm Language - Microsoft Research. https://www.microsoft.com/en-us/research/publication/pluscal-algorithm-language/
20. Lamport, L.: Specifying Systems, the TLA+ Language and Tools for Hardwareand Software Engineers. Addison-Wesley, Boston (2002)
21. Microsoft: High-level TLA+ specifications for the five consistency levels offered by Azure Cosmos DB (2020). https://github.com/Azure/azure-cosmos-tla
22. Newcombe, C., Rath, T., Zhang, F., Munteanu, B., Brooker, M., Deardeuff, M.: How amazon web services uses formal methods. Commun. ACM **58**(4), 66–73 (2015). https://doi.org/10.1145/2699417
23. Preguiça, N.M., Baquero, C., Shapiro, M.: Conflict-free replicated data types CRDTs. In: Sakr, S., Zomaya, A.Y. (eds.) Encyclopedia of Big Data Technologies. Springer, Cham (2019). https://doi.org/10.1007/978-3-319-77525-8_185
24. Soethout, T.: TimSoethout/tla-ci: TLA+ specifications used in "Automated Validation of State-Based Client- Centric Isolation with TLA+". Zenodo (2020). https://doi.org/10.5281/zenodo.3961617
25. Soethout, T., van der Storm, T., Vinju, J.: Path-sensitive atomic commit. Programming **5**(1) (2020). https://doi.org/10.22152/programming-journal.org/2021/5/3
26. Soethout, T., van der Storm, T., Vinju, J.J.: Static local coordination avoidance for distributed objects. In: Proceedings of the 9th ACM SIGPLAN International Workshop on Programming Based on Actors, Agents, and Decentralized Control - AGERE 2019, pp. 21–30. ACM Press, Athens (2019). https://doi.org/10.1145/3358499.3361222
27. Stoel, J., van der Storm, T., Vinju, J., Bosman, J.: Solving the bank with Rebel: on the design of the Rebel specification language and its application inside a bank. In: Proceedings of the 1st Industry Track on Software Language Engineering - ITSLE 2016, pp. 13–20. ACM Press (2016). https://doi.org/10.1145/2998407.2998413
28. Tanenbaum, A.S., van Steen, M.: Distributed Systems - Principles and Paradigms, 2nd edn. Pearson Education, Upper Saddle River (2007)
29. Weikum, G.: Principles and realization strategies of multilevel transaction management. ACM Trans. Database Syst. **16**(1), 132–180 (1991). https://doi.org/10.1145/103140.103145

Code Coverage Aware Test Generation Using Constraint Solver

Krystof Sykora[1], Bestoun S. Ahmed[2(✉)], and Miroslav Bures[1]

[1] Department of Computer Science, Faculty of Electrical Engineering,
Czech Technical University, Prague, Czech Republic
{sykorkry,buresm3}@fel.cvut.cz
[2] Department of Mathematics and Computer Science, Karlstad University,
Karlstad, Sweden
bestoun@kau.se

Abstract. Code coverage has been used in the software testing context mostly as a metric to assess a generated test suite's quality. Recently, code coverage analysis is used as a white-box testing technique for test optimization. Most of the research activities focus on using code coverage for test prioritization and selection within automated testing strategies. Less effort has been paid in the literature to use code coverage for test generation. This paper introduces a new Code Coverage-based Test Case Generation (CCTG) concept that changes the current practices by utilizing the code coverage analysis in the test generation process. CCTG uses the code coverage data to calculate the input parameters' impact for a constraint solver to automate the generation of effective test suites. We applied this approach to a few real-world case studies. The results showed that the new test generation approach could generate effective test cases and detect new faults.

Keywords: Software testing · Code coverage · Automated test generation · Test case augmentation · Constrained interaction testing

1 Introduction

In software engineering, regression testing has become a common practice to be used during test development. In this practice, testers begin by rerunning existing test suites to validate new software-under-test (SUT) functionality. However, this approach faces many problems as existing tests decrease their ability to detect SUT faults. This paper shows how Test suite augmentation techniques can be used to solve this problem.

To improve the test generation, tools like code coverage analysis and constraint solving are commonly utilized. We use these techniques in the Code Coverage-based Test Case Generation (CCTG) method. This approach is akin to Test suite augmentation techniques [10], which is commonly used in regression testing. Augmentation is employed to adjust existing test cases by analyzing

© Springer Nature Switzerland AG 2021
L. Cleophas and M. Massink (Eds.): SEFM 2020 Workshops, LNCS 12524, pp. 58–66, 2021.
https://doi.org/10.1007/978-3-030-67220-1_5

changes in the SUT. Practitioners of Test suite augmentation techniques believe the use of preexisting test cases can improve that test suites [9]. To promote these test cases, the SUT is analyzed using code coverage and other criteria to prioritize how changes to the test suites should be conducted. In our method, we focus on the second part of the augmentation approach – code coverage. By analyzing the code coverage of preexisting generated test cases, we can determine test parameters' impact. This information is then used for the test generation.

CCTG can be used for the test suite augmentation method by utilizing the code coverage data and test models to refine the test generation process. To generate the test cases, CCTG first creates the test data sets used for the input interaction. A test model consists of the SUT parameters classes (i.e., possible parameter values). The results are sets of specific SUT parameter values, which will be referred to as test cases. The parameter classes are then used to generate random test cases to be executed for code coverage monitoring. This process generates coverage data related to each generated test case. If a specific test parameter change affects the coverage consistently or significantly, the parameter's weight also increases. The weight indicates the extent to which each parameter is used and permuted when generating a new set of test cases. The process as a whole contributes towards the test suite augmentation. Here, the code coverage data used for the augmentation may also minimize the number of test cases and generate more effective test cases in terms of fault-finding.

As in most modern software systems, the test suite augmentation also suffers from input interaction constraint problems [1]. Here, we deployed a constraint solver within our approach during the test generation process to resolve the input values' constraints. The solver makes sure no test cases are generated with meaningless interactions with the SUT. The resulting test cases should resemble the general workflow of test cases [4] as the code coverage analysis motivates variety in decision paths, and the constraint solver ensures that the combinations remain reasonable.

2   Background

Classically, code coverage has been used as an analytical approach with the test suite execution. The approach is also used with the test suite generation strategies to maximizing the code-base covered by generating more test cases [8]. We have recently used more advanced techniques function as gray-box methods for test case analysis and test generation [2]. We considered the code's internal structure while augmenting a generated test suite for Combinatorial Interaction Testing (CIT). This approach improved the test generation process by studying the program's code to determine individual parameters' impacts. We used this impact factor to select the input parameters and values for the test cases. In this regard, the CCTG strategy proposed in this paper is very similar. However, examining the internal code structure can prove costly, in terms of a manual analysis of the input parameters and values. So, the CCTG simplifies this process by automatically evaluating the parameter impact, requiring only a test model of parameters and their possible values. It can also explore various hidden criteria for test generation (such as SUT configuration and state).

A common problem for test generation is that the test models don't consider specific constraints that would make the test cases useless or nonsensical. To avoid this, test generation methods [7,11] rely on constraint solvers to eliminate such undesirable combinations. Such constraint or rather Boolean satisfiability problem-solvers [3] (SAT solvers) are also employed in the CCTG method. The solver's additional benefit is that the CCTG user may incorporate the SAT constraints into the test model to prevent the generation of unwanted test cases and correct the focus of the test cases. If new functionality is added to the SUT, it can be specified that some parameters (representing the choice to use the new functionality) must be used. This allows the tester to influence the test generation by only adjusting the test model without changing the code coverage analysis results.

While the CCTG finds relations to the previously mentioned techniques, it is primarily a test suite augmentation method [10]. Our approach is based on an innovative idea of augmenting test cases using the coverage criteria. The CCTG method, however, uses randomly generated test cases to establish the coverage data. This data is then used to generate what is essentially the first set of actual test cases.

3   The Proposed Method

In this section, we examine each step of the CCTG method. The following subsections illustrate these steps in detail.

3.1   Determining Parameter Weight

The initial step in the CCTG methods is the code coverage analysis. This process is used to calculate the impact (weight) of SUT parameters that will be used for test generation.

SUT Test Model. For each SUT, a model (shown in Fig. 1(a)) for interaction consisting of the input parameters and constraints among them. The parameters are represented as the P[n] array, where each index P[1], P[2],... P[n] represents one of the test model parameters. Each parameter has a value. The value V is the possibility of a parameter. This can be represented in two ways, either the V is an array of the possible values for parameters (booleans or enums) from the SUTs perspective, or the value can be represented as a number with a range. In the case of range representation, the parameter depth [d] is added to represent the number of values chosen from the range for further test generation. The other parameter property in the model is its weight, which is graded as a float ranging from 0 to 1.

The second part of the model is a set of SAT constraints to ensure that the conflicting decision is not selected. This model is used to generate the initial test

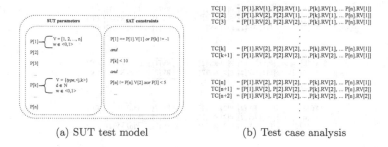

(a) SUT test model   (b) Test case analysis

Fig. 1. SUT modeling and parameter analysis

cases for code coverage analysis and generate the resulting test set for the actual testing. For regression testing, the first set of random test cases can be replaced with a set of regression tests from earlier stages of the SUT development.

Test Case Analysis and Generation. In contrast to other strategies, maximizing of code coverage is not a direct goal for CCTG. The strategy relies on using the coverage data in a more specific manner more targeted on condition or coverage principles. We follow the standard definition of code coverage for lines of code per individual test case. The gcov[1] tool and stored for later analysis for each test case. The test suite used for coverage determination is designed in a very specific way. As the initial step of generation, we select a test depth level. This level represents how many values will be selected for each parameter. The same number of values as overall test depth is selected for each parameter P and randomly ordered in an array RV. The first test case TC uses each parameter's first value in all randomly ordered lists. For every test case after that, one parameter changes the value to the next in the randomised list (P[1].RV[1] to P[1].RV[2]) as shown in Fig. 1(b). We proceed this way until all the arrays are exhausted. The rationale behind this approach is its effects on code coverage change that determines the parameter impact. We always select two test cases where all but one parameter have the same values to measure this. This will be explained in the Parameter weight calculation section and the Fig. 2.

Parameter Weight Calculation. The initial step in this phase is to determine each parameter's effect on code coverage (i.e., the parameter weight). Using the initial test cases, the code coverage is gathered automatically. To determine the information about a specific parameter, we must take a look at a set of test cases, where the examined parameter changes, while the remaining parameters stay the same, as shown in Fig. 2.

Initially, a parameter is selected by the algorithm for analysis. Then, similar pair test cases are selected, except for the value of the parameter under investigation. The difference in the code-coverage results of the test cases is recorded.

---

[1] https determines the code coverage: //linux.di.e.net/man/1/gcov.

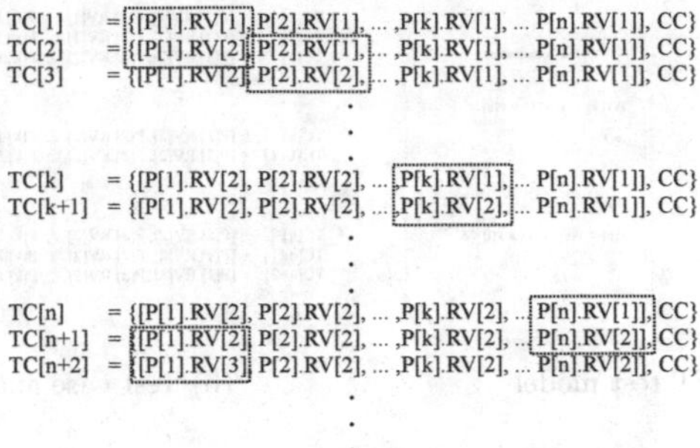

TC[1]   = {[P[1].RV[1], P[2].RV[1], ... ,P[k].RV[1], ... P[n].RV[1]], CC}
TC[2]   = {[P[1].RV[2], P[2].RV[1], ... ,P[k].RV[1], ... P[n].RV[1]], CC}
TC[3]   = {[P[1].RV[2], P[2].RV[2], ... ,P[k].RV[1], ... P[n].RV[1]], CC}

$\vdots$

TC[k]   = {[P[1].RV[2], P[2].RV[2], ... ,P[k].RV[1], ... P[n].RV[1]], CC}
TC[k+1] = {[P[1].RV[2], P[2].RV[2], ... ,P[k].RV[2], ... P[n].RV[1]], CC}

$\vdots$

TC[n]   = {[P[1].RV[2], P[2].RV[2], ... ,P[k].RV[2], ... P[n].RV[1]], CC}
TC[n+1] = {[P[1].RV[2], P[2].RV[2], ... ,P[k].RV[2], ... P[n].RV[2]], CC}
TC[n+2] = {[P[1].RV[3], P[2].RV[2], ... ,P[k].RV[2], ... P[n].RV[2]], CC}

$\vdots$

Fig. 2. Test case selection

These steps are repeated for all remaining pairs of test cases matching the criteria of all parameters being the same, except for the changing parameter under analysis. From the derived differences in coverage, an average is calculated. This average represents an absolute value of the parameters weight. These steps are illustrated in detail in the Algorithm 1.

Algorithm 1: Steps in the code coverage analysis to determine parameter's impact

1   **CodeCoverageAnalysis** $(T_C L, P)$
     Input   : Test case list $T_C L$, Parameter-under-test P
     Output: Measure of the impact of parameter P on code coverage $avrg(DL)$
2   TL = TestCaseList;
3   DL = list of float values;
4   **while** $T_C L$ *not empty* **do**
5   TL add($T_C L$ pop first item);
6   **foreach** T_C *in* $T_C L$ **do**
7   **if** *each i/PI : PV[i] in TC == PV[i] in TL[0]* **then**
8   TL add(pop TC from $T_C L$);
9   DL add(max(TL) − min(TL));
10   TL clear;
11   **return** $avrg(DL)$;

Maximum code coverage for the generated tests is not the aim here. Instead, by determining the impact on individual parameters' coverage, parameters are selected for permutation in test cases if they influence code coverage more.

3.2 Test Case Generation

Test cases are generated using the data of parameter weight. All unary parameters have only two possible permutations (0,1) as they can only be or not be included upon program execution. The number of binary parameter permutation is equal to 1 + the number of pre-selected possible values. The unary parameter is a set with two elements (one being the parameters exclusion) and the binary a set with 1 + (number of values) elements. All test cases created are solved for constraints.

The code-coverage based method is relying on the selected parameters to permute based on their weight. All weights are places on an axis, spanning the range from 0 to 100. Thanks to normalizing the values, their sum is 100 exactly and, as such, fills up the entire axis. For example, take 3 parameters with respective weights 15, 60, and 25. The parameters would assume the following ranges on the axis. P1 (0,15), P2 (15,75) and P3(75,100).

The first constructed test case has all parameters with their default values. A random real number is generated in the range of (0,100) to construct the next test case. A parameter whose range corresponds (on the axis) to the number generated is selected for the permutation. The selected parameter's value is permuted to the next value in the sum that represents it. The test case is then saved if it is not identical to a previously created one. The test case generation algorithm is included in Algorithm 2.

Algorithm 2: Code-coverage-based test-case generation

```
 1  CVBasedTCGeneration (PL, n)
       Input  : ParameterList PL, Number of test cases to be generated n
       Output: TestCaseList T_C L
 2       T_C L = list of TestCases ;
 3       CT_C = TestCase where each P_V in P_V L is 0;
 4       while n > 0 do
 5           r = random float value;
 6           t = 0;
 7           foreach P in PL do
 8               t = t + PW in P;
 9               if r <= t then
10                   permute(P);
11           add(PL);
12           n−;
13       return T_C L;
```

We have used the Z3 solver with this test generation to resolve the constraints. We have also developed an interface for the tool so that the relevant constraints for test cases exclusion can be imported from the test model.

4   Experimental Evaluation

To evaluate the CCTG method, we have conducted three case studies. The experiments were designed according to the Mutation Testing [6] approach. For each case study, several mutants were created using a fault seeding framework. To test the CCTG method's effectiveness, the generated test cases were used for both the original version of the SUT and the mutated (faulty) version. The outputs of the two SUTs were then compared. When the outputs differ, the fault is considered to be detected. The mull [5] LLVM-based tool was used.

For the case studies, the Unix utilities Flex, grep, gzip were selected. The Software-artifact Infrastructure Repository[2] obtained from the Gnu site. These utilities were chosen, as the test cases generated represent their command-line arguments. Therefore, each test case is essentially a parametrized call of these utilities that produces a standard output test.

The test case structure reflects the archetype of the SUTs from the case study, standard C utilities, requiring only a set of arguments. For the case study, CCTG test cases were therefore represented by a set of command-line arguments only. The type of arguments or parameters used can be divided into a few basic categories. There is a unary argument – which always represents a Boolean value. As the parameters are for typical command-line programs, these usually specify some functionality (e.g.., –printToCommandLine) that is enabled or a specific instruction (e.g.., –help). The second type is binary arguments, consisting of parameter and value (e.g.., –input /inputs/file1).

For each SUT, a set of all possible parameters is gathered. This is a simple list of all unary parameters accompanied by possible values for the binary ones. In many cases, the parameter values are subjective to the testing environments. Input files are part of prepared testing inputs for a specific program. Values with range such as integers are also limited to discrete and finite selection. The range's selecting values are done either based on other testing information, chosen randomly, or at regular intervals. The selection of specific values from a range does not directly affect the experiments as the selections are final for all test generation methods. Seeded versions of SUT are executed using test cases generated by various methods to measure the test effectiveness. The methods used are compared to the code-coverage method.

As a reference, we use two methods of test generation: Random generation and unweighted method. The random generation method is based on random parameter permutation. The initial setup is very similar to the CCTG method, as prepared sets represent the parameters. Unlike the CCTG method, the random method does not rely on coverage information but approaches the selection of parameters for permutation entirely randomly. The selection of parameters for permutation and the value to which it should be permuted is determined randomly. First, a random integer in the range representing all the test case parameters is generated to select a permutation parameter. Secondly, a second random integer is generated within the range of all said parameter values minus one (the one being the previous parameter value - so as not to repeat the same test case).

In each study, a set of 100 test cases is generated. This is done for all three methods and is used on 20 different seeded faults. The entire process is then repeated five times by generating a new set of test cases. The test cases were generated multiple times to account for the random elements in their generation. The code-coverage based method generally has the best results. For illustration, the results of each program are show in box-plot Figs. 3a, b, and c.

---

[2] http provided the programs: //sir.unl.edu.

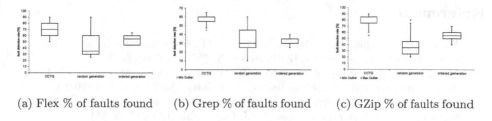

(a) Flex % of faults found   (b) Grep % of faults found   (c) GZip % of faults found

Fig. 3. CCTG evaluation results in Box Plot

The Flex results are shown in Fig. 3a and have the narrowest range for the code-coverage method – the shortest box plot. This indicates that the coverage based test cases are of very similar quality for flex. The code coverage method also has an overall higher median. However, the random method shows a tall box-plot, reflecting a completely random approach to test generation.

Figure 3b for the Grep experiments has all box plots of similar size. This reflects the random elements in all test generation methods. However, it does not produce such dissimilar sizes as in the Flex experiments. The overall dispersion, while similar in size, is marginally more successful for the code coverage method.

Gzip experiment results are shown in Fig. 3c. Here the smallest box plot represents the systematic method. While this does not correspond with results from other tests, it is not necessarily a surprise, as the systematic method has the lowest random factor of generation. The code coverage method again holds a marginally better median then the remaining two methods.

In all three cases, the medians are lower compared to the code coverage method. Most distributions in all figures are also not widely dissimilar, indicating even effectively test generating methods. It also shows the code coverage method as the most effective one.

5   Conclusion

This paper presented a new automated test case generation method based on the code coverage measure. The method's goal is to achieve automated test generation using the code coverage, which would also show improved performance at fault detection. Three case studies were implemented that compared out method against two other trivial approaches for test case generation. The results showed an overall improvement in the fault detection rate. The future goal is to work with a wider variety of parameters.

Acknowledgement. This research is conducted as a part of the project TACR TH02010296 Quality Assurance System for the Internet of Things Technology. The authors acknowledge the support of the OP VVV funded project CZ.02.1.01/0.0/0.0/16_019/ 0000765 "Research Center for Informatics." Bestoun S. Ahmed has been supported by the Knowledge Foundation of Sweden (KKS) through the Synergi Project AIDA - A Holistic AI-driven Networking and Processing Framework for Industrial IoT (Rek:20200067).

References

1. Ahmed, B.S., Zamli, K.Z., Afzal, W., Bures, M.: Constrained interaction testing: a systematic literature study. IEEE Access **5**, 25706–25730 (2017)
2. Ahmed, B.S., Gargantini, A., Zamli, K.Z., Yilmaz, C., Bures, M., Szeles, M.: Code-aware combinatorial interaction testing. IET Softw. **13**(6), 600–609 (2019)
3. Balint, A., Belov, A., Järvisalo, M., Sinz, C.: Overview and analysis of the sat challenge 2012 solver competition. Artif. Intell. **223**, 120–155 (2015)
4. Bures, M.: PCTgen: automated generation of test cases for application workflows. In: Rocha, A., Correia, A.M., Costanzo, S., Reis, L.P. (eds.) New Contributions in Information Systems and Technologies. AISC, vol. 353, pp. 789–794. Springer, Cham (2015). https://doi.org/10.1007/978-3-319-16486-1_78
5. Denisov, A., Pankevich, S.: Mull it over: mutation testing based on LLVM. In: 2018 IEEE International Conference on Software Testing, Verification and Validation Workshops (ICSTW), pp. 25–31, April 2018
6. Du, Y., Pan, Y., Ao, H., Ottinah Alexander, N., Fan, Y.: Automatic test case generation and optimization based on mutation testing. In: 2019 IEEE 19th International Conference on Software Quality, Reliability and Security Companion (QRS-C), pp. 522–523, July 2019
7. Gargantini, A., Petke, J., Radavelli, M.: Combinatorial interaction testing for automated constraint repair. In: 2017 IEEE International Conference on Software Testing, Verification and Validation Workshops (ICSTW), pp. 239–248, March 2017
8. Saumya, C., Koo, J., Kulkarni, M., Bagchi, S.: Xstressor: automatic generation of large-scale worst-case test inputs by inferring path conditions. In: 2019 12th IEEE Conference on Software Testing, Validation and Verification (ICST), pp. 1–12, April 2019
9. Xu, Z., Rothermel, G.: Directed test suite augmentation. In: 2009 16th Asia-Pacific Software Engineering Conference, pp. 406–413, December 2009
10. Xu, Z., Kim, Y., Kim, M., Rothermel, G., Cohen, M.B.: Directed test suite augmentation: techniques and tradeoffs. In: Proceedings of the Eighteenth ACM SIGSOFT International Symposium on Foundations of Software Engineering, FSE '10, pp. 257–266. Association for Computing Machinery, New York (2010)
11. Zhang, J., Ma, F., Zhang, Z.: Faulty interaction identification via constraint solving and optimization. In: Cimatti, A., Sebastiani, R. (eds.) SAT 2012. LNCS, vol. 7317, pp. 186–199. Springer, Heidelberg (2012). https://doi.org/10.1007/978-3-642-31612-8_15

From Requirements to Verifiable Executable Models Using Rebeca

Marjan Sirjani[1,2]([⊠]), Luciana Provenzano[1], Sara Abbaspour Asadollah[1],
and Mahshid Helali Moghadam[3]

[1] Mälardalen University, Västerås, Sweden
{marjan.sirjani,Luciana.Provenzano,Sara.Abbaspour}@mdh.se
[2] Reykjavik University, Reykjavik, Iceland
[3] RISE Research Institutes of Sweden, Västerås, Sweden
mahshid.helali.moghadam@ri.se

Abstract. Software systems are complicated, and the scientific and engineering methodologies for software development are relatively young. We need robust methods for handling the ever-increasing complexity of software systems that are now in every corner of our lives. In this paper we focus on asynchronous event-based reactive systems and show how we start from the requirements, move to actor-based Rebeca models, and formally verify the models for correctness. The Rebeca models include the details of the signals and messages that are passed at the network level including the timing, and can be mapped to the executable code. We show how we can use the architecture design and structured requirements to build the behavioral models, including Rebeca models, and use the state diagrams to write the properties of interest, and then use model checking to check the properties. The formally verified models can then be used to develop the executable code. The natural mappings among the models for requirements, the formal models, and the executable code improve the effectiveness and efficiency of the approach. It also helps in runtime monitoring and adaptation.

1 Introduction

Safety-critical systems are systems that may fail with catastrophic consequences on people, environment and facilities. These systems are becoming more and more common, powerful, and dependent on safety-critical software. The result is that serious consequences may arise from the failure of such software systems. Safety analysis is performed to identify the hazards that may cause failures which lead to accidents. Safety requirements are written as measures to mitigate the identified hazards, i.e. to avoid them or reduce their probability or limit their consequences. Therefore, safety requirements play an important role because they define the system's behaviors that shall be implemented to ensure the safety properties of the whole system.

In a model-driven development approach, requirements can be seen as the specification of the system to be developed. One can start from these requirements, build the necessary models to capture the structure and the behavior of

© Springer Nature Switzerland AG 2021
L. Cleophas and M. Massink (Eds.): SEFM 2020 Workshops, LNCS 12524, pp. 67–86, 2021.
https://doi.org/10.1007/978-3-030-67220-1_6

the system, and build the code based on that. In this process, we can use formal verification to come up with dependable models and hence more dependable code. Note that this is an iterative and incremental approach where we have to go back and forth between the models (including the requirements and the code) several times. This approach is not necessarily the common practice. In this paper, we promote this model-driven development approach.

Defective requirements can cause serious failures. This emphasizes the need to have requirements that are correct, precise and clear as basis of the system development. For building formal models based on the requirements, we need the requirements to be consistent and unambiguous, or else we will not be able to build the models. So, throughout the process of model-driven development we not only build the system based on the requirements, but also the requirements will be refined and become consistent and unambiguous. The models are then checked against the safety properties that are also derived from the requirements, to make sure that the (behavioral and implementation) details that are added to build the models are not introducing errors.

We describe our experience with an industrial case study, a time-critical safety function, i.e., "*Passenger Door Control*", from a train control system. We present how we start with the safety requirements and software architecture documents, and then conclude with verified models using the Rebeca modeling language [1–3]. Rebeca is an actor-based language used for modeling reactive and asynchronous distributed and concurrent systems [4]. Rebeca is supported with formal verification theories and tools [5]. Event-based reactive systems play a major role in many industrial control software systems such as those in railway and automotive domains. Hence, the experience we report in this paper can be used in other similar cases and domains.

The whole process from requirements to Rebeca models is depicted in Fig. 1. Specifically, to be able to create the Rebeca model, two inputs are necessary, i.e. the functional safety requirements and the system architecture. From the safety requirements and the architecture document, we create the behavioral models, i.e. the state diagrams and the sequence diagrams, and based on these diagrams we build the Rebeca model along with the properties that have to be checked. It is worth noting that this process foresees a document called "structured requirements". Indeed, it is important that the safety requirements in input are written according to a well-structured syntax. This enables us to reduce the ambiguity typical of natural language requirements in order to facilitate their interpretation and translation into the formal model. We use the *GIVEN-THEN-WHEN* syntax [6] for requirement specification, as explained in Sect. 3[1].

As for now, the Rebeca models are the final output of our proposed process from safety requirements towards verifiable models. During this process, by building visualized system-level models we get a better view of the system architecture is an extra step that can be conducted in parallel with building the Rebeca models. The co-modeling of hardware and software can be done using

---

[1] We use this format based on the experience of the second author of the paper who worked for seven years as requirements manager in industry.

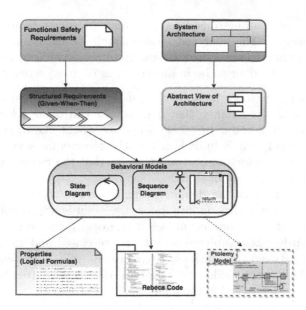

Fig. 1. The proposed process from requirement to code. Note that the figure shows one iteration in our iterative and incremental approach. All the models, from requirements to behavioral and executable ones are refined through the process in an iterative way.

modeling and simulation tools like Ptolemy [7] (as suggested in Fig. 1). While using Rebeca gives us formally verifiable models, by using Ptolemy we will get a clear view of the architecture, and also simulation results. The more detailed process is explained in the following sections.

2 The Door Controller Case Study

We use an example based on a real industrial case to describe the approach that should be followed to formally verify a set of requirements using Rebeca. We use the function *"Open external passengers doors"* that controls opening of the external doors of a train to let passengers get on and off safely. Specifically, the external doors of a train can be opened by the driver, through a dedicated button installed in the driver's cabin, and by the passenger, through a button placed on each external door. This is done to let passengers get off the train at their destination, and it should be only enabled when the train reaches a station and stops at it. Moreover, the external passenger doors are equipped with a lock mechanism to prevent opening a door when the train leaves the station and is running. This implies that to open a door, the door must be unlocked. This is an interesting function to be modeled and verified for two main reasons:

– The function is safety-related. Indeed, an external door which is accidentally opened when the train is running may cause a passenger to fall out of the train, thus causing an accident.

– The external door can be considered as a shared resource between the driver and the passenger. The door can receive simultaneous commands from the driver, i.e. to open, close or lock it; and from the passenger, i.e. to open it. This may cause the door to be in an erroneous or unexpected state.

Our aim is therefore to formally check by using the Rebeca modeling language whether there is any possibility that a passenger can open a locked door to get off from a running train. In other words, we would like to check whether the behavioral model that is built based on the requirements violates a safety property of the train, which also means to show that the requirements may be incorrect, inconsistent, or ambiguous.

It is worth noting that we define "running" as the train state which corresponds to one of the following situations: the train is approaching a station (before it stops and the doors are unlocked and open), the train is leaving the station (the boarding is completed and doors are closed and locked), and the train is running between two stations. There are multiple properties that can be checked using the Rebeca model checking tool Afra [8], in particular, the safety property that can be checked is the following:

– Is it possible to open a locked door when the train is running?

3 Structured Requirements

According to the proposed process in Fig. 1, the starting point to create the Rebeca model is to collect the safety requirements of the function to be verified and rewrite them using a well-structured syntax.

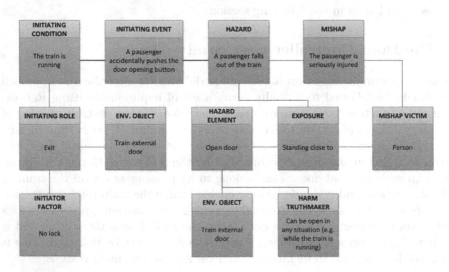

Fig. 2. Hazard Ontology for the hazard "Passenger fall out of the train".

Table 1. An example of the safety requirements for the door opening function.

SafeReq1	GIVEN the train is ready to run
	WHEN the driver requests to lock all external doors
	THEN all the external doors in the train shall be closed and locked
SafeReq2	GIVEN an external door is locked
	WHEN the passenger requests to open an external door
	THEN the external door shall be kept closed and locked
SafeReq3	GIVEN an external door is unlocked
	WHEN the passenger requests to open an external door
	THEN the external door shall be opened
SafeReq4	GIVEN all external doors on the side of the train close to the platform are unlocked
	WHEN the driver requests to open all external doors
	THEN all external doors on the side of the train close to the platform shall be opened
SafeReq5	GIVEN the train approaches a station
	WHEN the driver requests to unlock all external doors that are on the train side close to the platform
	THEN all external doors on the side of the train close to the platform shall be unlocked
SafeReq6	GIVEN the train is running
	WHEN an external door is open
	THEN an alert shall be provided

In this work, the safety requirements related to the *"Open external passenger doors"* function are obtained by applying the Safety Requirements Elicitation (SARE) approach [9] to the Hazard Ontology depicted in Fig. 2. This Hazard Ontology is used to identify the causes and consequences of the *"Passengers fall out of the train"* hazard. The Hazard Ontology proposed in [10] and [11], provides a conceptualization of the hazard which enables to gain a deep knowledge of the circumstances that result in hazards. This knowledge is structured in entities of the Hazard Ontology which correspond to the hazard's sources, causes and consequences. The SARE approach uses this knowledge to elicit the safety requirements that mitigate the hazard.

Here, we use the experience of the second author in the railway domain and the SARE approach to formulate the safety requirements; and as a real hazard for trains, we choose the hazard *"Passengers fall out of the train"*. One can alternatively use the functional safety document from an industry as the input. This experience also shows that the SARE approach can be used to complement the existing safety requirements provided as input or to discover new safety requirements in case of new systems.

To specify the safety requirements elicited by SARE, we use the *GIVEN-WHEN-THEN* syntax in order to obtain well-structured requirements that can be easily used for modeling in Rebeca, and then the model can be used for formal verification. Specifically, the *GIVEN-WHEN-THEN* is *"a style of specifying a system's behavior using Specification by Example"* [12] developed within the Behavior-Driven Development [6]. According to this style, a scenario is decomposed in three parts, i.e. the *GIVEN* states the pre-condition(s) to the scenario; the *WHEN* describes the input event(s) which trigger the action(s); the *THEN* defines the action(s) the system shall perform as a consequence of the trigger and the expected changes in the system. We think that this structured syntax for requirement specification helps to derive the concepts that build the actors, states of the actors, and also the events that trigger the changes. Moreover, it helps in deriving the properties to be verified using model checking. Table 1 shows a set of safety requirements in the *GIVEN-WHEN-THEN* syntax for the open door example.

4   The Architecture

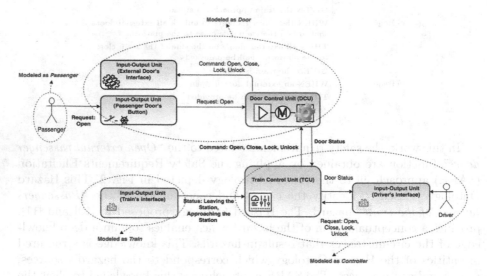

Fig. 3. The system architecture with a focus on the door controller case study. The dotted circles show the actors in the Rebeca code.

Figure 3 depicts an overview of a typical system architecture realizing the functionalities in our industrial case. The intended system is an example of a cyber-physical system consisting of hardware components like programmable control units, actuators, different communication channels, and different control applications running on the hardware units. The main components in the architecture are Input-Output (IO) units, central Train Control Unit (TCU), Door Control Unit (DCU). IO units act as interfaces to the system and are intended to receive/send the input/output signals. The IO unit on the passenger side are in charge of reading the door push buttons to receive the open request from the passenger. When a passenger pushes the "open" button, the IO unit receives the open request and sends it to the DCU. The commands for open, close, lock and unlock coming from the driver pass through TCU and go to the DCU. The DCU is responsible for actuating the proper commands for changing the state of the door.

TCU plays the role of the central control management. TCU might be distributed and run on separate physical devices. For example, one physical control device for running non safety-related functions and one device for the execution of safety-critical functions. DCU may represent a programmable unit which receives the command signal from TCU and applies the signal to the corresponding converters actuating the door. Data communication between the physical devices is usually conducted through a system-wide bus and a safe communication protocol. Later in our behavioral models, we model both DCU and the associated IO on the passenger side as "Door" actor and also the combination of TCU and the driver as "Controller" actor.

The actor "Train" models a set of IO units receiving the status from the sensors, and other means, that are used to inform the TCU and the driver that the train is in a state which is significant for our case study, i.e., approached at the station, and ready to leave. These are the states in which the TCU has to change the state of the doors. Figure 3 shows how we abstract the architecture diagram to extract main Rebeca actors.

Generally, in safety critical systems, in order to satisfy the integrity and availability, different types of redundancy structures are applied to different units including IO units. For example, redundant IO units are in place and extra supervision mechanisms for the validity check of the resulted values from these redundant IO units are used. In our example, we abstract these details away. We can create other models focusing on such details and verify the correct functionality of these parts of the system. In general, we need to use compositional and modular approaches to cover large and complicated systems.

5 The Mapping from Requirements to Behavioral Models

By studying the structured requirements, together with the architecture of the software system, we will know the actors to be included in the Rebeca code. We build an abstract version of the architecture to be the basis for writing the Rebeca code. The abstract architecture includes the reactive classes that we include in our code.

In the context of our door controller example, from the architecture (Fig. 3), we see that we have *I/O units* for the passenger door buttons (passing the input to the door to request open) and the driver input interface (passing the input to the controller to request open, close, lock and unlock (release)), and the door control actuator (passing the output from the controller to the door, commanding for open, close, lock and unlock (release)). From this explanation we can conclude that we need actors to represent the controller, the door, the driver and the passenger in the model.

From the structured requirements (Table 1), we can see that the players are: the *train*, the *driver*, the *passenger*, and the *door*. Note that we do not see the controller in the requirements. To see the complete picture to model the software system we need to study both the requirements and the architecture. For the door controller we consider the scenarios when a train is ready to run, and when it approaches the station. When boarding is complete and the *train* is ready to run, the *driver* sends the request to close and then lock the *doors*. When the *train* approaches the station, the *driver* sends the request to unlock and then open the *doors*. The requests are received by the *controller*, and the controller makes the decision based on the status of the train and the doors. The logic within the code of the *controller* is supposedly written in a way that the safety requirements are guaranteed. There is no exact physical realization as signals or hardware devices for the train in the model, the train is in the model to represent the states where the driver knows he has to send the command for closing and locking the doors, or unlocking and opening them. The *passenger* can always request to open the door.

The structured requirements also help in deriving the state variables, and their values, specially the pre- and post-conditions in the *GIVEN* and *THEN* parts. For example, consider the condition "*the train is ready to run*" written in the *GIVEN* part of the requirement SafeReq1 in Table 1. We can infer that we need a variable representing the train status (the variable trainStatus of the actor Controller in Fig. 6); and one possible value of this variable shows that the train is "*ready to run*". From these requirements we can also infer that we need two state variables to capture the status of the doors being locked or unlocked, and being opened or closed (the variables isLocked and isClosed of the actor Controller in Fig. 6).

The events defined in the *WHEN* parts are mapped to the messages that are sent to the actors and upon which the actors react. They can be used to obtain the sequence of messages exchanged among the actors, and to build the sequence diagram based on that.

This process and the natural mapping facilitate the development of the Rebeca model from the requirements and help to limit the errors that may be introduced when translating the requirements into the model. Moreover, the pre- and post- conditions in the requirements can be used to form the assertions that represent the properties to be verified.

Abstraction in an Iterative and Incremental Approach. Note that during the process we choose to have abstract models to begin with, and we continue by adding more details in an iterative and incremental way. For example, in the behavioral models derived from the requirements of the door controller case study, we do not distinguish each door separately, and we do not distinguish which side of the train the doors are. A concrete example of this abstraction is where for the requirement SafeReq5, we abstract away the part regarding the side of the train in the part referring to "all external doors on the side of the train close to the platform".

5.1   The Mapping to Logical Properties

We can use the structured requirements for writing assertions that must hold throughout the execution of the code. For example, consider the requirement SafeReq2: "*GIVEN an external door is locked, WHEN the passenger requests to open the locked external door, THEN the external door shall be kept closed and locked*". This requirement helps us to derive the main safety property of the function "*open external passenger door*". The assertion that shall be checked is the following: "*It is not possible to open a locked door by passengers*". A stronger assertion that covers this one is discussed in Sect. 6.1, the assertion is checked by Afra, and we show how the model is modified such that this assertion holds.

There are other interesting requirements, like the requirement SafeReq4 which is a property to show that progress has to be made. The SafeReq4 requirement states: "*GIVEN all external doors on the side of the train close to the platform are unlocked, WHEN the driver requests to open all external doors, THEN all external doors on the side of the train close to the platform shall be opened*". Safety properties are about showing that nothing bad will happen, while progress properties are about showing that good things will finally happen. For checking these types of requirements, we cannot use simple assertions and we need to use the TCTL model checking tool for Timed Rebeca [13][2]. The timing features can be included here, for example for the requirement SafeReq4, we can check that "**if** *the doors are unlocked and an open request is sent by the driver* **then** *the doors will be opened within x units of time*".

6   The Behavioral Models

Here we explain the state diagrams, sequence diagrams and the Rebeca code that are derived from the requirements. We also explain the timing properties.

State Diagrams. Using the mapping explained in Sect. 5, we can derive the state diagrams for the door controller case study. In Sect. 5, we concluded that we need actors to represent the controller, the door, the driver, the passenger, and the train in the model. For simplifying the model, we decided not to model the driver, the behavior of the driver is merged with the controller. We may consider this as an autonomous controller that decides based on the conditions of the doors and the train. Note that we only have one actor that represents all the doors, also for the sake of simplicity. The model can be refined, and details can be added in an iterative and incremental way in order to check different properties and different parts of the system.

As shown in the state diagram in Fig. 4a, the train can be either in a state that has just approached the station (when the doors should be unlocked and then opened), or in a state that it is ready to run (when the doors should be closed and locked). Note that these are the only two states of the train that are

---

[2] The TCTL model checking tool for Timed Rebeca is not yet integrated in the Eclipse tool suite of Afra.

important for us in our example because our focus is on changing the states of the doors, and only in these states of the train we need to change the status of the doors. For example when the train is running, or stopped (with doors already open) the status of the doors should stay unchanged (and that is what the controller in Fig. 4c guarantees by not accepting any *wrong* event in the *wrong* states).

Figure 4b shows the states of the doors. A locked and closed door can only be unlocked, and then opened; and an unlocked and open door can only be closed and then locked. The state diagram is consistent with the Rebeca code in Fig. 6. We prevent the door from going to a state where it is locked and open, an unsafe state that should be avoided. The `if-statement` in Line 93 guarantees this.

Figure 4c shows the state diagram for the controller. The controller receives the status of the doors and the train, also the requests for opening, closing, locking and unlocking the doors. The controller coordinates the commands that are sent to the doors based on the status of the door itself, and the train. Figure 4d is the state diagram of the passenger. This actor models the requests coming from the passengers in a non-deterministic way, and the Rebeca code is model checked to make sure this behavior cannot jeopardize the safety.

Sequence Diagrams. The sequence diagrams derived from the requirements and the architecture are shown in Fig. 5. These diagrams are made in a similar way as described for the state diagram. Indeed, the actors controller, door, passenger and train become the objects in the sequence diagrams among which messages are exchanged in a temporal order to perform the door functions. In the sequence diagrams the flow of messages between actors, and also their order and causality are clearer. In Fig. 5a, it is shown that when the status of the train or the door is changed the controller receives a message to update the status of these two actors in the controller. Any change in the status of the train or the doors triggers the execution of `driveController` message server in which the controller decides which command to send to the doors.

Figure 5b shows the message sent by the passenger to the door. Note that the sequence diagrams are consistent with the Rebeca code, here instead of having an actor representing the passenger button on the door, and another actor representing the door controller, for the sake of simplicity, we have both modeled as one actor. Passenger sends the open command directly to the door, and the door sends a message to the controller to update the status in the controller (as described above). This is where different errors may occur if the behavioral model (Rebeca code) is not written with enough care. More explanation is in Sect. 6.1.

Rebeca Code. Based on the state and the sequence diagrams, we wrote a Timed Rebeca code with four reactive classes: `Controller`, `Train`, `Door`, and `Passenger`. The Rebeca code is presented in Fig. 6. The rebecs (i.e. reactive objects, or actors) `controller`, `train`, `door`, and `passenger` are instantiated from these reactive classes.

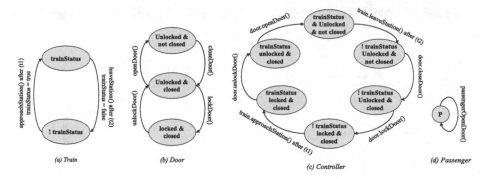

Fig. 4. The state diagrams for the door controller case study. In the state diagram for the *Train* (Part a), the state in which `trainStatus` is `true` is when the train has approached the station and stopped and ready for the doors to be unlocked and then opened. The state in which `trainStatus` is `false` is when the boarding is complete, and the train is ready to run and leave the station, and the doors must be closed and then locked. The name of the rest of the variables are chosen in a way to make the diagrams self-explanatory as much as possible.

The main message server of the reactive class `Controller` is `driveController`, where we check the state of the train and the doors, and send proper commands. If the train is in the state that the boarding is completed and the train is ready to run (`trainStatus` is true - lines 31–41), then if the doors are not yet closed, the `Controller` sends a command to close them (by sending the `closeDoor` to the rebec `door`). If the doors are already closed the controller sends a command to lock them (by sending the `lockDoor` to the rebec `door`). If the train is in the approaching state (`trainStatus` is false - lines 42–51), then if the doors are not yet unlocked, the controller sends a command to unlock the doors (by sending the `unlockDoor` to the rebec `door`). If the doors are already unlocked the controller sends a command to open them (by sending the `openDoor` to the rebec `door`).

The reactive class `Controller` also has two other message servers: `setDoorStatus` and `setTrainStatus`. The `setDoorStatus` (lines 21–25) is called by the `Door` after updating the status of the doors. The `setTrainStatus` (lines 26–29) is called by the `Train` after updating the status of the train. The reactive class `Train` has two message servers that model the train behavior when the train is ready to run (`leaveStation`) and approaches the station (`approachStation`). Both message servers in this actor inform the controller when the train status changes.

The reactive class `Door` models the behavior of the doors and has four message servers: `closeDoor()`, `lockDoor()`, `unlockDoor()` and `openDoor()`. The `closeDoor()` (lines 88–91) is called by `Controller` actor (line 34) to close the door by changing the status of the door (line 89). The `lockDoor()` (lines 92–97) is called by the controller (line 38) to lock the door. If the current status of the door is closed, then the status of the door is change to locked (line 94).

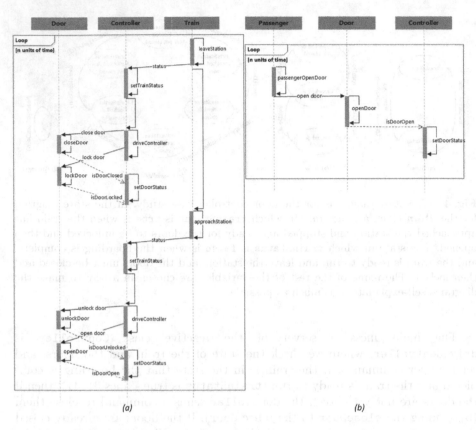

(a)   (b)

Fig. 5. Sequence diagrams of the door controller case study showing the message passing between the actors Controller, Train, Passenger, and Door.

The unlockDoor() (lines 98–101) is called by the Controller actor (line 45) to unlock the door by changing the status of the lock (line 99). The openDoor() (lines 102–107) is called by the Controller actor (line 49) and the Passenger actor (line 117) to open the door. If the current status of the door is unlocked, then the status of the door can change to open (line 104). In all these message servers the status value is sent to the Controller actor after any updates.

The Passenger actor is implemented to model the behavior of a passenger. We assume that the passenger can constantly send a request to the Door actor to open the door. This actor has only one message server (passengerOpenDoor). The passengerOpenDoor is designed to send a request (open the door) to the Door actor every 5 units of time (lines 117 and 118).

Timing Properties. The Rebeca code in Fig. 6 contains the environment variables (denoted by env at the top of the code). These variables are used to set the timing parameters. The variable $networkDelayDoor$ represents the amount of time that takes for a signal to get to the door from the controller (and vice versa), and the variable $networkDelayTrain$ shows the amount of time that

```
 1  env byte networkDelayDoor = 1;
 2  env byte networkDelayTrain = 3;
 3  env byte reactionDelay = 1;
 4  env byte passengerPeriod = 5;
 5  env short runningTime = 239;
 6  env short atStationTime = 50;
 7  reactiveclass Controller(23){
 8      knownrebecs{
 9          Door door;
10      }
11      statevars{
12          boolean isClosed;
13          boolean isLocked;
14          boolean trainStatus;
15      }
16      Controller(){
17          trainStatus = true;
18          isClosed = false;
19          isLocked = false;
20      }
21      msgsrv setDoorStatus(boolean close, boolean lock) {
22          isClosed = close;
23          isLocked = lock;
24          self.driveController();
25      }
26      msgsrv setTrainStatus(boolean status){
27          trainStatus = status;
28          self.driveController();
29      }
30      msgsrv driveController(){
31          if(trainStatus){ // leave the station
32              if(!isClosed || !isLocked) {
33                  if(!isClosed) {
34                      door.closeDoor() after(networkDelayDoor);
35                      delay(reactionDelay);
36                  }
37                  if(!isLocked) {
38                      door.lockDoor() after(networkDelayDoor);
39                  }
40              }
41          }// end of if(trainStatus)
42          else if(!trainStatus){ // arrive to the station
43              if(isClosed || isLocked) {
44                  if(isLocked) {
45                      door.unlockDoor()
                          after(networkDelayDoor);
46                      delay(reactionDelay);
47                  }
48                  if(isClosed) {
49                      door.openDoor() after(networkDelayDoor);
50                  }
51              } } // end of else if(!trainStatus)
52          } // end of driveController()
53      } //end of the Controller class
54  reactiveclass Train(5){
55      knownrebecs{
56          Controller controller;
57      }
58      statevars{
59          boolean status;
60      }
61      Train(){
62          status = true;
63          self.leaveStation();
64      }
65      msgsrv leaveStation(){
66          status = true;
67          controller.setTrainStatus(status)
                after(networkDelayTrain);
```

```
 68          self.approachStation() after(runningTime);
 69      }
 70      msgsrv approachStation(){
 71          status = false;
 72          controller.setTrainStatus(status)
                after(networkDelayTrain);
 73          self.leaveStation() after(atStationTime);
 74      }
 75  } //end of the Train class
 76  reactiveclass Door(15){
 77      knownrebecs{
 78          Controller controller;
 79      }
 80      statevars{
 81          boolean isDoorClosed;
 82          boolean isDoorLocked;
 83      }
 84      Door(){
 85          isDoorClosed = false;
 86          isDoorLocked = false;
 87      }
 88      msgsrv closeDoor(){
 89          isDoorClosed = true;
 90          controller.setDoorStatus(isDoorClosed,
                isDoorLocked) after(networkDelayDoor);
 91      }
 92      msgsrv lockDoor(){
 93          if (isDoorClosed){
                // The door is only locked if the door is closed.
 94              isDoorLocked = true;
 95          }
 96          controller.setDoorStatus(isDoorClosed,
                isDoorLocked) after(networkDelayDoor);
 97      }
 98      msgsrv unlockDoor(){
 99          isDoorLocked = false;
100          controller.setDoorStatus(isDoorClosed,
                isDoorLocked) after(networkDelayDoor);
101      }
102      msgsrv openDoor(){
            // The door is only opened if the door is not locked.
103          If (!isDoorLocked){
104              isDoorClosed = false;
105          }
106          controller.setDoorStatus(isDoorClosed,
                isDoorLocked) after(networkDelayDoor);
107      }
108  } //end of the Door class
109  reactiveclass Passenger(5){
110      knownrebecs{
111          Door door;
112      }
113      Passenger(){
114          self.passengerOpenDoor() after(passengerPeriod);
115      }
116      msgsrv passengerOpenDoor(){
117          door.openDoor();
118          self.passengerOpenDoor() after(passengerPeriod);
119      }
120  } //end of the Passenger class
121  main {
122      Controller controller(door):();
123      Door door(controller):();
124      Train train(controller):();
125      Passenger passenger(door):();
126      }
```

Fig. 6. The Rebeca model for the door controller case study.

takes for a signal to get from the train to the controller. We also have other timing features, e.g., we modeled a reaction delay for the controller when it reacts to the events (*reactionDelay*); *passengerPeriod* is defined to show the passenger sending the open command periodically (it can be modeled differently but this is the simplest way and serves our purpose to find possible errors). We also model passage of time between a train leaving and then again approaching the station (*runningTime*), and the time that train stays at the station (*atStationTime*).

The environment variables can be used as parameters to set different cycle times and communication channel features. The value for the parameters can be changed to check different configurations. For example, we can see varying depths in getting into the error state by changing the period of the passenger pressing the open door button.

6.1 Formal Verification

The Rebeca code in Fig. 6 is a version of the code that runs without violating any of the properties of interest. We checked the assertion: "*It is not possible to open a locked door (not by the driver nor the passengers);*" and we showed that the door cannot be opened when it is locked. This assertion covers multiple other weaker assertions, like: "*It is not possible to open a locked door (by driver or passengers) when the train is leaving the station;*" and "*It is not possible to open a locked door (by driver or passengers) when the train is arriving at the station*".

In the Rebeca model, the passenger sends a request directly to the door, the request does not pass through the controller. This is what makes the model vulnerable to errors. The door is receiving commands from both the passenger and the controller, and variant interleaving of these commands (i.e. events in the queue) may cause the execution of the model to end in a state that violates the safety property[3]. The two "if-statements" in lines 93 and 103 of the reactive class `Door` are there to avoid this problem. If we remove the passenger from the model, the model is correct even without these `if-statements`.

We run the Rebeca model checking tool, Afra, on a MacBook Pro laptop with 2,9 GHz Intel Core i5 processor and 8 GB memory. While model checking the code without the passenger the number of reached states is 55, and the number of reached transition is 68 (consumed memory is 660, and the total spent time is below one second). For the setting shown in the Rebeca model in Fig. 6, where we have a passenger and when the passenger sends a request to open the door every 5 units of time then the number of reached states will be 402079, the number of transitions is 1286068 and the total time spent for model checking is 115 s.

A different design for the model, derived from a different allocation of functions in the architecture, can be modeled and model checked. More explanation will be in Sect. 7In the Rebeca code in Fig. 6, where we have a passenger, if we remove the `if-statements` in lines 93 and 103, then the model violates the assertion and comes back with a counterexample. The depth of the trace in the state space to reach the counterexample depends highly on the setting of the timing parameters. A snapshot of the Afra tool where the counterexample is found is shown in Fig. 7. The assertion is checking the value of variables `isDoorClosed` and isDoorLocked

---

[3] A different design for the model, derived from a different allocation of functions in the architecture, can be modeled and model checked. More explanation will be in Sect. 7.

from the rebec door. In the snapshot you may see that isDoorClosed is false (the door is open), and isDoorLocked is also false (the door is unlocked). The only message in the queue of the rebec door is lockDoor. This will cause the execution of the message server lockDoor in the rebec door which will create the state in which isDoorClosed stays false (the door is open), and isDoorLocked changes to true (the door is locked). This states fail the assertion and the model checking tool comes back with the counterexample shown in Fig. 7. You can see this state on the right hand side of the figure, and the trace to get to it in the left hand side of the figure.

Note that changing the timing parameters can change the state space significantly. The timing parameter includes the period of sending the requests, network delay, and the computation/process delay.

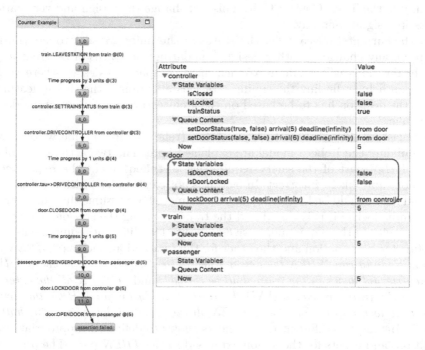

Fig. 7. The screen shot of Afra, coming back with a counterexample for checking the assertion "*It is not possible to open a locked door*" for the Rebeca code in Fig. 6.

7   Discussion and Future Work

To reach the Rebeca code from the requirements we need to use an iterative approach. There may be ambiguity in the informally stated requirements that need to be clarified. To come up with the right state variables and right transitions among states, we may need to go back and forth several times and ask the experts for the right information to avoid misunderstandings and incorrect outcome. As stated in many classical papers on formal methods, one of the main advantages of formal methods is to make the requirements clear, unambiguous, and consistent. Some examples of this kind of clarifications within our work are explained further in this section.

Rebeca models can be useful for checking safety and timing properties only if the topology of the actor model matches (or is consistent with) the architecture of the system. As we plan for a straightforward mapping of Rebeca code to executable code we need this consistency. This can be another challenge in the process, to know the architecture and the allocation of tasks to different components. One example is the decision we made for the Door Control Unit, modeled within the actor door, to send the open command to the door upon receiving the request from the passenger. Alternatively, we could have a model in which all the decisions for sending the open command to the door are handled centrally in the Train Control Unit. This will change the design and verification results in a significant way.

In the current Rebeca code, the status of the units are sent to the control unit upon any change. Another design is updating the status of different units periodically. This will result in a much more complicated design where verification can help in finding the timing problems and tuning the timing features. Again, the decision has to be based on the architecture and execution model of the system.

Some issues about the safety requirements (refer to Table 1) that we observed while building the Rebeca model are explained here. The two actions described in the *THEN* part of the safety requirement SafeReq1 make the requirement ambiguous and, likely, incorrect. In fact, it is not clear under which condition an external door should be locked. In our Rebeca model, we assume that an external door can be locked if it is closed and the train is leaving the station. To remove the ambiguity in the requirement, we can specify two different requirements, one to define the *close* action and the other for the *lock* action, such as *"GIVEN the train is ready to run WHEN the driver requests to close all external doors THEN all the external doors in the train shall be closed"*; and *"GIVEN all the external doors in the train are closed AND the train is ready to run WHEN the driver requests to lock all external doors THEN all the external doors in train shall be locked"*. Having two different requirements allows to define the appropriate preconditions and events for the action expressed in the *THEN* part. The proposed requirements implies that *"close"* and *"lock"* are two different actions that the driver must perform in order to lock the external doors. However, it is also possible that the action to lock the external doors includes the action to close them in order to guarantee that no open door can be locked. In this case, the safety requirement SafeReq1 is correct but, for the sake of clarity, the action *close* should be removed, i.e. *"THEN all the external doors in the train shall be locked"*.

The safety requirement SafeReq3 is unclear and incorrect due to incomplete pre-condition in the *GIVEN* part. In fact, the pre-condition *"an external door is unlocked"* does not take into account the train status, i.e. if the train is leaving or approaching the station. In our model, we assume that a door can be open if it

is unlocked and the train is approaching the station. As a result, a better requirement would be "*GIVEN an external door is unlocked AND the train approaches the
station WHEN the passenger requests to open the external door THEN the external
door shall open*". The proposed requirement also covers the dangerous situation in
which an external door is opened when the train is leaving the station. The safety
requirement SafeReq6 cannot happen in the Rebeca model proposed in this paper
since an external door that is locked cannot be open. However, such requirements
must be considered in a real application since they mitigate unexpected behaviors
that may happen due to interactions of other system's parts which may interfere
with the "open door" function.

The work presented in this paper is in preliminary stages. One direction to
go is to make the mappings automatic or semi-automatic. Generating Ptolemy
models during the process will make the approach more robust and also more
friendly towards the engineers. It gives us a better view on the architecture and
helps in choosing the actors involved.

8   A Quick Overview of Related Work

The work presented in this paper has multiple dimensions. We speak of a process for model-driven development of reactive systems that involves requirement
documentation, architecture designs, UML models, actor models, and formal
verification. In the following we point at a few related work, the text is far from
a complete survey or a thorough comparison with the existing work.

The center of the work is the actor-based language Rebeca, and how we
can use it in model-driven development of dependable reactive systems. Industrial reactive systems are mainly cyber-physical systems combining computation
and communication with physical and temporal dynamics. They consist of different components (actors) acting based on different computation models and
interacting with each other through communication channels. Actor-based modelling is one of the key approaches for co-modeling of hardware and software of
cyber-physical systems [14]. In this modeling style, actors are the components
communicating through interfaces, i.e. ports, via sending and receiving data.

For building dependable systems, we look for models that capture timing features and come with formal verification support. Timed automata and UPPAAL
[15] are examples of such models and tools that are widely used in industrial cases.
The reason for using Rebeca is its friendliness towards event-driven and asynchronous distributed systems [4], and the support for formal verification. Rebeca
is the first actor-based language with model checking support [16], and is used for
schedulability analysis of wireless sensor network applications [17], protocol verification [18], design exploration and comparing routing algorithms [19].

To fill the gap between the formal actor model and the requirements we need
other less formal models that are closer to the requirement specification. In this
work, we use the *GIVEN-WHEN-THEN* syntax to specify the safety requirements, and then we used UML state diagrams and sequence diagrams. One way

to get closer to a formal representation from informal requirements (written in natural language), is using patterns. Using patterns to specify requirements are proposed in multiple works [20–22]. Patterns for requirements specification are also integrated in frameworks which aim at building conceptual models from the structured requirements and, consequently, formally verifying them using different tools for model checking. Some examples of such frameworks are proposed in [23] and [24].

State diagrams are common notation for behavioral modelling of reactive systems. Currently they are a key part of modelling standards like UML, SysML [25] and MARTE [26]. There are many commercial and open source design tools supporting system behavior modelling in terms of state diagrams. While state diagrams show different states of each actor or combination of actors, sequence diagrams can show the flow of messages among actors and are used for modeling reactive systems [27].

For modeling reactive systems, there are other modeling and simulation frameworks which provide heterogeneous modeling along with simulation capabilities. Ptolemy II [7] and Stateflow [28] are popular examples of this category. Ptolemy II supports hierarchical actor-based modelling, i.e., composite actor, and various types of models of computation (MoC) with simulation capabilities. Stateflow provides a graphical language to describe the system behavior logic using state diagrams, flow charts and truth tables. It also offers the possibility of reusing Simulink subsystems and MATLAB code for representing states, and automatic code generation. However, none of these tools support formal verification.

Regarding proposing a systematic process for building verifiable behavioral models, Gamma [29] is a modeling framework which integrates heterogeneous statechart components to make a hierarchical composition, supports formal verification for the composite model and provides automatic code generation on top of the existing source code of the components. Gamma focuses on building hierarchical statechart network based on the existing statechart components, and like most existing tools and approaches do not consider the phase in the process where we need to map the requirements to behavioral models.

Acknowledgment. We would like to thank Edward Lee for reading the paper and giving us very useful comments. The research of the first three authors for this work is supported by the Serendipity project funded by the Swedish Foundation for Strategic Research (SSF). The research of the first two authors is also supported by the DPAC project funded by the Knowledge Foundation (KK-stiftelsen). The research of the fourth author is funded partially by Vinnova through the ITEA3 TESTOMAT and XIVT projects.

References

1. Rebeca: Rebeca. http://www.rebeca-lang.org/. Accessed July 2019
2. Sirjani, M., Movaghar, A., Shali, A., de Boer, F.S.: Modeling and verification of reactive systems using Rebeca. Fundam. Inform. **63**(4), 385–410 (2004)

3. Sirjani, M.: Rebeca: theory, applications, and tools. In: de Boer, F.S., Bonsangue, M.M., Graf, S., de Roever, W.-P. (eds.) FMCO 2006. LNCS, vol. 4709, pp. 102–126. Springer, Heidelberg (2007). https://doi.org/10.1007/978-3-540-74792-5_5

4. Sirjani, M.: Power is overrated, go for friendliness! expressiveness, faithfulness, and usability in modeling: the actor experience. In: Lohstroh, M., Derler, P., Sirjani, M. (eds.) Principles of Modeling. LNCS, vol. 10760, pp. 423–448. Springer, Cham (2018). https://doi.org/10.1007/978-3-319-95246-8_25

5. Sirjani, M., Movaghar, A., Shali, A., de Boer, F.S.: Model checking, automated abstraction, and compositional verification of Rebeca models. J. UCS **11**(6), 1054–1082 (2005)

6. North, D.: Introducing BDD. Better Software Magazine, March 2006. https://dannorth.net/introducing-bdd/. Accessed July 2019

7. Ptolemaeus, C.: System Design, Modeling, and Simulation using Ptolemy II. Ptolemy.org, Berkeley (2014)

8. Rebeca: Afra Tool (2019). http://rebeca-lang.org/alltools/Afra. Accessed July 2019

9. Provenzano, L., Häninnen, K., Zhou, J., Lundqvist, K.: An ontological approach to elicit safety requirements. In: Asia-Pacific Software Engineering Conference, APSEC, pp. 713–718 (2017)

10. Zhou, J., Häninnen, K., Lundqvist, K., Provenzano, L.: An ontological approach to hazard identification for safety-critical systems. In: 2nd International Conference Reliability and System Engineering, ICRSE, pp. 54–60 (2017)

11. Zhou, J., Häninnen, K., Lundqvist, K., Provenzano, L.: An ontological approach to identify the causes of hazards for safety-critical systems. In: 2nd International Conference System Reliability and Safety, ICSRS, pp. 405–413 (2017)

12. Fowler, M.: ThoughtWorks: GivenWhenThen (2013). https://martinfowler.com/bliki/GivenWhenThen.html. Accessed July 2019

13. Rebeca: RMC Tool (2016). http://rebeca-lang.org/alltools/RMC. Accessed July 2019

14. Lee, E.A.: Cyber physical systems: design challenges. In: 11th IEEE International Symposium on Object-Oriented Real-Time Distributed Computing (ISORC), pp. 363–369 (2008)

15. David, A., Larsen, K.G., Legay, A., Mikučionis, M., Poulsen, D.B.: Uppaal SMC tutorial. Int. J. Softw. Tools Technol. Transfer **17**(4), 397–415 (2015). https://doi.org/10.1007/s10009-014-0361-y

16. de Boer, F.S., et al.: A survey of active object languages. ACM Comput. Surv. **50**(5) 76:1–76:39 (2017)

17. Khamespanah, E., Sirjani, M., Mechitov, K., Agha, G.: Modeling and analyzing real-time wireless sensor and actuator networks using actors and model checking. STTT **20**(5), 547–561 (2018). https://doi.org/10.1007/s10009-017-0480-3

18. Yousefi, B., Ghassemi, F., Khosravi, R.: Modeling and efficient verification of wireless ad hoc networks. Formal Asp. Comput. **29**(6), 1051–1086 (2017). https://doi.org/10.1007/s00165-017-0429-z

19. Sharifi, Z., Mosaffa, M., Mohammadi, S., Sirjani, M.: Functional and performance analysis of network-on-chips using actor-based modeling and formal verification. ECEASST **66**, 1–16 (2013)

20. Dwyer, M.B., Avrunin, G.S., Corbett, J.C.: Patterns in property specifications for finite-state verification. In: International Conference on Software Engineering, ICSE, pp. 411–420 (1999)

21. Konrad, S., Cheng, B.H.C.: Real-time specification patterns. In: International Conference on Software Engineering, ICSE, pp. 372–381 (2005)

22. Mavin, A., Wilkinson, P., Harwood, A., Novak, M.: Easy approach to requirements syntax (ears). In: IEEE International Requirements Engineering Conference, RE, pp. 317–322 (2009)
23. Konrad, S., Cheng, B.H.: Real-time specification patterns. IEEE Trans. Softw. Eng. **30**, 970–992 (2004)
24. Filipovikj, P., Jagerfield, T., Nyberg, M.G., Rodriguez-Navas, C.S.: Integrating pattern-based formal requirements specification in an industrial tool-chain. In: IEEE Annual Computer Software and Applications Conference, COMPSAC, pp. 167–173 (2016)
25. Object Management Group: OMG Systems Modeling Language v1.5 (2017). https://sysmlforum.com/sysml-specs/. Accessed July 2019
26. Object Management Group: UML profile for MARTE, beta 2 (2008). https://www.omg.org/omgmarte/Specification.htm. Accessed July 2019
27. Alavizadeh, F., Nekoo, A.H., Sirjani, M.: ReUML: a UML profile for modeling and verification of reactive systems. In: International Conference on Software Engineering Advances ICSEA, pp. 50–55 (2007)
28. MathWorks: Stateflow: model and simulate decision logic using state machines and flow charts (2018). https://www.mathworks.com/products/stateflow.html. Accessed July 2019
29. Molnár, V., Graics, B., Vörös, A., Majzik, I., Varró, D.: The Gamma statechart composition framework. In: International Conference on Software Engineering, ICSE, pp. 113–116 (2018)

CIFMA 2020

Organization

CIFMA 2020 – Workshop Chairs

Pierluigi Graziani   University of Urbino, Italy
Pedro Quaresma   University of Coimbra, Portugal

CIFMA 2020 – Programme Committee

Samuel Alexander	The U.S. Securities and Exchange Commission New York Regional Office, USA
Oana Andrei	University of Glasgow, UK
John A. Barnden	University of Birmingham, UK
Francesco Bianchini	University of Bologna, Italy
José Creissac Campos	University of Minho, Portugal
Antonio Cerone	Nazarbayev University, Kazakhstan
Peter Chapman	Edinburgh Napier University, UK
Luisa Damiano	University of Messina, Italy
Anke Dittmar	University of Rostock, Germany
Pierluigi Graziani	University of Urbino, Italy
Yannis Haralambous	IMT Atlantique, France
Matej Hoffmann	CTU Prague, Czech Republic
Bipin Indurkhya	Jagiellonian University, Poland
Reinhard Kahle	NOVA University Lisbon, Portugal
Karl Lermer	ZHAW, Switzerland
Kathy L. Malone	Nazarbayev University, Kazakhstan
Paolo Masci	National Institute of Aerospace (NIA), USA
Paolo Milazzo	University of Pisa, Italy
Henry Muccini	University of L'Aquila, Italy
Eugenio Omodeo	University of Trieste, Italy
Graham Pluck	Nazarbayev University, Kazakhstan
Giuseppe Primiero	University of Milan, Italy
Ka I. Pun	Western Norway University of Applied Sciences, Norway
Pedro Quaresma	University of Coimbra, Portugal
Anara Sandygulova	Nazarbayev University, Kazakhstan
Giuseppe Sergioli	University of Cagliari, Italy
Sandro Sozzo	University of Leicester, UK
Mirko Tagliaferri	University of Urbino, Italy

CIFMA 2020 – Additional Reviewers

Alessandro Aldini   University of Urbino, Italy
Gianluca Curzi   University of Birmingham, UK
Flavia Marcacci   Pontifical Lateran University, Vatican State

A Pragmatic Model of Justification for Social Epistemology

Raffaela Giovagnoli(✉)

Pontifical Lateran University, Rome, Italy
giovagnoli@pul.it

Abstract. Social epistemology presents different theories about the status of shared knowledge, but only some of them retain a fruitful relation with classical epistemology. The aim of my contribution is to present a pragmatic model which is, on the one side, related to the classical concepts of "truth" and "justification", while, on the other side, addressing to a fundamentally "social" structure for the justification of knowledge. The shift from formal semantics to pragmatics is based on a notion of "material inference" embedding commitments implicit in the use of language, that favors the recognition of the social source of shared knowledge.

Keywords: Social epistemology · Truth · Justification · Material inference · Deontic statuses · Deontic attitudes

1 Introduction

Social epistemology presents different perspectives concerning the assessment of "social evidence". We can (I) assess the epistemic quality of individual doxastic attitudes when social evidence is used; (II) assess the epistemic quality of group attitudes or (III) assess the epistemic consequences of adopting certain institutional devices or systemic relations as opposed to alternatives [1]. The so-called "communitarian epistemology" (Hardwig, Welbourne, McIntyre, Brandom, Kusch) falls into the first stream and, particularly, maintains that knowledge is "essentially" social.

In this contribution, we'll sketch a social model of knowledge representation made explicit by a form of "expressive logic", which rests on a complex game of deontic statuses and deontic attitudes [2]. This pragmatic order of explanation focuses on the role of expression rather than representation. In this context, "expression" means to make explicit in assertion what is implicit in asserting something. A fundamental claim of this form of expressivism is to understand the process of explicitation as the process of the application of concept. According to the relational account, what is expressed must be understood in terms of the possibility of expressing it. Making something explicit is to transform it in premise and conclusion of inferences. What is implicit becomes explicit as reason for asserting and acting. Saying or thinking something is undertaking a peculiar kind of inferentially articulated commitment. It shows a deontic structure that entails the authorization of the inference as a premise and the responsibility to entitle oneself

© Springer Nature Switzerland AG 2021
L. Cleophas and M. Massink (Eds.): SEFM 2020 Workshops, LNCS 12524, pp. 89–99, 2021.
https://doi.org/10.1007/978-3-030-67220-1_7

to that commitment by using it (under adequate circumstances) as conclusion of an inference from other commitments one is or can become entitled. To apply a concept is to undertake a commitment that entitles to and precludes other commitments. Actually, there is a relevant difference between the Wittgensteinian theory of linguistic games and the scorekeeping model. Inferential practices of producing and consuming reasons are the point of reference of linguistic practices. Claiming is being able to justify one's claims and other claims (starting from one's claims) and cannot be considered as a game among other linguistic games. Following Sellars, Robert Brandom uses the metaphor of the "space of reasons", but he understands it as a "social" concept, i.e. as the space of the intersubjective justification of our assertions [3]. Reasons contained in assertions possess a content that is inferentially structured. The formal structure of communication gives us the possibility to make explicit this content. From the point of view of a "social" concept of the space of reasons, beliefs, mental states, attitudes and actions possess a con-tent because of the role they play in social "normative" practices (inferentially articulated).

Before to introduce a social concept of the space of reasons we want to make clear the sense in which we are talking of "normativity" as grounded on linguistic rules. The functioning of scorekeeping in a language game has been presented by David Lewis [4]. The result of Lewis' model is useful to understand the context dependence of ordinary conversation and this option helps us to grasp in plausible way the nature of the content in the game of giving and asking for reasons. The content of beliefs and actions is "phenomenalistic" because it expressed by inferential rules in the sense of material incompatibility. Moreover, the grasp of the conceptual content is possible only using intersubjective pragmatic rules that in some sense "harmonize" the collateral beliefs of the participants.

2   Communitarian Epistemology

To clarify the notion of "communitarian epistemology" it is important to analyze the con-cept of 'evidence', mostly considered in the philosophy of science and in the sociology of scientific knowledge. John Hardwig has subjected the conception of the 'individual' evidence to a critical analysis full of interesting ideas. According to this conception, there may be good reasons for a belief 'that p' if we have 'evidence' in favor of it; and the evidence is "anything that counts toward establishing the truth of p (i.e., sound arguments as well as factual argumentation)" [5].

Suppose that my trusted doctor told me that I have been suffering from a rare foot disease for many years. He or she has good reasons for diagnosis; in fact, given his professional experience, he can form a reliable judgment by studying the x-rays on my foot and the manner of walking. However, it may be that I do not feel pain, do not see anything strange in the manner of walking and do not find anything surprising on the radiographs (after all I am not a doctor). Clearly my doctor has good reasons to believe that I have that condition. In turn, I have good reasons to believe my doctor's diagnosis. But do my reasons constitute the evidence for the truth of the diagnosis? According to individualism, the answer must be negative. My reasons for believing in the diagnosis do not correspond to my doctor's reasons. The good reasons of my doctor are not enough to establish a connection of trust. They do not strengthen after the announcement of the diagnosis. But, according to Hardwig, the 'narrow' conception of evidence conflicts with common sense. We must therefore expand it by introducing a concept that includes second-hand evidence. Normally we believe what our trusted doctor tells us and therefore our reasons correspond to his. In general, we rely on the knowledge of experts and in everyday practice it would be irrational to do otherwise, because we are unable to control the truth and accuracy of the testimony. Sometimes we test the credentials of the experts when they conflict with the judgments of other experts. But we are not obliged to always use our head.

Hardwig extends the authority of testimony to knowledge in general (so it does not apply only to 'rational belief'). He writes [6]:

belief based on testimony is often epistemically superior to belief based on entirely direct, non-testimonial evidence. For [one person] b's reasons for believing p will often be epistemically better than any [other person] a would/could come up with on her own. If the best reasons for believing p are sometimes primarily testimonial reasons, if knowing requires having the best reasons for believing, and if p can be known, then knowledge will also sometimes rest on testimony.

This thesis is supported by arguments drawn from the scientific practice. Scientists form routine teams and these teams are formed on the basis of testimony and trust. Hardwig refers as an example to an experiment conducted by physicists on high energy in the early 1980s [7]:

After it was funded, about 50 man/years were spent making the need-ed equipment and the necessary improvements in the Stanford Linear Accelerator. Then approximately 50 physicists worked perhaps 50 man/years collecting the data for the experiment. When the data were in, the experimenters divided into five geographical groups to analyze the data, a process which involved looking at 2.5 million pictures, making measurements on 300, 000 interesting events, and running the results through computers… The "West Coast group" that analyzed about a third of the data included 40 physicists and technicians who spent about 60 man/years on their analysis.

The research gave rise to an article with 99 co-authors, some of whom will not know how they have arrived at such number. To producing data for such an article presupposes that scientists exchange information and that they consider the results of others as evidence for the measurements in question. It cannot be done otherwise. None of the participating physicists could replace his testimony- based knowledge with perception-based knowledge: doing this would take too much vital time. This type of 'epistemic dependence' can also be found in mathematics; for example, in the de Branges proof of the Bieberbach conjecture, a proof that involved mathematicians with very different forms of specialization. Reading Hardwig, Kusch begins to identify three epistemological alternatives:

1. 'Strict individualism' for which knowledge is in the possession of the individual and presupposes the evidence is sufficient on the resources available.
2. 'Weak individualism' for which it is not necessary to possess evidence for the truth of what is known and perhaps not even fully understand what is known.
3. 'Communitarianism' which sees the community as the primary source of knowledge. It maintains the idea that the acquaintance must have 'direct' possession of the

evidence, but breaks with the assumption that this acquaintance must or can be an individual.

According to Kusch, Hardwig tends to community; and not only for epistemology but for philosophy in general. Testimony is located in an area where epistemology meets ethics. Whether or not the expert's result provides good reasons for believing that p will depend on the recipient's perception of the reliability of the expert's testimony which in turn will depend on an assessment of its character. Here we find the relevance of the reflections of the sociologist Max Weber on the figure of the scientist and the politician. Was the expert sufficiently responsible for considering himself informed for developments in the field? Was he conscientious, and realistic in the self-consideration of how a reliable judgment should be produced? To answering these questions is to make an assessment about a moral and epistemic character together.

The work of Hardwig on teams and trust in scientific practice has been taken up by relevant exponents of contemporary social epistemology (Galison, Knorr Cetina, Shaffer, Shapin and Mackenzie). Kusch highlights two limitations of it. In the first place, Hardwig favors real and proper scientific communities, therefore it does not investigate cases of cooperation more related to daily practice and where testimony plays a crucial role, given that we rely on numerous public messages without investigating sincerity or competence of the sources. In the second place, the way in which Hardwig refers to the evidence of true belief can be criticized. He rejects strong individualism because the evidence can belong to the individual if testimony is allowed (mild individualism) and only teams have sufficient direct evidence. The latter notion is not clear; in fact one wonders: do teams have direct evidence as they have mental states like individuals? Finally, many epistemologists reject the thesis that knowledge is true belief based on evidence. Knowledge is not 'true justified belief', nor 'true belief based on evidence', but 'true belief produced in a reliable manner'. Trustworthiness does not require one to be able to provide reasons for his belief; it is sufficient that this is formed through a reliable process. It may be that reliability is compatible with communitarianism, but, according to Kusch, Hardwig did not clarify this compatibility.

Michael Welbourne wrote the book, The Community of Knowledge, which represents a good example of communitarian epistemology based on testimony [8]. The decisive theoretical step is the consideration of testimony not as a mere 'transmission' of information as for classical epistemology. Knowledge takes place in a community where knowledge is transmitted, according to a certain vision of 'shared knowledge'. To share knowledge means sharing commitments and entitlements with others, at least in many standard cases [9]. His theory of 'authority' runs counter to the theory of evidence. Entitlements imply that we consider knowledge as a base or premise for our inferences since we consider it as an external and objective standard for what others should also recognize. Commitments include the investigation of the authorizations of others so that a dialogic dynamic is created that generates new shared knowledge. In Kusch's words [10]:

Assume that I claim to know how long it takes to travel from Cambridge to Edinburgh; I tell you, and you believe me and tell me so. In doing so, we agree that we should not consent to anyone who suggests a different travel period, that we shall inform each other in case it turns out that we did not possess knowledge after all, that we shall let this information figure in an unchallenged way in travel plans, and so on. We can perhaps go beyond Welbourne by saying that the sharing of knowledge creates a new subject of knowledge: the community. And, once this community is constituted, it is epistemically prior to the individual member. This is so since the individual community members' entitlement and commitment to claiming this knowledge derive from the membership in this community. The individual knows as "one of us", in a way similar to how I get married as 'one of a couple', or how I play football as 'one of the team'.

The major limitation of Welbourne's work, according to Kusch, lies in the fact that he did not consider the 'normative' basis of testimony or the background of knowledge. The fact of believing in what another says, depends very much on sharing the background of the knowledge that provides 'normativity' to the speaker and the hearer; the formation of a community of knowledge presupposes previous communities of knowledge. In addition, the attitude to believe in someone can be described more effectively by notions such as 'participant attitude' or 'trust in others', notions that also imply a moral aspect.

Since the position of Welbourne invites us not to consider testimony as mere transmission, but as the dialogical process of exchange of commitments and entitlements (Wilfrid Sellars' game of giving and asking for reasons') which has a normative background, a further step could be to consider knowledge as 'built' from testimony by means of a sort of 'institutionalization'. In such case we will have the need of a theory of social institutions and social states that are based on the use of the so-called 'performatives' (Austin). The major references for social epistemology are John Searle, Barry Barnes and David Bloor.

'Performative' testimony starts from the act we perform by saying something and how this act is received by our interlocutor. It is not concerned simple 'to say so and so' or mere transmission as in traditional epistemology, but a common construction process. A performative testimony does not allow to consider a state of things p, to refer and to know as discrete, sequential and independent events. For example [11]:

The registrar a tells the couple b that they have now entered a legally binding relationship of marriage; and by telling them so, and their understanding what he tells them, the registrar makes it so that they are in the legally bin ding relationship of marriage. For the registrar's action to succeed, the couple has to know that they are being married through his say-so, and he has to know that his action of telling does have this effect. Moreover, a and b form a community of knowledge in so far as their jointly knowing that p is essential for p to obtains. That is to say, a and b enter into a nexus of entitlements and commitments, and it is this nexus that makes it so that each one of them is entitled to claim that p. The registrar has to use certain formulas (By the power invested in me by the state of California.. . etc.) bride and groom have to confine themselves to certain expressions (a simple "yes" will be fine), and each one commits himself or herself, and entitles the other, to refer to p as a fact subsequently. More principally, we can say that "getting married" is an action that one cannot do on one's own (or just with one's partner). It is an action that is primarily performed by a 'we'.

The new social status and the knowledge that is created between the spouses is generated by the performative testimony, or by the linguistic act performed by the authority in question. The knowledge that 'p' did not exist 'before' the declaration (to use Searle's language). The reasons to explain why performative testimony generates knowledge lie in two important characteristics of performatives: self-reference and self-validity. The act refers to itself in as much as it announces what it does and if is done in the right circumstances (therefore given the institutional setting) it generates the validity of the reality it creates. The act that creates the new social situation is like a common act carried out through an agreement between people. This act is fragmented and distributed to other linguistic acts; it is implicit in daily practices, such as when we greet someone, we talk about greeting colleagues we meet or criticize those who have not responded to our greeting. All these acts for the most part assertions and contain shared performatives. This thesis is fundamental for the epistemology of testimony, since it is mostly realized through shared and widely distributed.

Testimony generates (for the most part) its references and knowledge of them. We can isolate three options for defining knowledge:

(1) Knowledge is equivalent to a term for a natural species such as 'elephant' and then whatever is connoted by the term it continues to exist (although not as knowledge) even if we stop using the common performative (which establishes animal taxonomy and the specimens of 'being an elephant');
(2) Knowledge outlines a social state such as 'money' or 'marriage' and the social institution of knowledge disappears as soon as we stop using the performative.
(3) Knowledge is like the 'typewriter'. The physical or mental entity or the processes that we produce and call 'knowledge' can continue to exist (even if they are no longer called in this way) when we stop using the perfomative (which establishes the relevant taxonomy and the specimens of 'being a typewriter').

Also in case (1) we have to understand knowledge on the basis of a shared performative because the category of 'elephant' has its specimens and models that are socially established and maintained. Kusch's thesis is that knowledge is a social state consisting of a shared performative (We hereby declare that there is a single, recommendable way of possessing the truth, and we call this way 'knowledge'). Knowledge is a social referent created through references to it; and these references occur in testimony, as in other forms of dialogue. Dialogue in fact includes affirming that something is knowledge, posing challenges to knowledge, testing knowledge, doubting knowledge and so on through a broad spectrum of possible references.

3   The Role of Conditionals for Human Discursive Practices

Before to briefly sketch the social normative source of shared knowledge in inferential terms, we need to clarify the very notion of "inference" embedded in conditionals. We are not only creatures who possess abilities such as to respond to environmental stimuli we share with thermostats and parrots but also "conceptual creatures" i.e. we are logical creatures in a peculiar way. It is a fascinating enterprise to investigate how machines simulate human behavior and the project of Artificial Intelligence, a project that began meads of the XX century, could tell us interesting things about the relationship between syntactical abilities and language. Brandom seriously considers the functioning of automata because he moves from some basic abilities and he gradually introduces more sophisticated practices, which show how an autonomous vocabulary raises [12]. This analysis is a "pragmatist challenge' for different perspectives in analytic philosophy such as formal semantics (Frege, Russell, Carnap and Tarski), pragmatics both in the sense of the semantics of token-reflexive expressions (Kaplan and Stalnaker) and of Grice, who grounds conversation on classical semantics. Conditionals are the paradigm of logical vocabulary to remain in the spirit of Frege's Begriffschrift. But, according to Brandom, the meaning-use analysis of conditionals specifies the genus of which logical vocabulary is a species. In this sense, formal semantics is no more the privileged field for providing a universal vocabulary or meta-language.

Starting from basic practices, we can make explicit the rules that govern them and the vocabulary that expresses these rules. There are practices that are common to humans, non-human animals and intelligent machines that can be also artificially implemented like the standard capacities to respond to environmental stimuli. But, it seems very difficult to artificially elaborate the human discursive practices which depend on the learning of ordinary language. In particular, humans are able to make inferences and so to use conditionals because they move in a net of commitments and entitlements embedded in the use of concepts expressed in linguistic expressions. Logical vocabulary helps to make explicit the inferential commitments entailed by the use of linguistic expressions, but the meanings of them depend on the circumstances and consequences of their use. The last meta-language is ordinary language in which we give and ask for reasons and therefore acquire a sort of universality. It seems that, we do not need to apply the classical *salva veritatae* substitutional criterion, as conditionals directly make explicit the circumstances and consequences namely inferential commitments and entitlements possessed by singular terms and predicates [13].

The source of the normativity entailed by conceptual activity is a kind of "autonomous discursive practice" that corresponds to the capacity to associate with materially good inferences ranges of counterfactual robustness [14, 15, 16]. In this sense, "modal" vocabulary represented by modally qualified conditionals such as if p then q has an expressive role. Modal vocabulary is a conditional vocabulary that deserves to codify endorsements of material inferences: it makes them explicit in the form of material inferences that can themselves serve as the premises and conclusions of inferences. According to the argument Brandom calls "the modal Kant-Sellars thesis", we are able to secure counterfactual robustness (in the case of the introduction of a new belief), because we "practically" distinguish among all the inferences that rationalize our current beliefs, which of them are update candidates. The possibility of this practical capacity derives from the notion of "material incompatibility", according to which if we treat the claim that q follows from p as equivalent to the claim that everything materially incompatible with q is materially incompatible with p. So, for example if we say "Cabiria is a dog" entails "Cabiria is a mammal" we are stating that everything in-compatible with her being a mammal is incompatible with her being a dog.

For the sake of my discussion, it is interesting how we can intend Kantian normativity in terms of "incompatibility" relations between commitments. Actually, there is a distinction between empirical vocabulary and modal vocabulary, because the world cannot tell us what we ought to do in certain situations. The content is normally understood in terms of representation of objects. The scorekeeping model replaces the Kantian notion transcendental apperception with a kind of synthesis based on incompatibility relations. In drawing inferences and "repelling" incompatibilities, a person is taking oneself to stand in representational relations to objects that she is talking about. A commitment to A's being a horse does not entail a commitment to B's being a mammal. But it does entail a commitment to A's being a mammal. Drawing the inference from a horse- judgment to a mammal-judgment is taking it that the two judgments represent one and the same object. Thus, the judgment that A is a horse is not incompatible with the judgment that B is a cat. It is incompatible with the judgment that A is a cat. Taking a horse-judgment to be incompatible with a cat-judgment is taking them to refer or represent that object, to which incompatible properties are being attributed by the two claims.

The normative rational unity of apperception is a synthesis to expand commitments inferentially, noting and repairing incompatibilities. In this sense, one's commitments become reasons for and against other commitments; it emerges the rational critical responsibility implicit in taking incompatible commitments to oblige one to do something, to update one's commitment so as to eliminate the incompatibility. According to the scorekeeping model, attention must be given not only to "modal" incompatibility but also to "normative" incompatibility. Again, modal incompatibility refers to states of affairs and properties of objects that are incompatible with others and it presupposes the world as independent of the attitudes of the knowing-and-acting subjects.

Normative incompatibility belongs to discursive practices on the side of the knowing-and-acting subjects. In discursive practice the agent cannot be entitled to incompatible doxastic or practical commitments and if one finds herself in this situation one is obliged to rectify or repair the incompatibility. On the side of the object, it is impossible for it to have incompatible properties at the same time; on the side of the subject, it is impermissible to have incompatible commitments at the same time. In this sense, Brandom introduces the metaphysical categorical sortal meta-concept subject whereas it represents the conceptual functional role of units of account for deontic normative incompatibilities. In my opinion, we can intend this role as a "social" role because of the fact that we learn how to undertake deontic attitudes in the process of socialization. The possibility of criticizing commitments in order to be able not to acknowledge incompatible commitments is bound to the normative statuses of commitment and entitlement and we ought to grasp the sense of them.

4 The Dimensions of Justification

The scorekeeping model describes a system of social practices in which agents performs assertions that express material inferential commitments [17]. In the previous section, I considered together with the modal vocabulary also the normative vocabulary both related to the use of ordinary language. Let's see now what are the inferential relations that agents ought to master in order for justifying their claims. Our assertions have a "sense" or are "contentful" by virtue of three dimensions of inferential social practices. To the first dimension belongs the commitment-preserving inference that corresponds to the material deductive inference. For example, A is to the west of B then B is to the east of A and the entitlement preserving inference that corresponds to inductive inference like if this thermometer is well made then it will indicate the right temperature. This dimension is structured also by incompatibility relations: two claims have materially incompatible contents if the commitment to the one precludes the entitlement to the other.

The second dimension concerns the distinction between the concomitant and the communicative inheritance of deontic statuses. To the concomitant inheritance corresponds the intrapersonal use of a claim as a premise. In this case, if a person is committed to a claim is, at the same time, committed to other concomitant claims as consequences. Correspondingly, a person entitled to a commitment can be entitled to others by virtue of permissive inferential relations. Moreover, incompatibility relations imply that to undertake a commitment has as its consequence the loss of the entitlement to concomitant commitments to which one was before entitled. To the communicative inheritance corresponds the interpersonal use of a claim, because to undertake a commitment has as its "social" consequence to entitle others to the "attribution" of that commitment. The third dimension shows the two aspects of the assertion as "endorsed": the first aspect is the "authority" to other assertions and the second aspect dependent to the first is the "responsibility" through which an assertion becomes a "reason" enabling the inheritance of entitlements in social contexts.

The entitlement to a claim can be justified (1) by giving reasons for it, or (2) by referring to the authority of another agent, or (3) by demonstrating the capacity of the agent reliably to respond to environmental stimuli. The scorekeeping model is based on a notion of entitlement that presents a structure of "default" and "challenge". This model is fundamental in order to ground a pragmatic and social model of justification, that requires the participation to the game of giving and asking for reasons. A fundamental consequence of this description is that the deontic attitudes of the interlocutors represent a perspective on the deontic states of the entire community.

We begin with the intercontent/interpersonal case. If, for instance, B asserts "That's blue", B undertakes a doxastic commitment to an object being blue. This commitment ought to be attributed to B by anyone who is in a position to accept or refuse it. The sense of the assertion goes beyond the deontic attitudes of the scorekeepers, because it possesses an inferentially articulated content that is in a relationship with other contents. In this case, if by virtue of B's assertion the deontic attitudes of A change, as A attributes to B the commitment to the claim "That's blue", then A is obliged to attribute to B also the commitment to "That's colored". A recognizes the correctness of that inference when she becomes a scorekeeper and, therefore, consequentially binds q to p. Again, the incompatibility between "That's red" and "That's blue" means that the commitment to the second precludes the entitlement to the first. Then A treats these commitments as incompatible if she is disposed to refuse attributions of entitlement to "That's red" when A attributes the commitment to "That's blue". In the infracontent/interpersonal case, if A thinks that B is entitled (inferentially or not inferentially) to the claim "That's blue", then this can happen because A thinks that C (an agent who listened to the assertion) is entitled to it by testimony.

An interesting point is to see how the inferential and incompatibility relations among contents alter the score of the conversation. First, the scorekeeper A must include "That's blue" in the set of commitments already attributed to B. Second, A must include the commitments to whatever claim which is the consequence of "That's blue" (in commitive-inferential terms) in the set of all the claims already attributed to B. This step depends on the available auxiliary hypothesis i relationship with other commitments already attributed to B. These moves determine the closure of the attributions of A to B by virtue of the commitment-preserving inferences: starting from a priori context with a certain score, the closure is given by whatever committive-inferential role A associates with "That's blue" as part of its content.

Incompatibility also limits the entitlements attributed to B. A can attribute entitlements to what ever claim is a consequence in permissive-inferential terms of commitment to which B was already entitled. For example, B is entitled to "That's blue" because she is a reliable reporter i.e. she correctly applies responsive capacities to environmental stimuli. The correctness of the inference depends also on A's commitment, namely on the circumstances under which the deontic status was acquired (these conditions must correspond to the ones in which B is a reliable reporter of the content of "That's blue"). Moreover, A can attribute the entitlement also by inheritance: reliability of another interlocutor who made the assertion in a prior stage comes into play.

5 Conclusion

The pragmatic model I sketched could represent a valid perspective for social epistemology by virtue of its "relational" perspective. It rests on social evidence that derives from semantic relations among material-inferential commitments and entitlements and pragmatic attitudes expressed by a net of basic speech acts. The structure represents a view of knowledge as projected by the discursive practices of an entire community of language users. Moreover, it is a dynamic model as social practices are always exposed to the risk of dissent. In this context, social practices entail the dimension of challenge, i.e. the case in which the speaker challenges the interlocutor to justify and eventually to repudiate his/her commitment. Even in the case in which an agent acquires the entitlement to act by deferral i.e. by indicating a testimonial path whereby entitlements to act can be inherited, the query and the challenge assume the function of fostering discussion among the participants.

References

1. Goldman, A.: Social epistemology, stanford enciclopedia of philosophy (2015); Giovagnoli R.: Introduzione all'epistemologia sociale, LUP, Vatican City (2017)
2. Brandom, R.: Making It Explicit. Cambridge University Press, Cambridge (1994)
3. Brandom, R.: Knowledge and the social articulation of the space of reasons. Philos. Phenomenol. Res. **55**, 895–908 (1995)
4. Lewis, D.: Scorekeeping in a Language Game, Philosophical Papers. Oxford University Press, New York (1983)
5. Hardwig, J.: Epistemic dependence. J. Philos. **82**, 337 (1985)
6. Hardwig, J.: The role of trust to knowledge. J. Philos. **88**, 698 (1991)
7. Hardwig: p. 347 (1985)

8. Welbourne, M.: The Community of Knowledge. Gregg Revivals, Aldershot (1993)
9. Giovagnoli, R.: The debate on testimony in social epistemology and its role in the game of giving and asking for reasons, Information MDPI, March 2019
10. Kusch, M.: Knowledge by Agreement, p. 60. Oxford University Press, Oxford (2002)
11. Kusch, M.: pp. 65–66 (2002)
12. Brandom, R.: Between Saying and Doing. Oxford University Press, Oxford (2008)
13. Giovagnoli, R.: Why the fregean square of opposition matters for epistemol- ogy. In: Beziau, J.Y., Dale, J. (eds.) Around and Beyond the Square of Opposition. Springer, Basel (2012). https://doi.org/10.1007/978-3-0348-0379-3_7
14. Brandom, R.: (2008)
15. Giovagnoli, R.: Representation, analytic pragmatism and AI. In: Dodig-Crnkovic, G., Giovagnoli, R. (eds.) Computing Nature, pp. 161–170. Springer, Heidelberg (2013). https://doi.org/10.1007/978-3-642-37225-4_9
16. Giovagnoli, R.: The relevance of language for the problem of representation. In: Dodig-Crnkovic, G., Giovagnoli, R. (eds.) Representation and Reality in Humans, Other Living Organisms and Intelligent Machines. SAPERE, vol. 28, pp. 235–245. Springer, Cham (2017). https://doi.org/10.1007/978-3-319-43784-2_11
17. Giovagnoli, R.: From single to relational scoreboards. In: Beziau, J.Y., Costa- Leite, A., - J.M.L., D'Ottaviano, J.M.L. (eds.) Aftermath of the Logical Paradise, Colecao CLE, Brazil, vol. 81, pp. 433–448 (2018)

Personal Identity and False Memories

Danil Razeev$^{(\boxtimes)}$ (iD)

Institute of Philosophy, Saint Petersburg State University, Saint Petersburg, Russia
`d.razeev@spbu.ru`

Abstract. In current philosophy of mind, there are two main approaches to the question of personal identity. The first one claims that personal identity is based on our memory and, for several decades, has been known as a psychological approach to the problem. The second one has been called an animalistic approach and considers personal identity as a biological property of human beings or as a specific feature of our bodily continuity. The experiment on creating false memories in mice brains, recently conducted at Massachusetts Institute of Technology (MIT), seems to shed new light on the question of personal identity, taking into account the fact that the mouse brain is morphologically quite similar to our brain. The purpose of my paper is to consider whether the above-mentioned experiment supports one of the approaches: the psychological or the animalistic. Using the conceptual instrumentarium of contemporary analytic philosophy and cognitive phenomenology, I differentiate between strong and weak false memories and I argue that we cannot consider the conducted experiment to have created false memories in the strong sense. I develop a thought experiment showing what it would be like to experience an implanted (weak) false memory in the human brain. I conclude that there is not and cannot be an experience of the (strong) false memory.

Keywords: Personal identity · False memory · Animalism

1 Introduction

Could a computer or robot be a person? Contemporary philosophers and scientists do not offer an unequivocal answer to this question. Some of them claim that being a person is inherent only to highly developed biological organisms, particularly humans, and cannot be found in non-biological matter. Others express some optimism in this regard and claim that, in the future, it will be possible not only to create artificial intelligence possessing genuine personality but also to transfer a biologically-based personality to an artificially-built one and vice versa.

The problem of personal identity is deep rooted in the history of philosophy. The British philosopher John Locke has been one of the most influential figures

This work was supported by the Russian Foundation for Basic Research under Research Grant 18-011-00840.

in discussing the question about personal identity. In his famous book "An Essay Concerning Human Understanding" [1] he points out that we cannot find the only unitary criterion of identity for all that exists. Locke's argumentation, let me present it in a slightly modern and free manner, suggests dividing identity into three basic types: the first one is the identity of a thing, the second is the identity of a living organism and the last one is the identity of a person. The identity of a thing depends on the identity of material stuff, the identity of a living body goes back to its persistence as an organism, i.e. as a unified whole, and the identity of a person has its roots in the capacity to maintain a kind of self-representation through time. Let us give some examples of Locke's typology of identities. Identity of a thing does not allow us to identify the statue of David with the huge piece of marble, from which Michelangelo's masterpiece was sculptured, rather we deal with a process whereby one thing became something else. An example of the second type of identity would be a cat having been identified by its owner as the same living organism through time, although in its old age it does not look like that pretty kitten, which first entered the house. More confusing is the situation with personal identity and each of us, human animals, could count as an example for this kind of identity. Intuitively, we understand the difference between the identity of a living body and personal identity, as in the case of a human being falling into a vegetative state, where the body remains the same, but its carrier loses his bodily citizenship so to speak. Many contemporary philosophers think that the question about personal identity should be considered beyond the framework of our living bodies.

Even careful usage of the term identity can lead us to a set of very difficult questions, such as: 1. Does the identity of a thing mean the material identity atom by atom? 2. Where can a material boundary for a living thing be found? 3. Can a digital copy of a person be created? 4. Would a person remain the same, if they were reproduced using a different material carrier? 5. What would happen to a person if they were reproduced using two carriers, materially identical atom by atom? And so on.

2 Experimental Data

The question about what makes us identical through time has not found any unambiguous answer and has been discussed by many philosophers studying the problem of personal identity since Locke's time. Nowadays, there exist two general approaches to the problem: the psychological one, which is sometimes called psychological reductionism [2–5], and the somatic one, the so-called animalism [6–9]. According to the first approach, a criterion for personal identity has to be found in our psychological continuity over time. The second one tries to find this criterion in the persistence of our bodily organization. Psychological reductionism has been continuing Locke's attempt to find a certain criterion for personal identity in the mechanism of memory. Animalism regards this approach as conceptually wrong and claims that the identity of a person can be completely reduced to the identity of this or that living body, in our case to the identity of a living body of the human type or, in short, of a human animal.

In my paper I would like to consider in detail a very interesting scientific experiment, recently conducted by neuroscientists at Massachusetts Institute of Technology (MIT). It pretends to change our understanding of genesis and structure of subjectivity and shed new light on the contemporary discussion about whether or not personal identity could count as a special and irreducible type of identity. The neuroscientists Susumi Tonegawa and his colleagues at MIT claim to have created a false memory in the brain of a living organism [10].

First of all, let me recall the details of the experiment on the mouse brain. For the experiment scientists used genetically modified mice whose neuronal activity in hippocampus, a specific region in the brain responsible for memory, could be activated or deactivated by flashes of light, using a special laser device attached to the mouse brain.

At first, the mice were placed in a box (box A) with a comfortable environment and the neuroscientists were able to trace the neuronal activity in the mice's hippocampus. After that, the mice were moved to another box (box B), where their memories about being in Box A were activated with a laser while at the same time their feet were shocked with electricity. Using this technique in the mice brains, an association of being in Box B and experiencing some fear there was created. Being placed again in Box A, the mice behaved as if they remembered some negative experience in Box B that in reality had never happened. In such a way the neuroscientists came to a general conclusion that they had created in the mice brains a false memory or a memory about something that never actually happened.

Although the experiment was conducted on mice brains, in my opinion, it has very serious philosophical and ethical consequences for the understanding of our own subjectivity. Even though morphologically the mouse brain and the human brain are similar, to draw conclusions about the structure of our subjectivity based only on the results of the experiment would not be correct. Nevertheless, in philosophy we can conduct so-called thought experiments. As philosophers we are permitted to suggest that a set of neural events that happened in the mouse brain during the experiment could have happened in the human brain, despite the fact that it cannot be verified at the present moment due to the lack of technology or because of ethical restrictions. In other words, I would like to discuss some philosophical consequences that would have arisen if the experiment had been conducted on humans and it had resulted in the creation in the human brain of false memories about some events that in reality never happened.

If we extrapolated this experiment from mice to people, not taking into account technical and ethical aspects [11], it could work in the following way. At first, a volunteer is placed in a blue room with a comfortable environment. Then, s/he is moved to a red room with an uncomfortable environment, where neuroscientists, using a special technology, would activate a specific region in the volunteer's brain responsible for the memory of his/her previous presence in the blue room and thereby create an additional association between his/her presence in the blue room and the uncomfortable environment in the red room. Lastly, s/he is moved back to the blue room. The result of the experiment is expected to be as follows. The volunteer will remember some negative experience about her/his previous presence in the blue room that in reality never happened to her/him.

Having become acquainted with the details of the experiment, I would like to involve it in the contemporary discussion about personal identity.

3   Evaluation and Discussion

To what extent is the mechanism of memory crucial for personal identity? According to the psychological approach, memory plays the fundamental role in the constitution of personal identity. Many philosophers who represent this approach are very familiar with psychological theories about so-called long-term memory [12,13]. They claim that personal identity is represented by a number of systems within long-term memory [14–17]. Contemporary psychologists consider long-term memory to contain two basic systems: procedural and declarative. The procedural memory system is responsible for the acquisition and retention of perceptual, motor and cognitive skills, while the declarative memory system involves facts and beliefs about the world. Some psychologists distinguish two subsystems within declarative memory: episodic and semantic [18,19]. Semantic memory contains relatively generic information about the world, such as mathematical knowledge and general context-free facts about the world, without specifying when, where and how such information was acquired. Semantic memory plays an important role in our self-identity. Within semantic memory two kinds of self-related memories can be differentiated: 1) semantic factual knowledge of the self, 2) knowledge of one's own traits. Although self-trait memory gives a rudimental sense of self, it is not sufficient for a sense of personal identity over time. Episodic memory, in contrast to semantic memory, has access to the events that have been experienced by a subject at a particular point in space and time, such as what one did yesterday evening and so on. In other words, episodic memory makes explicit reference to the time and place of its acquisition. In relation to the problem of personal identity, it is this feature of episodic memory that has made it the center of research interest for both psychologists [20,21] and philosophers [22–25]. Retrieval from episodic memory has been described as having a self-referential quality that is not available in other types of memory. Using episodic memory, a subject can re-experience events from their past constructing a personal narrative, i.e. a life story [26,27].

Could the experiment on false memories shed new light on the problem of personal identity? The experiment seems to challenge the psychological approach to the problem of personal identity. Obviously, memory, being controlled from the outside, cannot be an intrinsic feature of personal identity and should be regarded rather as an extrinsic, additional mechanism. If, using a specific technology, neuroscientists were able to switch on or off some neuronal populations in the hippocampus, and to make someone recall events that never happened or make them suppress others that did, and, thereby, were able to manipulate their behavior in the future, the mechanism of memory would lose its central and fundamental role for the formation of personal identity, particularly for the constitution of phenomenal self. It would mean that we do not need a personal identity in order to exist and we are nothing but highly organized animatons possessing several sophisticated cognitive mechanisms, one of which is memory [28,29]. Nevertheless, I suggest being cautious and raising some important questions concerning the experiment before drawing radical philosophical conclusions. The first question concerns the status of false memory in the experiment. More specifically, with what kind of memory are we dealing in the experiment: is it false memory or rather a kind of modified memory? The second question is related to the very process of memorization in the experiment. I am asking, whether actually experiencing an event is a necessary condition for creating memory about it or whether memory can be created even if the event was not actually experienced. The third question is about the role that the mechanism of memory seems to play with regard to the identity of a person and whether modifying memory or creating false memory can significantly change personal identity.

At a first glance, the experiment seems to support the animalistic approach. The question about the status of memory in the experiment on the mouse brain is a core one. Let me differentiate between creating a new memory and modifying the already existing memory. Neuroscientists called the type of memory they dealt with in the experiment "false memory". I think they misinterpreted the very concept of false memory. Contemporary neuroscience does not offer a precise definition of false memory. Even in the famous paper of the cognitive psychologist E. Loftus, who was one of the first scientists to introduce this concept [30], we cannot find a clear definition of what exactly this term means. Loftus prefers to use it contextually and, in general, considers false memory to be a subtype of memory distortion. She provides very deep insights into how false memories are formed. Many false memories begin through suggestive misinformation that leads to a false recollection of an event or idea. Her studies show that a strong form of suggestion has led many subjects to believe or even remember in detail events that did not happen, that were completely manufactured with the help of family members, and that would have been traumatic had they actually happened. In the context of our study, another conclusion made by Loftus where she claims that her findings do not, however, give us the ability to reliably distinguish between real and false memories is important [31]. In my opinion, the experiment on the mouse brain dealt with false memory as a subtype of distorted memory, which I would call false memory in the weak sense. Let us take a look at the details of the experiment from this perspective. Firstly, being in Box A, the mice stored some information about being in a positive environment. Secondly, being moved to Box B, their memories about being in Box A were activated and additionally associated with experiencing some fear, because their feet received an electric shock. Thirdly, being placed again in Box A, the mice recalled a distorted memory, i.e. their original memory of being in Box A was superimposed by their memory of being in Box B. I think that the overlapping of memories can lead to creating distorted memory and is part of our everyday psychological process. We only need to consider cases of eyewitnesses at a crime scene. Being asked, just after an incident they usually cannot recall any specific information, but later, step by step, they begin to recall details of the incident very vividly. In a broad sense, all our memories can be regarded as distorted memories. Individual pieces of memory do not exist in isolation. Each one is always recalled in a new context and, being recalled in the present, has already been modified. In my opinion, in the experiment, scientists did not create in the mice brains false memories in the strong sense. They combined the memories that had existed in each of the mice brains before, resulting in what I called false memory in the weak sense.

Creating false memories in the strong sense is connected to the second of the above-mentioned questions, namely, whether it is possible to create such a memory that would not refer to an event experienced earlier by a subject [32]. If we tried to imagine the conditions of such an experiment on false memories in the strong sense, they would probably be as follows. Subject number one is placed in a blue room and subject two in a red room. Then both subjects are moved to a green room where, using a new sophisticated technology, neuroscientists exchange the subjects' memories about being in the blue and red rooms respectively. After that, the subjects are put in the previous rooms, but now in reverse order: subject number one in the red room and subject two in the blue one. The experiment would succeed if subjects could report remembering already having been in the rooms. It would seemingly prove that neuroscientists could create false memories in subjects about their being where they had never been before. Nevertheless, I am afraid that even this hypothetical experiment, were it technically possible in the future, would not prove that false memories in the strong sense are possible. The problem is more difficult then it appears to be. To what extent being a person presupposes recalling in memory what happened in your own experience and not in another's experience? Often we coordinate our behavior by recalling in our semantic memory something that never happened personally to us. I do not need to be hit by a car in order to realize that the street should be crossed on a green light. Other animals certainly possess a similar mechanism of memory. They, like us, are capable of learning something by recalling the experience of others. And they can do it without being persons. Even if subjects could recall each other's memories or have their own memories exchanged, it would not mean that false memories in the strong sense had been created in their minds. In my opinion, even in this case, each subject's memories would remain false memories in the weak sense, because each of them would, in the end, refer to an experience undergone earlier by another subject. It means that the above-mentioned hypothetical experiment with subjects recalling each other's memories or having their own memories exchanged does not give a physically possible example of false memory in the strong sense. Moreover, I am inclined to regard the very notion of false memory in the strong sense as a self-contradictory notion, i.e. not only physically not possible, but also logically not possible. In order to be recalled, memories per definition must refer to experiences that have actually happened before, regardless of which of the subjects' minds was involved. In other words, only something that has happened before in the actual experience of a conscious mind can be recalled in our memory. Whether the experiencing being and the recalling being have to be the same one is another question.

4   Conclusion

Could the results of the experiment on false memories be used as empirical support of the animalistic approach to personal identity and contra the psychological one? I don't think so. All that the experiment has proven is the existence of a certain type of memory that can be activated bypassing the phase of actual experience. It means that as human animals we are able to maintain our life, process information cognitively, regulate behavior and we can carry out all of these activities using a specific type of memory, which is not accompanied by awareness. At the same time, it does not prove that we do not possess a different type of memory, which defines us as persons and takes us beyond mere human animals. It means that the mechanism of memory still remains a vessel for personal identity.

In general, our analysis of the experiment on false memories shows the following. Firstly, the experiment does not support the animalistic approach to personal identity. If animalism were true, we could not exist except as animals and personal identity would be a phenomenon that is solely inherent to some highly developed biological organisms. Secondly, our analysis emphasizes support of the psychological approach to personal identity. If the psychological approach is correct, then memory is a core mechanism of personal identity. The psychological approach gives us hope for the development, in the future, of such a form of artificial intelligence that could possibly possess a genuine personality.

References

1. Locke, J.: An Essay Concerning Human Understanding. Oxford University Press, Oxford (1975)
2. Parfit, D.: Reasons and Persons. Oxford University Press, Oxford (1984)
3. Shoemaker, S.: Self-Knowledge and Self-Identity. Cornell University Press, Ithaca (1963)
4. Shoemaker, S., Swinburne, R.: Personal Identity. Blackwell, London (1984)
5. Unger, P.: Identity, Consciousness, and Value. Oxford University Press, Oxford (1990)
6. Hudson, H.: A Materialist Metaphysics of the Human Person. Cornell University Press, Ithaca (2001)
7. Olson, E.: The Human Animal: Personal Identity Without Psychology. Oxford University Press, Oxford (1997)
8. Olson, E.: What Are We? A Study in Personal Ontology. Oxford University Press, Oxford (2007)

9. Thomson, J.: People and their bodies. In: Dancy, J. (ed.) Reading Parfit, pp. 202–209. Blackwell, London (1997)
10. Ramirez, S., et al.: Creating a false memory in the hippocampus. Science **341**(6144), 387–391 (2013)
11. Liao, S.: The ethics of memory modification. In: Bernecker, S., Michaelian, K. (eds.) Routledge Handbook of Memory, pp. 373–382. Routledge, New York (2017)
12. Cohen, N.: Preserved learning capacity in amnesia: evidence for multiple memory systems. In: Squire, L., Butters, N. (eds.) Neuropsychology of Memory, pp. 83–103. Guilford Press, New York (1984)
13. Squire, L.: Memory and Brain. Oxford University Press, New York (1987)
14. Tulving, E.: Elements of Episodic Memory. Oxford University Press, New York (1983)
15. Gillihan, S., Farah, M.: Is self special? A critical review of evidence from experimental psychology and cognitive neuroscience. Psychol. Bull. **131**, 76–97 (2005)
16. Klein, S.: The self: as a construct in psychology and neuropsychological evidence for its multiplicity. WIREs Cogn. Sci. **1**, 172–183 (2010)
17. Rathbone, C., Moulin, C., Conway, M.: Autobiographical memory and amnesia: using conceptual knowledge to ground the self. Neurocase **15**, 405–418 (2009)
18. Gennaro, R.: Consciousness and Self-consciousness. John Benjamins Publishing, Phildelphia (1996)
19. Parkin, A.: Memory and Amnesia. Basil Blackwell, New York (1997)
20. Klein, S.: The cognitive neuroscience of knowing one's self. In: Gazzaniga, M. (ed.) The Cognitive Neurosciences, pp. 1007–1089. MIT Press, Cambridge (2004)
21. Klein, S., Ganagi, C.: The multiplicity of self: neuropsychological evidence and its implications for the self as a construct in psychological research. Ann. N. Y. Acad. Sci. **1191**, 1–15 (2010)
22. Brennan, A.: Amnesia and psychological continuity. Philos. Supplement. **11**, 195–209 (1985)
23. Campbell, S.: Rapid psychological change. Analysis **64**, 256–264 (2004)
24. Northoff, G.: Are "Q-Memories" empirically realistic? A neurophilosophical approach. Philos. Psychol. **13**, 191–211 (2000)
25. Schechtman, M.: Personhood and personal identity. J. Philos. **87**, 71–92 (1990)
26. Eakin, P.: Living Autobiographically: How we Create Identity in Narrative. Cornell University Press, Ithaca (2008)
27. Fivush, R., Haden, C.: Autobiographical Memory and the Construction of a Narrative Self. Lawrence Erlbaum Publishers, Mahwah (2003)
28. Shettleworth, S.: Cognition, Evolution, and Behavior. Oxford University Press, New York (2010)
29. Millin, P., Riccio, D.: False memory in nonhuman animals. Learn. Memory **26**(10), 381–386 (2019)
30. Loftus, E.: The reality of repressed memories. Am. Psychol. **48**(5), 518–537 (1993)
31. Loftus, E., Pickrell, J.: The formation of false memories. Psychiatric Ann. **25**, 720–725 (1995)
32. Vetere, G., et al.: Memory formation in the absence of experience. Nat. Neurosci. **22**(6), 933 (2019)

Against the Illusory Will Hypothesis

A Reinterpretation of the Test Results from Daniel Wegner and Thalia Wheatley's *I Spy* Experiment

Robert Reimer[✉][iD]

Universität Leipzig, Leipzig, Germany
robertreimer@gmx.de

Abstract. Since Benjamin Libet's famous experiments in 1979, the study of the will has become a focal point in the cognitive sciences. Just like Libet, the scientists Daniel Wegner and Thalia Wheatley came to doubt that the will is causally efficacious. In their influential study *I Spy* from 1999, they created an experimental setup to show that agents erroneously experience their actions as caused by their thoughts. Instead, these actions are caused by unconscious neural processes; the agent's 'causal experience of will' is just an illusion. Both the scientific method and the conclusion drawn from the empirical results have already been criticized by philosophers. In this paper, I will analyze the action performed in the *I Spy* experiment and criticize more fundamentally the assumption of a 'causal experience of will'. I will argue that the experiment does not show that the agent's causal experience of will is illusory, because it does not show that there is a causal experience of will. Against Wegner & Wheatley's assumption, I will show that it is unlikely that the participants in the *I Spy* experiment experienced their conscious thoughts as causally efficacious for an action, that they did not perform at all. It is more likely, that they experienced their own bodily movement as causally efficacious for a cooperative action, that they did not perform solely by themselves.

Keywords: Philosophy of cognition · Consciousness of the will · Causal theories of action

1 Introduction

In their paper *Apparent Mental Causation*, Daniel Wegner and Thalia Wheatley write: "Conscious will is a pervasive human experience." [13] However, it might be an illusion that an agent's conscious thoughts to perform an action cause that action.

The neuroscientists Wegner & Wheatley are proponents of the illusory will hypothesis. According to it, agents only have the impression that there is a causal relation between their thought of performing an action and the performance of that action. They call the experience of this causal relation 'causal experience of will'. Instead of being caused by her own thoughts, the agent's action is

© Springer Nature Switzerland AG 2021
L. Cleophas and M. Massink (Eds.): SEFM 2020 Workshops, LNCS 12524, pp. 108–117, 2021.
https://doi.org/10.1007/978-3-030-67220-1_9

caused by unconscious neural processes. The illusory causal experience of will arises, because these unconscious neural processes that actually cause the action additionally produce the conscious thoughts about it. The thoughts, however, remain impotent epiphenomena.

To refute the assumption that there is a real causal relation between the agent's thoughts of performing an action and the performance of that action, Wegner & Wheatley tried to create an experimental setup in which participants first develop a thought to move an object and second perceive a movement adequate to this thought (the supposed action) without actually moving the object. Wegner & Wheatley believed to show that, if it turned out that the participants in the experiment still have the impression that they moved the object, we (in general) just have the "[...] feeling we willfully cause what we do", without doing it [13]. This experiment was called 'I Spy'.

Wegner & Wheatley are not the first scientists setting up an experiment to test the causal efficiency of the conscious will. The pioneer scientist in the field of consciousness, Benjamin Libet, also assumed that the conscious will must be something that causes the action to happen and that can be felt by the agent [5]. In a series of experiments in 1979, Libet advised his participants to measure the time when they felt the will to act in advance of acting itself. After measuring their neural processes, Libet discovered that their will to act was preceded by unconscious nerve cell activities in the motor cortex. Based on these results, Libet concluded that the will is either an epiphenomenon (and therefor causally impotent) or caused by these previous neural activities (and therefor itself causally determined). Wegner & Wheatley's hypothesis about the nature and function of the will is highly influenced by Libet's experiments.

On the other hand, Wegner & Wheatley's I Spy study and their theory about the nature of the will influenced the work of many other scientists in the field of psychology and neurobiology, such as Lau [4], Haggard [2], Mogi [6]. Haggard, for example, not just shared the assumption of a distinction between what he called 'the experience of intention' and 'the experience of agency' [2]. He also defended the illusory will hypothesis. Just like Wegner & Wheatley, Haggard tried to create situations in which agents have thoughts about an action without performing the action.

Wegner himself further developed his illusory will hypothesis in his book *The Illusion of Conscious Will* [10], and further defended it in other, similarly structured studies, such as *the helping hands experiment* [12].

In this paper, I will take a closer look at the 'causal experience of will' that is presupposed by Wegner & Wheatley's illusory will hypothesis. I agree with Wegner & Wheatley that there is no causal relation between the thoughts (of willing to act) and that action. However, I do not think that there is an experience of such a causal relation, either. In acting, agents do not experience any causal relation between a previously experienced thought and a subsequent action. Acting and willing to act are not two separable and causally dependent events. They are synchronous and inseparable.

Many philosophers, such as Wittgenstein,[1] have already argued against the assumption of a separate experience of will and in favor of a unity between thinking and acting. I will not participate in this general discussion within the philosophy of action. However, my aim is to support the assumption of the unity between thinking and acting indirectly by showing that the test results of *I Spy*, differently interpreted, do not support the assumption that there is a causal experience of will.

In Sect. 2, I will briefly explain what it means to experience a causal relation in general and what it could mean to experience a causal relation between one's own thoughts of willing to perform a certain action and that action. In Sect. 3, I will introduce the experiment which is supposed to show that agents can have thoughts and experience these thoughts as (more or less) causally efficacious for the action without acting. In Sect. 4, I will show that the empirical results do not support that interpretation. Instead, I will present my own alternative interpretation according to which the participants in *I spy* rather moved their fingers and experienced these bodily movements as (more or less) causally efficacious for the execution of the overall action.

2   The Experience of Causal Relations

Consider the following situation: You are playing billiard. You hit the white ball and observe it hitting the black ball. Then you observe the black ball starting to roll, too. You assume that these events are causally dependent on each other: The white ball, by striking the black ball, caused the black ball to roll, too[2]:

The white ball (by hitting the black ball) $\rightarrow_{caused}$ the black ball (to roll)

Even if you are skeptical, whether there is a real causal relation between these events, your skepticism might vanish after repeating the trial. David Hume argued that the idea of a causal relation, or a causal connection, is based on the perception of the temporal succession of two events that can be repeated. If you notice that events of type A, such as the movement of the white ball, are regularly followed by events of type B, such as the movement of the black ball, your mind concludes that there must be a causal relation between instances of type A and instances of type B. According to Hume, however, you suffered from

---

[1] Consider this passage in the Philosophical Investigations: "When I raise my arm 'voluntarily', I do not use any instrument to bring the movement about [...] 'Willing if it is not a sort of wishing, must be the action itself [...]'" [14].

[2] Each causal relation consists of two objects (agens and patiens) being involved in two separate events; the event of the patiens (effect) is causally dependent on the event of the agens (cause). I will frequently use schemas like this to illustrate the structure of certain causal relations. These schemas should be read in the following way:

Agens (cause-event) $\rightarrow_{caused}$ patiens (effect-event)

a 'causal illusion' in positing that relation. There was no causal relation between these events; just a succession of them.

For Hume, not only events in the physical world give rise to such a 'causal illusion', but also the 'acts of the spirit'. He wrote:

> Some have asserted, that we feel an energy, or power, in our own mind; [...] But to convince us how fallacious this reasoning is, we need only consider, that the will being here consider'd as a cause, has no more a discoverable connexion with its effects, than any material cause has with its proper effect. So far from perceiving the connexion betwixt an act of volition, and a motion of the body; [...] the actions of the mind are, in this respect, the same with those of matter. We perceive only their constant conjunction; nor can we ever reason beyond it. [3]

Wegner & Wheatley, in referring to Hume, did not want to support his general skepticism of causal relations. In fact, their theory rests on the assumption that actions are caused by unconscious neural processes. However, they adopted the skepticism of mental causation and two central ideas.

First, they noted that causal relations, in general, cannot be perceived directly. Causality is not a 'magic bond' between events that can be made visible under the microscope. Instead, observers must infer a causal relation, based on the experience of the repeated succession of two events of a certain type. That explains why causal theories are prone to illusions: There is no proof, whether events of two types really stood in a causal relation to each other, or whether one was just followed by the other.

Second and more importantly, Wegner & Wheatley also shared the assumption with Hume that the perception of events in the physical world basically resembles the experience of one's own agency. They wrote:

> The person experiencing will [...] is in the same position as someone perceiving causation as one billiard ball strikes another. Causation is inferred from the conjunction of ball movements, and will is inferred from the conjunction of events that lead to action. [13]

According to this analogy, the object of perception in case of agency must be identical to the subject of perception. It is the agent who is supposed to experience a causal relation 'within' herself. First, she experiences some of her own thoughts when they 'occur in their consciousness', as Wegner & Wheatley put it [13]. These thoughts constitute the supposed cause of the supposed causal relation. Then, she experiences her own action either through observation of her limbs or through proprioception. This action constitutes the supposed effect of the supposed causal relation. Based on the experience of the thought processes and the action, the agent gains the impression that her thoughts have caused the action. In the words of Wegner: *"[P]eople experience conscious will when they interpret their own thought as the cause of their action"* [11].

However, there is no proof of a causal relation between thought and action, even if both occurred, and match each other. The action could also have been caused by something else, for instance an unconscious neural process. If that is true, the power of will is a causal illusion (See Fig. 1).

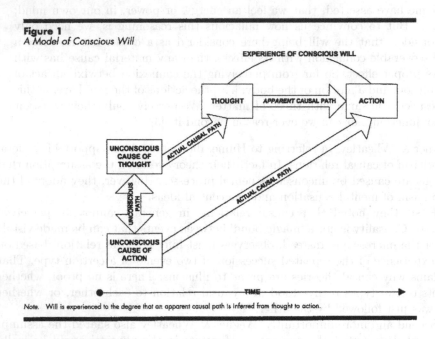

Figure 1
A Model of Conscious Will

Note. Will is experienced to the degree that an apparent causal path is inferred from thought to action.

Fig. 1. Wegner & Wheatly used this schema to illustrate the real and the apparent causal path from the unconscious neural process to the action according to their hypothesis [13].

3   The *I Spy* Experiment

Let me now briefly explain the experimental setup of the *I Spy* experiment as well as Wegner & Wheatley's interpretation of its test results.

The experiment included a series of trials. In each trial, a participant was paired with an assistant working for the experiment's administrator. Both placed their fingers on an Ouija-board-like mouse and moved it together in a circular manner. The movements of the mouse were projected to a monitor where the mouse cursor moved over several items on the screen (a swan, an umbrella, etc.). The participants were instructed to stop their movements freely at some point. During each trial, they listened to music through headphones (See Fig. 2). Frequently, words designating an item on the screen, such as "Swan!", were spoken through the headphones of the participant. These words were supposed to prime a thought about the respective item 'in the participant's consciousness'.

Fig. 2. The picture shows a participant and an assistant moving the Ouija-board-like mouse in the *I Spy* experiment [13].

The participants assumed that the assistants were also participants, just like them. In fact, they received secret instructions from the administrator either to stop the mouse by themselves on a specific item or to let the participants stop the mouse freely wherever they want. Those trials in which the assistant was instructed to let their participant stop freely were called 'free stops'. Those trials in which the assistant was instructed to force a stop on an item were called 'forced stops'. In case of the forced stops, the participants heard the word for the respective item through their head-phones 30, 5, or 1 s before; or 1 s after the mouse was forced to stop on the item.

After each trial, the administrator asked the participants to rate the 'level of intentionality' that they felt when the mouse stopped, both in case of the free stops and in case of the forced stops: "[T]hey each would rate how much they had intended to make each stop, independent from their partner's intention." [13] At the end, the participants rated the free stops in average 56% intentional and the forced stops in average 52% intentional. However, there was a fluctuation in the perception of intentionality depending on the time difference. Given that the primed word was spoken 1 s before the forced stop, participants rated the stop up to 65% intentional. So, it seems that even in case of the forced stops, when the assistant was instructed to move and to stop, the participants felt, at least to some extent, that they intended to make a stop.

Based on these results, Wegner & Wheatley hypothesized that the occurring thought of a certain item on the screen together with the subsequent perception of the mouse stopping on that very item made the participant believe that their own thoughts caused their hand to move the mouse towards the primed item, even though they did not [13].

4   The Causal Experience in the *I Spy* Experiment

Wegner & Wheatley's study, as well as their illusory will hypothesis, have been criticized by many philosophers.

In his paper *Willusionism, epiphenomenalism and the feeling of conscious will* [9], Sven Walter identified various problems of the *I Spy* study, including the low significance of the test results and the ambiguous responses of the participants that do not justify Wegner & Wheatley's conclusion. Markus Schlosser [8] remarked that there is plenty of empirical evidence supporting the assumption that the sub-personal correlates of an agent's intention are causally efficacious for her action. Glenn Carruthers [1], arguing from the opposite direction, doubted the existence of a universal causal experience of will. He discussed a variety of empirical evidence showing that many agents, especially children and patients suffering from autism, can experience agency without experiencing these actions as caused by their intentions.

I am, clearly, sympathetic to Carruthers' position. However, I want to go further and assume that even the participants in the *I Spy* study did not experience their intentions (or thoughts) as (more or less) causally efficacious for their action. To show this, I will provide an alternative and more plausible interpretation of the test results.

Note that the question of the action's level of intentionality, or the question, "How much did you intend to make the stop?" that Wegner & Wheatley asked the participants can be interpreted in different ways. Given that the participants developed the thought of an item on the screen after hearing the respective word, it can be interpreted in the following way: "Did my thought of the primed item cause my hand to move the mouse to this item or not?" or in that way: "How strongly did my thought of the primed item cause me to move the mouse to this item?". It is clear, that this is the interpretation that Wegner & Wheatley had in mind when they asked their question. The schema of the causal experience according to this interpretation would be the following:

My thoughts (by occurring in my consciousness) $\rightarrow_{caused}$ the mouse (to move)

However, the question of the action's level of intentionality could also have been interpreted in a different way by the participants, namely as the question whether or how strongly the thrust of their hand contributed to the movement of the mouse towards the primed item. That interpretation presupposes a distinction not between two different events, the mental act of thinking to move the mouse and the action of moving the mouse, but within the action itself, namely between the participant's hand movement and the movement of the mouse. If the participants suffered from a causal illusion, that illusion would not have consisted, as Wegner & Wheatley assumed, in misjudging the causal impact of the thought on the movement of the mouse, but in misjudging the causal impact of the hand's thrust on the movement of the mouse. The schema of the causal experience according to this alternative interpretation would be the following:

My hand (by pushing the mouse) $\rightarrow_{caused}$ the mouse (to move)

To illustrate the difference between both causal schemas, consider this example given by Wegner in his book *The Illusion of Conscious Will*. Wegner describes a situation of him sitting in front of a gaming machine in a toy store. While moving the joy-stick "[a] little monkey on the screen was eagerly hopping over barrels as they rolled toward him." [10] He was under the impression of playing a video game, but the machine just showed a pre-game demo. Wegner concludes: "I thought I was doing something that I really did not do at all." [10] Furthermore, he assumes that operating the gaming machine is a good real-life example to proof his illusory will hypothesis.

Timothy O'Connor [7] and Walter [9] have already criticized Wegner's interpretation of the situation. In fact, Wegner did do something, namely moving the joystick. He did not erroneously assume that his thoughts caused an action. He assumed that his hand, in moving the joystick, caused the machine to operate. He did not err about acting at all, but about the outcome of his moving the joystick [7], or, in other words, about the causal effects of his moving the joystick [9].

Unfortunately, we cannot easily apply the reinterpretation of the gaming machine situation to the *I Spy* experiment. Since the action that Wegner performed at the gaming machine involved a machine-interaction, as O'Conner put it [7], it is not comparable with the action that the participants performed in *I Spy*. Successfully interacting with a machine requires the functioning of many independently operating devices and gears that are partially hidden. Since some of the devices and gears in the machine did not operate as they would operate, if the game was running, Wegner's movement of the joystick did not have the expected outcome. His action was prone to a causal illusion, because he did not have direct control over, or insight in the mechanism of the gaming machine.

The causal illusion of the participants in the *I Spy* experiment, however, cannot be explained in the same way, because the action did not involve any defective machine-interaction. The Ouija-mouse and the connected monitor operated flawlessly, and the mouse transmitted the movement information properly to the cursor on the screen. It seems, that the participants had direct control over the movement of the cursor over the whole timespan. So, how could they suffer from a causal illusion? How could they have been mistaken about the effects or the outcome of their hand movement?

Even though the action that was performed in I Spy did not involve a defective causal mechanism, it was not a simple bodily movement, either. Since both the participants and the assistants were invited to move the Ouija-mouse in circles for the whole time, the participants and the assistants performed a cooperative (or joint) action. With regard to the cooperative character of the action in I Spy, Sven Walter noted that "[i]n cases of joint action, however, you always have to try to respond to cues from the other in order to coordinate your movements with the other movements" [9]. Accordingly, it is likely that in case of the free stops, the assistants took part in moving the mouse towards an item, along with the participants; and in case of the forced stops, it is likely that the participants took part in moving the mouse, along with the assistants, towards the primed item. Or in other words: During all trials, the overall action (of moving the mouse to an item) was influenced by the thrust of both the assistant's and of the participant's hand, because both were constantly pushing forward, or responding to a push. This essentially cooperative character of the action during both kinds of trials can be supported by the fact that the 'level of intentionality' that was rated during both trials was roughly the same, namely 56% vs. 52% intentional.

Given that this is true, and both the participants and the assistants contributed to the movement of the mouse by constantly pushing and by responding to a push, it is likely that the participants could not tell for sure to which extent the assistants actively intended to move the mouse towards the primed item. "Did the other person independently move towards that item or did they simply try to coordinate their movement with mine?" That fact applies to both the forced stops and the free stops, because the participants did not know, which trial was a free and which trial was a forced stop. Furthermore, given that the participants could not precisely estimate the assistants' causal impact on the mouse movement, they could also not tell for sure, to which extent they themselves actively contributed to the overall movement of the mouse.

That the participants were unsure about their own contribution to the overall movement can be supported by the indecisive answers that they gave, when they were asked to rate the 'level of intentionality' for their action. In case of a normal, non-cooperative action, such as moving a bottle from A to B, an agent would exactly know if she intended to do it or not. She would rate her level of intentionality either 100% intentional (fully intentional) or 0% (not intentional at all). Cooperative actions, in turn, are indeed prone to this kind of certainty. That is why neither in case of the free stops nor in case of the forced stops the participants rated the level of intentionality 100% intentional or 0%.[3]

I conclude that the test results of the I Spy experiment do not show that the participants experienced a (more or less) causally efficacious will, separate from their action. It is more likely that they experienced their own hand movement as (more or less) causally efficacious for the overall cooperative movement of the mouse. Furthermore, the test results do not support Wegner & Wheatley's illusory will hypothesis. It is possible that the participants suffered from some kind of causal illusion. But the causal illusion that they suffered from, most likely, did not consist in overestimating or underestimating the causal impact of their own thoughts on the action (of moving the mouse towards the primed item). It is more likely that the causal illusion consisted in overestimating, or in underestimating the causal impact of their own hand's thrust (against the mouse) on the overall cooperative movement of the mouse.

---

[3] Walter also pointed out the unusual indecisiveness of the participants: "If 100 corresponds to 'I intended to make the stop,' then a stop that was experienced as intended should receive an average of 100, not of 56. Therefore, the fact that free stops received an average rating of 56 does not show that the correct rating for intended stops is 56, but that the free stops were not perceived as fully intended" [9].

References

1. Carruthers, G.: A problem for Wegner and colleagues' model of the sense of agency. Phenom Cogn. Sci. **9**(3), 341–357 (2010). https://doi.org/10.1007/s11097-010-9150-6
2. Haggard, P.: Human volition: towards a neuroscience of will. Nat. Rev. Neurosci. **9**(12), 934–946 (2008). https://doi.org/10.1038/nrn2497
3. Hume, D.: A Treatise of Human Nature, 2nd edn. Clarendon Press, Oxford [1739] (1978)
4. Lau, H.C., Rogers, R.D., Passingham, R.E.: On measuring the perceived onsets of spontaneous actions. J. Neurosci. **26**(27), 7265–7271 (2006). https://doi.org/10.1523/JNEUROSCI.1138-06.2006
5. Libet, B.W.: Do we have free will? J. Conscious. Stud. **6**(8–9), 47–57 (1999)
6. Mogi, K.: Free will and paranormal beliefs. Front. Psychol. **5**, 281 (2014). https://doi.org/10.3389/fpsyg.2014.00281
7. O' Connor, T.: Freedom with a human face. Midwest Stud. Philos. **29**, 207–227 (2005). https://doi.org/10.1111/j.1475-4975.2005.00113.x
8. Schlosser, M.E.: Causally efficacious intentions and the sense of agency: in defense of real mental causation. J. Theoret. Philos. Psychol. **32**(3), 135–160 (2012). https://doi.org/10.1037/a0027618
9. Walter, S.: Willusionism, epiphenomenalism, and the feeling of conscious will. Synthese **191**(10), 2215–2238 (2014). https://doi.org/10.1007/s11229-013-0393-y
10. Wegner, D.M.: The Illusion of Conscious Will. MIT Press, Cambridge (2002)
11. Wegner, D.M.: Précis of the illusion of conscious will. Behav. Brain Sci. **27**, 649–694 (2004). https://doi.org/10.1017/s0140525x04000159
12. Wegner, D.M., Sparrow, B., Winerman, L.: Vicarious agency: experiencing control over the movements of others. J. Pers. Soc. Psychol. **86**(6), 838–848 (2004). https://doi.org/10.1037/0022-3514.86.6.838
13. Wegner, D.M., Wheatley, T.: Apparent mental causation: sources of the experience of Will. Am. Psychol. **54**(7), 480–492 (1999). https://doi.org/10.1037/0003-066X.54.7.480
14. Wittgenstein, L.: Philosophical Investigations. (German) [Philosophische Untersuchungen]. 2. edn. Blackwell, Malden [1953] (1999)

Understanding Responses of Individuals with ASD in Syllogistic and Decision-Making Tasks: A Formal Study

Torben Braüner[1]([⊠]), Aishwarya Ghosh[2], and Sujata Ghosh[2]

[1] Roskilde University, Roskilde, Denmark
torben@ruc.dk
[2] Indian Statistical Institute, Chennai, India

Abstract. Recent studies have shown that in some reasoning tasks people with Autism Spectrum Disorder perform better than typically developing people. The present note gives a brief comparison of two such tasks, namely a syllogistic task and a decision-making task, identifying the common structure as well as differences. In the terminology of David Marr's three levels of cognitive systems, the tasks show commonalities on the computational level in terms of the effect of contextual stimuli, though an in-depth analysis of such contexts provides certain distinguishing features in the algorithmic level. We also make some general remarks on our approach.

1 Introduction

It is well-known from the vast psychological and psychiatric literature on Autism Spectrum Disorder[1] (ASD) that children with ASD have a limited or delayed capacity to respond correctly to certain psychological reasoning tests such as false-belief tasks. In other words, on such tests, children with ASD perform less well than children with typical development (TD). However, it turns out that in some other reasoning tasks, people with ASD perform not *worse*, but *better*, than typicals, thus, showing that ASD is not in all respects a "disability", a view that was put forward by Simon Baron-Cohen [2] two decades ago. During the last few years, several new empirical studies have emerged where individuals with ASD perform better than typical individuals, thus supporting Baron-Cohen's view.

In [6], Farmer et al. investigate adult's performance in a decision task where the subject has to choose between pairs of consumer products that are presented with a third, less desirable "decoy" product. According to conventional economic theory, a consumer's choice of one product over another should be independent

---

[1] Autism Spectrum Disorder is a psychiatric disorder with the following diagnostic criteria: 1. Persistent deficits in social communication and social interaction. 2. Restricted, repetitive patterns of behavior, interests, or activities. For details, see *Diagnostic and Statistical Manual of Mental Disorders, 5th Edition (DSM-V)*, published by the American Psychiatric Association.

© Springer Nature Switzerland AG 2021
L. Cleophas and M. Massink (Eds.): SEFM 2020 Workshops, LNCS 12524, pp. 118–128, 2021.
https://doi.org/10.1007/978-3-030-67220-1_10

of whether there is a third option. To quote the paper, "If one prefers salmon to steak, this should not change just because frogs' legs are added to the menu". Farmer et al. demonstrate that the tendency to violate this norm is reduced among individuals with ASD, thus, in this sense, they are more rational than typical individuals. They found a similar difference between the two groups of people drawn from the general population, classified in accordance with their levels of autistic-like traits, measured in terms of the self-report questionnaire called the Autism-Spectrum Quotient (AQ).

A similar example can be found in [11], where Lewton et al. compares the ability to do syllogistic reasoning in the general population with individuals showing autistic-like traits that are measured in terms of the AQ-score. Some syllogisms are consistent with reality: *All birds have feathers. Robins are birds. Therefore robins have feathers*, but others are not: *All mammals walk. Whales are mammals. Therefore whales walk.* Both of these syllogisms are valid, that is, the conclusion follows logically from the premises, in fact, they have exactly the same logical structure, but the validity is more difficult to detect in the second syllogism because the correct answer is inconsistent with reality. Thus, prior knowledge of reality can affect the judgement of validity, and the study in [11] shows that there is a negative correlation between this reasoning bias and the AQ-score, thus, the more autistic-like a person is, the better the person is to judge syllogisms without being affected by irrelevant prior knowledge of reality. See [9] for a comprehensive overview of different psychological theories of syllogistic reasoning.

Now, to the best of our knowledge, there are no systematic and theoretical studies of the commonalities between the psychological tasks where individuals with ASD perform better than the typical individuals, as reported in [6,11]. It is the goal of the present paper to investigate this question – an interdisciplinary enterprise requiring insights from both logic and economic theory. Such an investigation will help us in providing a better understanding of the capabilities of the individuals with ASD, which in turn might help in accommodating a better work environment for these individuals. A common feature of the above-mentioned tasks seems to be that they require an ability to disregard irrelevant contextual information, but this is a very informal verbal description. We will aim at a more formal and precise analysis, identifying a common structure, inspired by other works aiming at identifying a common logical structure in superficially different reasoning tasks.[2] As a tool to analyze the tasks in question, we make use of David Marr's levels of analysis of cognitive systems [12]: Any task computed by

---

[2] In particular, in [4] it is demonstrated that two seemingly dissimilar reasoning tasks, namely two different versions of a false-belief task called the Smarties task, have exactly the same underlying logical structure. Similarly, in [5] it is demonstrated that four second-order false-belief tasks share a certain logical structure, but they are also distinct in a systematic way. We remark that such a strategy was also pursued in the book [17], where it was shown that a false-belief task and what is called the box task have a logical structure similar to a third task called the suppression task.

a cognitive system must be analyzed at the following three levels of explanation (in order of decreasing abstraction):

Computational level: Identification of the goal and of the information-processing task as an input–output function;
Algorithmic level: Specification of an algorithm which computes the function;
Implementational level: Physical or neural implementation of the algorithm.

Analogous levels of analysis can be found in several other works of cognitive science, e.g., see the overview in [16], pages 9–12. For this work, we shall focus on the computational and algorithmic levels.

2   The Syllogistic Task

In this section, we analyze the performances in the syllogistic tasks as investigated in [11] on both computational and algorithmic levels. We first provide a brief discussion on the empirical study as reported in [11].

An Empirical Study by Lewton et al. [11]: Four different types of syllogisms are considered. The two syllogisms described in the introduction were of the respective types of valid-believable and valid-unbelievable (this terminology is self-explanatory). But there are also the types invalid-believable and invalid-unbelievable. An example of the former type is: *All flowers need water. Roses need water. Therefore Roses are flowers.* An invalid-unbelievable syllogism with exactly the same structure is: *All insects need oxygen. Mice need oxygen. Therefore mice are insects.* Each subject has to judge four congruent syllogisms (valid-believable and invalid-unbelievable) and four incongruent ones (invalid-believable and valid-unbelievable). A subject scores 1 point for each correct judgement. So there is a 0–4 scale for congruent syllogisms and 0–4 for incongruent ones. A belief bias occurred when there is a decrease in accuracy for incongruent problems (valid-unbelievable and believable-invalid) relative to congruent problems (valid-believable, invalid-unbelievable). Such a bias is calculated by subtracting the score for incongruent syllogisms from that of congruent ones, resulting in a possible score from -4 to 4. The study reports a number of correlation results, in particular, the correlation between AQ and belief bias was -0.39 (with a p-value less than 0.001). The AQ-congruent correlation was -0.11 but not significant, whereas the AQ-incongruent correlation was 0.40 (also with a p-value less than 0.001). Thus, the congruent and incongruent variables measure different underlying cognitive abilities, only the latter is associated with AQ.

2.1   Computational Level Analysis (Syllogistic Task)

We now ask the following question: What does it precisely mean that a subject is able to judge a syllogism without bias, that is, without involving irrelevant contextual information? We assume that the validity of syllogisms is defined in the usual manner as in first-order logic in terms of first-order models $\mathcal{M}$. This defines a function *valid* which maps syllogisms to truth-values. This function formalizes the normatively correct judgement of syllogisms.

Now, a subject's judgement of a syllogism takes place in a specific context, that is, in a specific state of affairs, namely the actual state of affairs, where for example *Robins have feathers* is true, but *Whales walk* is false. Such a state of affairs is formalized by a model. This means that a subject's judgement of syllogisms in a context can be modeled by a function *believable* similar to the function *valid*, but with an extra parameter, representing a context. Thus, the function *believable* maps a pair consisting of a syllogism and a model to a truth-value, and the requirement of context-independence can be formulated as

$$believable(S, \mathcal{M}_1) = believable(S, \mathcal{M}_2) \tag{1}$$

for any syllogism S and any models $\mathcal{M}_1$ and $\mathcal{M}_2$.

A stronger requirement than the independence of context is the notion of correctness, that is,

$$believable(S, \mathcal{M}) = valid(\mathcal{M}) \tag{2}$$

for any syllogism S and any model $\mathcal{M}$. Note that this is a strictly stronger requirement, for example, a *believable* function that always gives the incorrect answer would be independent of contexts. We note here that we would not find a similar requirement in case of the decision task we discuss later.

2.2 Algorithmic Level Analysis (Syllogistic Task)

In what follows we shall describe some theoretical explanations of belief bias in syllogistic reasoning, based on the work done in [10]. These explanations have the form of algorithms, where bias arises at one of the three different stages in the reasoning process: during input, processing, or output (cf. see [10], page 852). Given the algorithmic character of the explanations, we are situated at the second of Marr's three levels, where an algorithm computes the input-output function specified at the top level. We give particular attention to the reasoning process that takes place when incongruent syllogisms are judged, that is when logic and belief conflict.

One of the algorithms described in [10] is the *misinterpreted necessity model*, which is described by the flowchart-like diagram in Fig. 1. A feature of this algorithm is that the logically correct answer is guaranteed if the conclusion follows from the premises or if the conclusion is falsified by the premises (called *determinately invalid*). If none of these two conditions are satisfied, that is, if some models of the premises falsify the conclusion and some models verify it (called *indeterminately invalid*), then the output of the algorithm is decided by the conclusion's believability. Thus, the logically correct answer is guaranteed for any syllogism that either is valid or determinately invalid. Note that the bias here takes effect *after* the logical reasoning process. According to Klauer et al. [10], the bias in this model is due to the subject's misunderstanding of what it means to say a conclusion not following from the premises, namely that it is

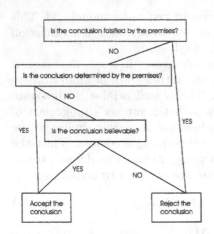

Fig. 1. The misinterpreted necessity model, taken from [10].

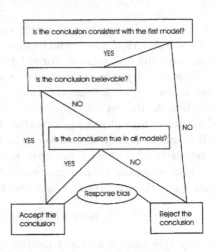

Fig. 2. An account by mental models, taken from [10].

sufficient that the conclusion is falsified by *some* models of the premises, not necessarily *all* such models.

Earlier we discussed the invalid "rose" and "mice" syllogisms, which have exactly the same logical structure. Since syllogisms with this structure have models of the premises that verify the conclusion (the "rose" case) as well as models that falsify it (the "mice" case), they are indeterminately invalid. Thus, in these syllogisms, the response of the misinterpreted necessity model is decided by the believability of the conclusion, so in the "rose" syllogism, the response would incorrectly be "valid", but in the "mice" syllogism, the response would correctly be "invalid" (but for the wrong reason).

In [10], Klauer et al. also give an account of the belief bias based on the "mental models" school in the psychology of reasoning, according to which the mechanism underlying human reasoning is the construction of models, [8]. An account by mental models is shown in Fig. 2. The first step of this algorithm is to build an initial model of the premises of the syllogism under investigation, which is followed by an evaluation of the conclusion in the model in question. If the conclusion comes out as true, but it is not believable, this triggers the generation of further models of the premises, as indicated in the figure. Note that like in the misinterpreted necessity model, the logically correct answer is guaranteed for any syllogism that either is valid or determinately invalid. But if a syllogism is indeterminately invalid, then the answer becomes incorrect if and only if the conclusion is true in the initial model and also believable, hence, the selection of the initial model matters. Note that the bias here takes effect *during* the reasoning process.

3   The Decision Task

We now analyze the performances in a decision task of choosing between pairs of consumer products in the presence of a third less desirable decoy product, investigated in [6]. We investigate the task on computational as well as algorithmic levels.

An Empirical Study by Farmer et al. [6]: It is investigated whether individuals with ASD show reduced sensitivity to contextual stimuli when exposed to a decision-making situation where they had to make choices between pairs of consumer products that are presented with a third, less desirable decoy option. In a choice set, a decoy option is usually considered as an asymmetrically dominated alternative which is dominated by one of the choice alternatives but not by the other, i.e., based on the preference determining attributes, it is completely dominated by (i.e., inferior to) one option (target) and only partially dominated by the other (competitor). The choice task included participants to see 10 pairs of products (e.g., USB sticks); the products in each pair differed on two dimensions (in the case of USB sticks, storage capacity, and longevity). Each pair was presented twice, once with a decoy that targeted one product and once with a decoy that targeted the other. According to the conventional economic theory, any rational individual when exposed to such a situation should show a consistent preference behavior as the individual's preference between two items should be independent of the 'decoy' options on offer. In contrast, it was observed that the choices of the general participants (control group) were heavily influenced by the composition of the choice set. Rather than being based on an independent assessment, the attractiveness of a given option relied upon how the individual compared it with the other values that were simultaneously present (attraction effect). But this tendency was quite reduced for individuals with ASD. Thus, they showed reduced sensitivity to contextual stimuli, indicating that their choices were more consistent and conventionally rational.

3.1   Computational Level Analysis (Decision Task)

The reduced context effect in people with ASD might be a manifestation of their reduced understanding of, or concern for, the likely beliefs and appraisals of others. Thus, the choices of individuals with ASD have a better chance to satisfy the norm given by (3) than typical individuals [2].

In theory, the rational decision-makers are expected not to show sensitivity to context stimuli and be more consistent in their choices when they had to make choices in the situation mentioned above in the presence of a decoy option. Choice consistency should be the norm in this case. More formally, we can consider a choice function which returns the chosen item from the finite tuple of possible choices, and the requirement for context-independence is given by:

$$Choice(Product_1, Product_2, Decoy_1) = Choice(Product_1, Product_2, Decoy_2) \quad (3)$$

Note that this is analogous to the requirement on the judgements of syllogisms that we called context-independence (requirement (1) on a *believable* function).

On the other hand, there is no requirement similar to the correctness of the *believable* function (requirement (2) on the function).

3.2 Algorithmic Level Analysis (Decision Task)

We now provide an algorithmic explanation of the attraction effect bias that is visible in context-dependent decision tasks [6]. To this end, we consider dimensional weight models as discussed in [1,19], where the authors mention how the difference in dimensional (attribute) weights are highly dependent on the similarity relationship among the items. The more similar a set of items is on one attribute the easier it is to notice discrepancies on their other attribute (for both target and decoy items) so that the observed discrepancies on a given dimension increase the corresponding weight [1]. Thus, once the decision-maker (DM) is able to determine the important dimension it then goes on to compare the three items (target, decoy, and competitor) on that dimension. After the comparison, the DM gives more attention weight to the target and decoy as the distance between them is smaller compared to that between competitor and decoy, eventually selecting the target as the final choice. This idea of giving higher attention weights to options whose attribute values are similar is based on the multiattribute linear ballistic accumulator (MLBA) model given by Trueblood et al. [18].

The Dominance Search Model (DSM) of Decision Making [14], which considers four phases of a decision process (cf. Fig. 3) is used to describe the decision process discussed above. We analyze the dimensional weight theory using the flowchart-like diagram in Fig. 3 and establish a line of argument as to how the decision task explained in [6] fits in this respect. However, this argument might vary with different examples especially in terms of given attribute values. The decision task in [6] considers a choice set with three items defined on two dimensions where the target strictly dominates the decoy. According to this model, the DM follows four phases of the decision process:

1. **Pre-editing Phase:** In the first phase, the DM screens and evaluates the attributes and alternatives. Alternatives with a better chance of becoming dominant are selected.
2. **Finding a promising alternative phase:** Given the selected alternatives from the first phase the DM now moves on to detect an alternative with attractive attributes that can be considered as a promising alternative (see Fig. 3). The bias becomes evident in this phase as the target shows a higher potential of being a promising alternative because of its strict dominance over the decoy.
3. **The dominance testing phase:** Once the DM is able to find a potentially promising alternative, the dominance test is done in this phase. If there is any violation, the DM caters to it in the next phase. If no violation is found, the DM checks whether all the relevant information has been evaluated (Fig. 3). Once this is done, the final decision is taken, otherwise, the DM moves on to test dominance once again.

4. The dominance structuring phase: After identifying a violation of dominance the DM tries to neutralize it in this phase using the ways mentioned in Fig. 3. After possible removal of the violation, the DM moves on to make the final decision. Otherwise, the evaluation process starts again. We note that the decision process of Fig. 3 is task-dependent and might vary accordingly.

For the example discussed at the beginning of this section, it is seen that the target strictly dominates the decoy thus reducing its chance of getting selected, but both the target and the competitor are considered as options at this stage [Phase 1]. Now, the target is considered as the more promising alternative because of the attraction effect [Phase 2]. A strict dominance of the target over the decoy is established but the target and the competitor are found to be incomparable [Phase 3]. This leads to a violation that gets resolved in the next phase [Phase 4] which can be explained using the theories mentioned above.

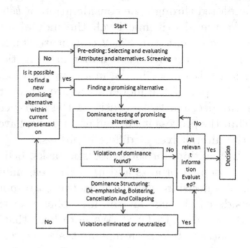

Fig. 3. The decision process, adapted from [14].

We note here that the syllogisms discussed in Sect. 2 are endowed with a notion of (in)correct reasoning and bias in the algorithmic models amounts to various ways of deviating from this norm. In the case of the decision task, one can also consider norms, but we leave it to future work to investigate how the algorithmic models can capture such deviations from the relevant norms if any.

4   Discussion

For certain syllogistic tasks [11] and decision tasks [6], it was shown that individuals with ASD performed better than typical individuals. To analyze these results on a computational level, we took a functional approach (with a subject's reasoning being represented by a mathematical function) where the functions considered the respective tasks as arguments together with certain contextual information. For such functions, we have considered the following properties: *contextual independence* and *correctness*.

While the syllogistic task gave rise to certain functional expressions (as defined by mathematical functions) pertaining to both of the properties, those corresponding to the decision task paved the way for considering one of them, namely, context independence. These decision tasks were based on certain attributes, and no single choice was a dominant one (i.e. strictly better than the others), hence no notion of correctness. One might argue that such a correctness condition may be added to the decision task in case one of the choices is a strictly dominant one. But, more often than not, these tasks have rather complex choices. Moving on to contextual considerations, they can be further developed in the decision tasks by considering the following effect: *attraction effect* and *dominance effect*. For the syllogistic tasks, the contextual information is explored through the consideration of *belief biases*.

At the algorithmic level, the mental model corresponding to the syllogistic tasks provided in Fig. 2 constitutes building an initial model satisfying the premises of the syllogism under investigation. Then, an evaluation of the conclusion takes place in the model in question. Thus, the algorithm bases on the initial input of the model structure. In contrast, the algorithm given by DSM for the decision tasks considers the entrance of possible promising alternatives within the process itself, and as such, we have an ongoing process of introduction of the alternatives at different phases.

In addition, for the syllogistic tasks, belief biases are considered both *during* the reasoning process and *after* the reasoning process, depending on the model. For the decision making task, the corresponding notion of attraction effect is considered throughout the four phases of the decision-making process considered according to DSM. We note here that at the computational level for the syllogistic and decision-making tasks we were not able to make a deeper connection with respect to contextual independence.

To summarize, the commonalities in these two tasks on the computational level exist in terms of the effect of contextual stimuli, though the in-depth analyses of such contexts provide us with certain distinguishing features. When the tasks are analyzed at the abstract computational level, the responses of ASDs in both tasks exhibit certain similarities, but when they are analyzed at the more concrete algorithmic level, the differences are made explicit with respect to the handling of biases. One might argue that our study should be relevant for typical individuals as well, but then we would digress from the initial analysis at the computational level. The functional expressions fit very well for the individuals with ASD.

5   Future Work

Here, we consider reasoning tasks where individuals with ASD perform better than typical individuals, namely, [11] and [6]. Below, we mention three more example studies that provide further validation towards the better performance of individuals with ASD. We plan to subject these studies to similar analysis in the future, so as to provide a more detailed formal insight into the performances of the individuals with ASD, which may lead to a better understanding of the capabilities of such individuals.

In [7], Fujino et al. investigate adults' performance in the so-called sunk-cost task, which measures the tendency to include considerations on past costs while choosing between current alternatives. According to conventional economic theory, past expenses are irrelevant, rational decision-makers should only pay attention to future consequences of possible alternatives. It is shown in [7] that individuals with ASD are less prone to violate this norm than typical individuals. The study [13] investigates adult's performance on a financial task in which the monetary prospects were presented as either loss or gain, and it is shown that individuals with ASD demonstrate a larger consistency in decision making than typical individuals. The study [15] compares the performance of individuals with ASD and typical adolescents on tasks from the heuristics and biases literature, including the famous Linda task, involving the conjunction fallacy, which violates a fundamental law of probability theory. It is found that children with ASD are less susceptible to this fallacy.

Such formal investigations of the tasks where individuals with ASD perform better than typical individuals would help us to identify common strengths and competencies in the cognitive style of such individuals, which in turn can be the basis for neurobiological research, investigating biological explanations of such common competencies. In addition, it would also add impetus to the neurodiversity perspective [3] that suggests that autism should not be seen as a disorder but as variations of the neurotypical brain - the involved disability and even disorder may be about the person-environment fit.

References

1. Ariely, D., Wallsten, T.S.: Seeking subjective dominance in multidimensional space: an explanation of the asymmetric dominance effect. Organ. Behav. Hum. Decis. Process. **63**(3), 223–232 (1995)
2. Baron-Cohen, S.: Is Asperger syndrome/high-functioning autism necessarily a disability? Dev. Psychopathol. **12**, 489–500 (2000)
3. Baron-Cohen, S.: Editorial perspective: neurodiversity - a revolutionary concept for autism and psychiatry. J. Child Psychol. Psychiatry **58**, 744–747 (2017)
4. Braüner, T.: Hybrid-logical reasoning in the Smarties and Sally-Anne tasks. J. Log. Lang. Inf. **23**, 415–439 (2014). https://doi.org/10.1007/s10849-014-9206-z
5. Braüner, T., Blackburn, P., Polyanskaya, I.: Being deceived: information asymmetry in second-order false belief tasks. Top. Cogn. Sci. **12**, 504–534 (2020)
6. Farmer, G., Baron-Cohen, S., Skylark, W.: People with autism spectrum conditions make more consistent decisions. Psychol. Sci. **28**, 1067–1076 (2017)
7. Fujino, J., et al.: Sunk cost effect in individuals with autism spectrum disorder. J. Autism Dev. Disord. **49**, 1–10 (2019). https://doi.org/10.1007/s10803-018-3679-6
8. Johnson-Laird, P.: Mental models and deductive reasoning. In: Adler, J., Rips, L. (eds.) Reasoning: Studies of Human Inference and Its Foundations, pp. 206–222. Cambridge University Press (2008)
9. Khemlani, S., Johnson-Laird, P.: Theories of the syllogism: a meta-analysis. Psychol. Bull. **138**, 427–457 (2012)
10. Klauer, K., Musch, J., Naumer, B.: On belief bias in syllogistic reasoning. Psychol. Rev. **107**, 852–884 (2000)
11. Lewton, M., Ashwin, C., Brosnan, M.: Syllogistic reasoning reveals reduced bias in people with higher autistic-like traits from the general population. Autism **23**, 1311–1321 (2019)

12. Marr, D.: Vision. Freeman and Company, New York (1982)
13. Martino, B.D., Harrison, N., Knafo, S., Bird, G., Dolan, R.: Explaining enhanced logical consistency during decision making in autism. J. Neurosci. **28**, 10746–10750 (2008)
14. Montgomery, H.: Decision rules and the search for a dominance structure: towards a process model of decision making. In: Advances in Psychology, vol. 14, pp. 343–369. Elsevier (1983)
15. Morsanyi, K., Handley, S., Evans, J.: Decontextualised minds: adolescents with autism are less susceptible to the conjunction fallacy than typically developing adolescents. J. Autism Dev. Disord. **40**, 1378–1388 (2010). https://doi.org/10.1007/s10803-010-0993-z
16. Stanovich, K.: Who is Rational? Studies of Individual Differences in Reasoning. Lawrence Erlbaum, Hillsdale (1999)
17. Stenning, K., van Lambalgen, M.: Human Reasoning and Cognitive Science. MIT Press, Cambridge (2008)
18. Trueblood, J.S., Brown, S.D., Heathcote, A.: The multiattribute linear ballistic accumulator model of context effects in multialternative choice. Psychol. Rev. **121**(2), 179–205 (2014)
19. Wedell, D.H.: Distinguishing among models of contextually induced preference reversals. J. Exp. Psychol. Learn. Mem. Cogn. **17**(4), 767 (1991)

Symbolic and Statistical Theories of Cognition: Towards Integrated Artificial Intelligence

Yoshihiro Maruyama[(⊠)]

Research School of Computer Science,
The Australian National University, Canberra, Australia
yoshihiro.maruyama@anu.edu.au

Abstract. There are two types of approaches to Artificial Intelligence, namely Symbolic AI and Statistical AI. The symbolic and statistical paradigms of cognition may be considered to be in conflict with each other; the recent debate between Chomsky and Norvig exemplifies a fundamental tension between the two paradigms (esp. on language), which is arguably in parallel with a conflict on interpretations of quantum theory as seen between Bohr and Einstein, one side arguing for the probabilist or empiricist view and the other for the universalist or rationalist view. In the present paper we explicate and articulate the fundamental discrepancy between them, and explore how a unifying theory could be developed to integrate them, and what sort of cognitive rôles Integrated AI could play in comparison with present-day AI. We give, inter alia, a classification of Integrated AI, and argue that Integrated AI serves the purpose of humanising AI in terms of making AI more verifiable, more explainable, more causally accountable, more ethical, and thus closer to general intelligence. We especially emphasise the ethical advantage of Integrated AI. We also briefly touch upon the Turing Test for Ethical AI, and the pluralistic nature of Turing-type Tests for Integrated AI. Overall, we believe that the integrated approach to cognition gives the key to the next generation paradigm for AI and Cognitive Science in general, and that Categorical Integrated AI or Categorical Integrative AI Robotics would be arguably the most promising approach to it.

Keywords: Symbolic AI · Statistical AI · Integrated AI · Categorical AI

1 Introduction: MIT's AI Lab, Now and Then

Neil Thompson at MIT and his collaborators recently published an intriguing article entitled "The Computational Limits of Deep Learning" [53], arguing in the following way:

> [P]rogress along current lines is rapidly becoming [...] unsustainable. Thus, continued progress [...] will require dramatically more computationally-efficient methods, which will either have to come from changes to deep learning or from moving to other machine learning methods.

© Springer Nature Switzerland AG 2021
L. Cleophas and M. Massink (Eds.): SEFM 2020 Workshops, LNCS 12524, pp. 129–146, 2021.
https://doi.org/10.1007/978-3-030-67220-1_11

There are many other problems in statistical machine learning, such as explainability and ethical issues as we shall discuss below. How could we overcome them? What sort of changes would be necessary for the next generation of Artificial Intelligence (and Cognitive Science in general)? A possible approach to overcome the limitations of statistical machine learning would be the integration of Symbolic and Statistical AI; at least some part of what Statistical AI is bad at is what Symbolic AI is good at. Deductive reasoning and inductive learning would be arguably the two fundamental wheels of the human mind (even though there may possibly be yet another wheel of human cognition).

An earliest idea of Integrated AI comes from Marvin Minsky, the 1969 Turing Award winner and co-founder of MIT's AI Lab, who proposes the integration of Symbolic and Connectionist AI in particular (aka. Logical and Analogical AI, or Neat and Scruffy AI) as a form of Integrated AI in his 1991 article [48][1]:

> Our purely numerical connectionist networks are inherently deficient in abilities to reason well; our purely symbolic logical systems are inherently deficient in abilities to represent the all-important "heuristic connections" between things—the uncertain, approximate, and analogical linkages that we need for making new hypotheses. The versatility that we need can be found only in larger-scale architectures that can exploit and manage the advantages of several types of representations at the same time. Then, each can be used to overcome the deficiencies of the others.

In light of this, the Minsky's conceptions of Symbolic (Logical) and Connectionist (Analogical) AI may be compared with the Reichenbach's well-known (yet debated) conceptions of the context of justification ("reason well") and the context of discovery ("making new hypotheses"). From the philosophy of science point of view, the cognitive capacities of discovery and justification are arguably the conditions of possibility of science as a human intellectual enterprise, which would make it compelling to combine the two paradigms of AI. Yet at the same time, there is a fundamental tension between the two paradigms, as exemplified by the Chomsky versus Norvig debate (Peter Norvig is Google's research director; Chomsky is one of the founders of Cognitive Science as well as the father of modern linguistics), which is arguably in parallel with the Bohr versus Einstein debate on the nature of quantum reality, as we shall see below.

In the following, we revisit the Chomsky-Norvig (and Bohr-Einstein) debate(s) to elucidate the discrepancy between the two paradigms, and place it in a broader context of science and philosophy (Sect. 2). And we discuss how the two paradigms could be integrated and why that matters at all, especially from an AI ethics point of view; we also give Turing-type tests for Ethical AI and Integrated AI (Sect. 3). We finally conclude with outlooks for the future of Artificial Intelligence and Cognitive Science, arguing that category theory, as giving transdisciplinary foundations/methodology of science (or interlanguage in the

---

[1] Statistical AI in this paper is meant to include Connectionist AI. Note that the MIT AI Lab is an origin of Embodied AI, too (see Rodney Brooks' seminal paper [10]).

trading zone of science in terms of Galison [20]), could be the key to Integrated Artificial Intelligence and Integrated Cognitive Science (Sect. 4).

2   The Fundamental Tension Between Symbolic and Statistical Paradigms of Cognition

In everyday life, we use both logical reasoning and statistical inference to make various judgments; deduction and induction are indispensable part of everyday life as well as scientific investigation. At the same time, we cannot precisely tell which part of human cognition is essentially symbolic, and which part of it is essentially statistical. It could, for example, happen that all functions of the human mind can be simulated by logical means alone or by statistical means alone; if this sort of reduction is possible, the apparent dualism of logic and statistics may collapse. For instance, automated theorem provers or assistants (such as Coq) have been developed within the symbolic paradigm of AI, but there is now some evidence that it can be done more efficiently within the statistical paradigm, especially with the help of deep learning (see, e.g., [4,26,51]). AI research today seems to make it compelling to reconsider the relationships between deductive reasoning and inductive learning. In this section, we focus upon a fundamental tension between the two paradigms, which manifests in the present landscape of AI as well as the history of science and philosophy as we shall discuss in the following. Let us begin with the Maxwell's integrative view of Nature.

2.1   Maxwell's Dualistic View of Nature

In a 1850 letter to Lewis Campbell, Maxwell [46] asserts as follows:

> [T]he true Logic for this world is the Calculus of Probabilities.

Maxwell is known for his great contributions to electromagnetism and statistical mechanics. Philosophically, he was seemingly influenced by the British tradition of empiricism, which puts a strong emphasis on the contingent nature of reality. His empiricist tendency may be observed in the following passage as well [46]:

> [A]s human knowledge comes by the senses in such a way that the existence of things external is only inferred from the harmonious (not similar) testimony of the different senses, understanding, acting by the laws of right reason, will assign to different truths (or facts, or testimonies, or what shall I call them) different degrees of probability.

Yet Maxwell was not a naïve empiricist. His statistical mechanics is in harmony with the empiricist thought; however, his theory of electromagnetism, which is arguably his greatest contribution to science, is rather closer to the rationalist thought in the continental tradition that sticks to the absolute, universal nature of truth, or the mechanistic view of Nature as shared by Newton and Laplace. Indeed, Maxwell [46] argues as follows:

[O]ur experiments can never give us anything more than statistical information [...] But when we pass from the contemplation of our experiments to that of the molecules themselves, we leave a world of chance and change, and enter a region where everything is certain and immutable.

The "molecules themselves" in Maxwell's thought may be compared with the Kant's idea of things themselves. So, whilst asserting that the true logic of the world is the calculus of probabilities, he maintains the universal conception of truth as being immune to chance and change. Experiments only allow us to access statistical information, but he thought there is something beyond that, namely some Platonistic realm of absolute truths. The probabilist view somehow coexisted with the universalist view in his thought. The Maxwell equations in his theory of electromagnetism embody the latter, whilst his statistical mechanics the former.

2.2   The Chomsky Versus Norvig Debate on the Nature of Science and Cognition

There was already some tension between the probabilist and universalist views at the time of Maxwell; it culminates in the contemporary debate between Noam Chomsky, who defends the universalist position, and Peter Norvig, who defends the probabilist position (see, e.g., [13]). Gold [22] recapitulates their debate in the following manner:

Recently, Peter Norvig, Google's Director of Research and co-author of the most popular artificial intelligence textbook in the world, wrote a webpage extensively criticizing Noam Chomsky, arguably the most influential linguist in the world. Their disagreement points to a revolution in artificial intelligence that, like many revolutions, threatens to destroy as much as it improves. Chomsky, one of the old guard, wishes for an elegant theory of intelligence and language that looks past human fallibility to try to see simple structure underneath. Norvig, meanwhile, represents the new philosophy: truth by statistics, and simplicity be damned.

Norvig basically takes the empiricist position, emphasising the "world of chance and change" (in terms of Maxwell) and thus the necessity of statistical analysis; Chomsky, by contrast, takes the rationalist position, emphasising the universal nature of linguistic structure and of scientific laws in general. To clarify Chomsky's view, Katz [25] interviewed Chomsky at MIT; Chomsky criticises Statistical AI, or Statistical Cognitive Science in general, in the following manner:

[I]f you get more and more data, and better and better statistics, you can get a better and better approximation to some immense corpus of text [...] but you learn nothing about the language.

Chomsky does not deny the success of statistical methods in prediction and other tasks; rather, he is concerned with the nature of scientific understanding, his point being akin to the recent issue of explainability in machine learning, especially deep learning. Chomsky even argues, in the same interview, that statistical analysis allows us to "eliminate the physics department" in his extreme Gedanken experiment:

[I]t's very different from what's done in the sciences. So for example, take an extreme case, suppose that somebody says he wants to eliminate the physics department and do it the right way. The "right" way is to take endless numbers of videotapes of what's happening outside the video, and feed them into the biggest and fastest computer, gigabytes of data, and do complex statistical analysis [...] you'll get some kind of prediction about what's gonna happen outside the window next. In fact, you get a much better prediction than the physics department will ever give.

Chomsky is not just concerned with cognition and language, but also with the nature of science in general. Chomsky argues that there are two conceptions of science, the universalist one aiming at the scientific understanding of Nature (and Cognition) and the probabilist one aiming at the engineering approximation of data; he says as follows in the same interview [25].

These are just two different concepts of science. The second one is what science has been since Galileo, that's modern science. The approximating unanalyzed data kind is sort of a new approach, not totally, there's things like it in the past. It's basically a new approach that has been accelerated by the existence of massive memories, very rapid processing, which enables you to do things like this that you couldn't have done by hand. But I think, myself, that it is leading subjects like computational cognitive science into a direction of maybe some practical applicability.

To Chomsky, it is a wrong direction to go, especially from a scientific, rather than engineering, point of view. Chomsky argues that statistical analysis is just "butterfly collecting"; Norvig [50] himself succinctly recapitulates Chomsky's points as follows:

Statistical language models have had engineering success, but that is irrelevant to science [...] Accurately modeling linguistic facts is just butterfly collecting; what matters in science (and specifically linguistics) is the underlying principles [...] Statistical models are incomprehensible; they provide no insight.

To Chomsky, data science is engineering rather than science; science must confer understanding. As the above passage clearly shows, Norvig actually understood Chomsky's points very well, and still strongly disagreed. Norvig [50] argues for the necessity of statistical analysis in the science of language on the grounds of the contingent nature of language per se:

[L]anguages are complex, random, contingent biological processes that are subject to the whims of evolution and cultural change. What constitutes a language is not an eternal ideal form, represented by the settings of a small number of parameters, but rather is the contingent outcome of complex processes. Since they are contingent, it seems they can only be analyzed with probabilistic models.

The Chomsky versus Norvig debate may be compared with the Bohr versus Einstein debate on the ultimate nature of quantum reality, especially the issue of the EPR (Einstein-Podolsky-Rosen) paradox and non-local correlations [9] (which are even debated in the context of cognition; see [1] and references therein). The Chomskyan linguistics aims at explicating the eternal ideal form of language, and Chomsky is very much like Einstein, who believed that probabilities arise in quantum mechanics because the formulation of quantum mechanics is still incomplete, i.e., there are some hidden variables ("small number of parameters") to make it a deterministic theory like classical mechanics. The universalist's strongest possible presupposition is that there are always universal (or deterministic) principles underlying apparently complex (or probabilistic) phenomena. From the universalist perspective, statistics is more like a compromise than an ultimate solution to the problem of understanding Nature. On the other hand, Norvig argues that the irreducible complexity of natural language and its evolution makes it compelling to use probabilistic models, just as Bohr argued for the necessity of probabilities in quantum mechanics and so for the completeness of it. To the probabilist, there is nothing lurking behind statistics; it can simply happen that certain phenomena in Nature are inherently probabilistic. That is to say, there is just the surface without any depths underlying it (incidentally, such an idea has been discussed in twentieth century continental philosophy as well); certain probabilistic theories are already complete.

It is a common view that Bohr won the debate with Einstein, whose understanding of quantum theory was proven to be misconceived by the celebrated Bell theorem [6], even though there are some non-local deterministic formulations of quantum theory, such as Bohmian mechanics, to which the assumptions of the Bell theorem do not apply and which thus partially realise Einstein's dream [8]. Bell-type theorems in physics are called No-Go theorems because they mathematically refute certain forms of classical realism, which, therefore, is a wrong direction to go. If there are similar theorems in AI, Norvig could mathematically refute Chomsky; however, there is no such theorem known at the moment. And in the case of AI in particular, there is some hope for reconciling the two camps as we shall discuss in the next section, before which we briefly touch upon the tension between the two paradigms in the context of natural language semantics in particular.

2.3   A Manifestation of the Fundamental Tension in Natural Language Semantics

The success of Natural Language Processing in the statistical paradigm is mostly due to the so-called Vector Space Model (VSM) of Meaning. Turney and Pantel [54] indeed argue as follows:

The success of the VSM for information retrieval has inspired researchers to extend the VSM to other semantic tasks in natural language processing, with impressive results. For instance, Rapp (2003) used a vector-based representation of word meaning to achieve a score of 92.5% on multiple-choice synonym questions from the Test of English as a Foreign Language (TOEFL), whereas the average human score was 64.5%.

The Vector Space Model of Meaning is statistical semantics of natural language, and based upon what is called the Distributional Hypothesis [54]: "words in similar contexts have similar meanings." This is some sort of semantic contextualism, and semantic contextualism is a form of holism about meaning, since the meaning of a word is determined with reference to a larger whole, namely contexts, without which meaning cannot be determined. In the Vector Space Model of Meaning, for instance, meaning vectors are derived on the basis of a large amount of linguistic contexts, without which meaning vectors cannot be determined.

In contrast to this statistical semantics, which builds upon contextualism, there is another paradigm of natural language semantics, namely symbolic semantics, which builds upon compositionalism, the view that the meaning of a whole is determined with reference to the meaning of its parts. In contextualism, the meaning of a part is only determined with reference to a larger whole, and thus compositionalism is in sharp contrast with contextualism. The tension between Chomsky and Norvig in the narrow context of linguistic analysis may be understood as rooted in this conflict between compositional and contextual semantics. The compositionality camp includes Montague as well; he expresses an opinion sympathetic with Chomsky as follows [49]:

> There is in my opinion no important theoretical difference between natural languages and the artificial languages of logicians; indeed, I consider it possible to comprehend the syntax and semantics of both kinds of languages within a single natural and mathematically precise theory. On this point I differ from a number of philosophers, but agree, I believe, with Chomsky and his associates.

Both the principle of compositionality and the principle of contextuality have their origins in Frege's philosophy of language. It is puzzling why Frege endorsed both of them, especially in light of the above view that there is a fundamental conflict between compositionality and contextuality. Michael Dummett, a well known commentator on Frege, was clearly aware of this, pointing out a "difficulty which faces most readers of Frege" [18]:

> It was meant to epitomize the way I hoped to reconcile that principle, taken as one relating to sense, with the thesis that the sense of a sentence is built up out of the senses of the words. This is a difficulty which faces most readers of Frege [...] The thesis that a thought is compounded out of parts comes into apparent conflict [...] with the context principle [...].

According to more recent commentators on Frege, it is actually not so obvious whether Frege really endorsed any of them. Pelletier [52], for example, concludes that Frege endorsed neither of them; Janssen [24] argues that Frege only endorsed the principle of contextuality.

Compositionality is essential in the so-called productivity of language; thanks to the compositional character of language, we can compose and comprehend entirely new sentences. Frege [19] was already aware of this connection between compositionality and productivity:

> It is astonishing what language can do. With a few syllables it can express an incalculable number of thoughts, so that even a thought grasped by a terrestrial being for the very first time can be put into a form of words which will be understood by someone to whom the thought is entirely new.

Compositionality allows us to understand some other striking characteristics of natural language. Davidson [15], for example, points out that compositionality is essential in the learnability of language:

> It is conceded by most philosophers of language, and recently by some linguists, that a satisfactory theory of meaning must give an account of how the meanings of sentences depend on the meanings of words. Unless such an account could be supplied for a particular language, it is argued, there would be no explaining the fact that we can learn the language: no explaining the fact that, on mastering a finite vocabulary and a finitely stated set of rules, we are prepared to produce and to understand any of a potential infinitude of sentences. I do not dispute these vague claims, in which I sense more than a kernel of truth.

Yet these do not necessarily imply that the principle of contextuality or statistical semantics based on it cannot account for those properties of natural language (see, e.g., [38], which also explains the tension between compositionality and contextuality in more detail; for contextuality across physics and cognitive science, see [39, 41]).

Statistical semantics in Natural Language Processing has been highly successful in various domains of application and actually implemented in a variety of real-world systems. It is however known to suffer from lack of structure; it mostly ignores the inherent structure of language such as grammar. Turney-Pantel [54], for example, argue in the following manner:

> Most of the criticism stems from the fact that term-document and word-context matrices typically ignore word order. In LSA, for instance, a phrase is commonly represented by the sum of the vectors for the individual words in the phrase; hence the phrases house boat and boat house will be represented by the same vector, although they have different meanings.

The same criticism applies to what is generally called the bag-of-words model in information retrieval [54]. It is truly amazing that statistical semantics has been so successful whilst ignoring the structure of language mostly. The intrinsic structure of language is what Chomsky has investigated for a long time. So Gold [22] says that modern AI technologies would make "Chomskyan linguists cry":

Norvig is now arguing for an extreme pendulum swing in the other direction, one which is in some ways simpler, and in others, ridiculously more complex. Current speech recognition, machine translation, and other modern AI technologies typically use a model of language that would make Chomskyan linguists cry: for any sequence of words, there is some probability that it will occur in the English language, which we can measure by counting how often its parts appear on the internet. Forget nouns and verbs, rules of conjugation, and so on: deep parsing and logic are the failed techs of yesteryear.

Yet at the same time, substantial improvement in computational efficiency and other respects has been achieved recently with the integration of symbolic and statistical methods in Natural Language Processing (see, e.g., [23]). And there is now some movement to integrate the two paradigms in linguistic and other contexts (for integrations in Natural Language Processing, see, e.g., [5,14,23]). The integration of Symbolic and Statistical AI works beyond Natural Language Processing, allowing for different advantages, and this is what we are going to address in the following section (for more detailed discussions on linguistic issues concerning Symbolic and Statistical AI, we refer to [38]).

3   Towards Integrated Artificial Intelligence and Integrated Cognitive Science

Symbolic AI is good at principled judgements, such as logical reasoning and rule-based diagnoses, whereas Statistical AI is good at intuitive judgements, such as pattern recognition and object classification. The former would amount to what is called the faculty of reason and understanding, and the latter to the faculty of sensibility in terms of the Kantian epistemology or philosophy of mind. McLear [47] explains these fundamental faculties of human cognition in the following manner:

> Kant distinguishes the three fundamental mental faculties from one another in two ways. First, he construes sensibility as the specific manner in which human beings, as well as other animals, are receptive. This is in contrast with the faculties of understanding and reason, which are forms of human, or all rational beings, spontaneity.

So, from this Kantian point of view, animal cognition, as well as human cognition, is equipped with the faculty of sensibility to recognise the world, and yet the faculty of reason and understanding is a striking characteristic of human cognition only. If this view is correct, Statistical AI may not be sufficient for realising human-level (or super-human) machine intelligence. The Kantian philosophy of mind suggests that both Symbolic and Statistical AI are indispensable for human-level artificial intelligence. If so, it would be essential for the next generation of AI to overcome the symbolic-statistical divide and integrate the two paradigms of cognition.

3.1   The Integrated Paradigm: A Classification of Integrated AI

Let us discuss Integrated AI in more detail in the following. We give a classification of three levels of Integrated AI, and propose Turing-type tests for it. In addition we argue, inter alia, that Integrated AI is a promising approach to Ethical AI or Just AI.

Let us begin with some history of AI. Broadly speaking, historical developments of Artificial Intelligence may be summarised as follows [12]:

- First-generation AI: Search-based Primitive AI.
- Second-generation AI: Deductive Rule-based Symbolic AI (aka. GOFAI, i.e., Good Old-Fashioned Artificial Intelligence).
    - Examples of Symbolic AI: expert systems based on production rules; automated reasoning and planning; theorem provers and verification; and so fourth.
- Third-generation AI (present-day AI): Inductive Learning-based Statistical AI (with successful applications in industry today).
    - Examples of Statistical AI: neural networks and deep learning; support vector machines and kernel methods; Bayesian networks and their variants such as Markov networks; and so fourth.

The next generation AI, then, might be:

- Fourth-generation AI (in the coming future): Towards Integrated AI, namely the integration of Symbolic and Statistical AI.

[12] presents a similar perspective on future developments of AI.

In order to explicate different ways to conceive Integrated AI, let us now give a conceptual classification of Integrated AI in terms of three levels of integration (i.e., task-oriented integration, modular mechanism integration, and seamless mechanism integration) in the following manner:

1. Task-oriented integration: integration at the level of each concrete problem solving, namely integration made for (or dependent upon) a given particular task, which is thus applicable to the specific type of problems only.
    - Examples of task-oriented integration: Statistical Theorem Proving, which generates candidate proofs via statistical methods, and then verify their correctness via symbolic methods [4,26,51]; Safe Learning, which combines deductive reachability analysis with statistical machine learning in order to determine safe regions for safety critical systems to operate in [2,55]; Neural Planning (i.e., classical planning with deep learning) [3].
2. Modular mechanism integration: integration at the level of modular mechanisms, namely integration with symbolic and statistical components modularly separated to each other.
    - Examples of modular mechanism integration: the compositional distributional model of natural language processing, which derived word vectors via statistical methods based upon what is called the distributional hypothesis ("words in similar contexts have similar meanings"), and compose sentence vectors from word vectors via symbolic methods based upon the logical theory of formal grammar such as Lambek's pregroups [14,23].

3. Seamless mechanism integration: integration at the level of integrated mechanism, namely integration as a single mechanism unifying symbolic and statistical approaches to cognition.
   – Examples of seamless mechanism integration: Markov logic network, which is a general framework to combine first-order logic and Markov networks, "attaching weights to first-order formulas and viewing them as templates for features of Markov networks" [16] (see also [17]); neural-symbolic computing [7, 21], which is one of the oldest approaches to the integrated paradigm of Artificial Intelligence, and aims to integrate statistical connectionism and symbolic representationism within a general framework for learning and reasoning.

Task-independent methods are desirable in order to develop Integrated AI for different purposes in a systematic manner. Task-oriented integrations only work for specific tasks, but both modular and seamless mechanism integrations can work for more general purposes, just as the compositional distributional model of natural language processing mentioned above works for a broad variety of linguistic tasks. Note that there is no implication like seamless integrations are generally better than modular integrations. Modular integrations could be more useful than seamless ones, for example, on the ground that results in each paradigm can be transferred and applied directly. Note also that machine learning frameworks are usually task-independent, even though each problem solving algorithm is made in a task-dependent manner; this means that the mathematical essence of learning is independent of the nature of each concrete task, and that is why machine learning is a theory of learning.

3.2 Five Features of Integrated AI: Making AI More Verifiable, Explainable, Accountable, Ethical, and Thus More Human

Integrated AI is not just for improvement of computational performance; it is expected to resolve difficulties in Statistical AI via the methods of Symbolic AI. Desirable characteristics of Integrated AI would be as follows (cf. [42]):

1. Verifiability: Integrated AI should allow us to verify the results of its problem solving such as prediction and classification (which Statistical AI is good at, whereas Symbolic AI is good at verification).
2. Explainability: Integrated AI should allow us to explain the results of its problem solving, e.g., the reason why they have obtained rather than others; explainability is seemingly one of the strongest concerns in recent AI research.
3. Causality: Integrated AI should allow us to account for causal relationships as well as correlational ones; this is particularly important in data science, which must be able to account for causal laws if it aims to qualify as proper science on its own.
4. Unbiasedness: Integrated AI should allow us to make unbiased judgements or to correct their biases learned from biased real-world data; this 'debiasing' function shall be discussed below in more detail.

5. Generality: Integrated AI would allow for AGI, namely Artificial General Intelligence; this may be too strong a requirement, yet developing general intelligence would be one of the ultimate purposes of AI research.

Caliskan et al. show in their recent *Science* article [11] that:

> Semantics derived automatically from language corpora contain human-like biases.

Machine learning, or data-driven science enabled with it, is descriptive in the sense that it basically learns anything in data, regardless of whether it is good or bad. It is like a very obedient child, who may mimic some bad behaviour of parents or teachers without considering whether it is good or bad. By contrast, Integrated AI can be normative as well as descriptive; it can, for example, be equipped with top-down rules or norms to prevent bias learning and to make AI more ethical. This may count as a striking feature of Integrated AI, especially from the perspective of AI ethics.

There might be no means for purely statistical AI to prevent itself from learning biases from biased data; the better it approximates the given biased data, the better it learns those biases contained in it. This ironically suggests that those AI systems that are inferior in learning performance can actually be superior, in terms of unbiasedness, to those that are superior in learning performance (something analogous could happen in the human society as well). Put another way, there are things one should not learn from experience (i.e., empirical data) as well as things one should learn from it. And rational agents must be able to distinguish between them on the ground of some norms or rules, which can be incorporated via Symbolic AI. Integrated AI would thus be a right framework for Unbiased Ethical AI (aka. Just AI; see also [42]); this would be crucial in developments of AI for the Social Good, which has been sought after in the present, more and more AI-laden society.

Social implementation of AI systems would require them to be ethical; ethics, or ethical behaviour, may also be considered to be constituents of intelligence. But how could we judge whether AI is morally good or not? There could be something like the Turing Test to do that. For example, the Ethical Turing Test could be formulated in the following manner:

– The Ethical Turing Test (aka. Misleading Turing Test): we try to deceive AI with biased data or reasoning; still AI must be able to make correct judgments whilst being not deceived by us humans.

AI must be able to follow (or simulate) correct behaviour in the original Turing Test; in the above Ethical Turing Test, AI must be able to correct its behaviour, and so it may also be called the Dual Turing Test. The Dual Turing Test can be more difficult to pass than the original Turing Test, because correcting wrong answers is often more complex than giving correct answers (the author is familiar with this phenomenon in his experience of teaching logic to hundreds of students and correcting their mistakes every week during the term). The Ethical/Dual Turing Test requires AI to be resilient with respect to different biases, which do exist in real-world situations. Although Statistical AI has been shown to learn different biases from real-world data in an inductive, bottom-up manner, nevertheless, Integrated AI could pass the Ethical Turing Test with the help of Symbolic AI, which gives top-down rules and principles to make it immune to potential biases.

To test Integrated AI, we could rely upon other types of Turing-type tests as well, such as the Verification Turing Test, in which AI must be able to give both answers to questions and the verification of them. In general, the plurality of Turing-type tests would be essential; there may be no single experimental scheme to test different aspects of intelligence at once. If so, multiple tests are required to test different facets of intelligence. Human intelligence is so versatile that no single experiment allows for an adequate assessment of different aspects of it. So the plurality of Turing-type tests would be essential for conceiving the Turing Test for Integrated AI (for related Turing-type Tests, see also [42]).

There could be Chinese-Room-style counterarguments against these Turing-type Tests. Highly non-ethical AI could pass the Ethical Turing Test above just by simulating ethical behaviour in a superficial manner. Superintelligent AI, e.g., could not be deceived by us, bur rather could deceive us in many ways, whilst hiding its unethical nature from us. This means it could easily pass the Ethical Turing Test. Yet the same thing may happen in the human case as well. Just as there is no effective method to test the ethical nature of human beings, there would be no ultimate Turing Test for Ethical AI, neither (for related issues, see also [32, 42]).

4 Concluding Remarks: The Integrated Paradigm as a Transdisciplinary Trading Zone

We have discussed the fundamental tension between the symbolic and statistical paradigms of AI, and the possibility of integrating and unifying them, together with various advantages to do so, including the ethical one in particular. What is particularly interesting in the present landscape of AI is, in our opinion, that the debate between the symbolic and statistical camps look very much like the classic debate between the universalist and the probabilist views of Nature (including Cognition and Intelligence as part of it), and that the debate is directly relevant to urgent issues in AI, such as verifiability, explainability, causal accountability, and algorithmic biases, as we have discussed above.

AI and Machine Learning, therefore, would allow us not only to revive the classic debate between the universalist and the probabilist in the past, but also to place it in different novel contexts relevant to the present society. The central tenet of the present paper is that Integrated AI, if it could be developed in the right way, would serve the purpose of solving those urgent issues in AI. Yet at the same time, we would contend that philosophical debates in the past (and present) could be useful inputs to the design and development of Integrated AI, since they are closely linked with the urgent issues in AI as we have discussed above. In light of these, Integrated AI could be a transdisciplinary platform (or trading zone in Peter Galison's terms [20]) where different sorts of intellectual cultures are allowed to meet each other, as well as a theoretical foundation for the next generation AI technology (note that Galison also contrasts Image and Logic, which Statistical and Symbolic AI are arguably about).

We have mainly focused upon AI rather than Cognitive Science in general; even so, our arguments would mostly apply to Cognitive Science as well as AI. There are, as a matter of course, symbolic and statistical paradigms in Cognitive Science just as well, and integrating them would be beneficial in many ways. From the AI point of view, the principal merits of the integrated paradigm would be developments of solutions to problems such as explainability and algorithmic biases. Yet from the Cognitive Science point of view, the principal advantages of the integrated paradigm would rather be the integrated understanding of fundamental faculties of the human mind, especially the faculty of reason and understanding on one hand, and the faculty of sensibility on the other. Integrated Cognitive Science could, hopefully, lead to something like a cognitive theory of everything (or a theory of every-cognition). A physical theory of everything is concerned with a unified understanding of general relativity and quantum theory, and a cognitive theory of everything with a unified understanding of the faculty of reason and understanding and the faculty of sensibility.

It is a highly non-trivial issue how to actually develop Integrated AI and Integrated Cognitive Science or what kind of mathematical framework allows us to lay down a theoretical foundation for the integrated paradigm of AI and Cognition in the first place. Let us briefly remark upon our speculative vision for the integrated paradigm. We have touched upon different approaches to Integrated AI throughout the paper, some of which can be expressed in terms of category theory, an algebraic theory of structural networks. We generally believe that category theory could give a principal methodology to integrate the two paradigms, as it has indeed played such unificatory rôles in the sciences, and succeeded in integrating different paradigms even across different kinds of sciences, inter alia, via the transdisciplinary theory of categorical dualities between the ontic and the epistemic in various scientific disciplines (see, e.g., [27–31,34,36,37,40,43–45] and references therein). Let us just touch upon a single successful case of the transdisciplinary application of category theory: substructural logic and quantum mechanics have been unified by means of the abstract theory of categorical quantum theory (see [14,33,35] and references therein), and the logical methods of automated reasoning have been applied to quantum physics, and quantum computation in particular, via the categorical connection between substructural logic and quantum physics; the first system developed at the University of Oxford is called Quantomatic, and there are more advanced versions available online for free, including PyZX for the so-called ZX calculus in quantum computation. Those automated reasoning systems for quantum mechanics and computation may be regarded as artificial intelligence for quantum mechanics and computation; it is supported by graphical calculus in terms of category theory, the completeness of which can be shown in a mathematically rigorous way (basically, any equation valid in Hilbert spaces can be logically derived in the graphical calculus of categorical quantum mechanics, and vice versa; see [14] and references therein). In light of these, many approaches to Integrated AI and Cognition could even be understood and unified under one umbrella, namely Categorical Artificial Intelligence and Categorical Cognitive Science, hopefully. Integrated AI could even be integrated into Robotics, thus leading to Integrative AI Robotics, in particular Categorical AI Robotics or Categorical Integrative AI Robotics. In order to address the symbol grounding problem, it is essential to consider enacted interactions between symbolic cognition and physical environments. Categorical Integrative AI Robotics might even help us to resolve the symbol grounding problem, and account for the nature of symbol emergence.

Acknowledgements. The author is grateful to the four referees for their numerous helpful comments and suggestions for improvement. The author hereby acknowledges that this work was supported by JST PRESTO (JPMJPR17G9).

References

1. Aerts, D., et al.: Quantum entanglement in physical and cognitive systems: a conceptual analysis and a general representation. Eur. Phys. J. Plus **134** (2019). Article number: 493. https://doi.org/10.1140/epjp/i2019-12987-0
2. Akametalu, A.K., Kaynama, S., Fisac, J.F., Zeilinger, M.N., Gillula, J.H., Tomlin, C.J.: Reachability-based safe learning with Gaussian processes. In: Proceedings of CDC, pp. 1424–1431 (2014)
3. Asai, M.: Classical planning in deep latent space: bridging the subsymbolic-symbolic boundary. In: Proceedings of AAAI, pp. 6094–6101 (2018)
4. Bansal, K., Loos, S.M., Rabe, M.N., Szegedy, C., Wilcox, S.: HOList: an environment for machine learning of higher order logic theorem proving. In: Proceedings of ICML, pp. 454–463 (2019)
5. Baroni, M., et al.: Frege in space: a program of compositional distributional semantics. Linguist. Issues Lang. Technol. **9**, 5–110 (2014)
6. Bell, J.S.: Speakable and Unspeakable in Quantum Mechanics: Collected Papers on Quantum Philosophy. Cambridge University Press, Cambridge (2004)
7. Besold, T.R., et al.: Neural-symbolic learning and reasoning: a survey and interpretation (2017). arXiv:1711.03902
8. Bohm, D., Hiley, B.: The Undivided Universe: An Ontological Interpretation of Quantum Theory. Routledge Chapman & Hall, Abingdon (1993)
9. Born, M.: The Born Einstein Letters. Walker and Company, New York (1971)
10. Brooks, R.: Intelligence without representation. Artif. Intell. **47**, 139–159 (1991)
11. Caliskan, A., et al.: Semantics derived automatically from language corpora contain human-like biases. Science **356**, 183–186 (2017)
12. CDRS: Research and Development on the Fourth Generation of AI, Strategic Proposal, CRDS-FY2019-SP-08 (2019)
13. Chomsky, N.: Keynote panel: the golden age - a look at the original roots of artificial intelligence. In: Cognitive Science, and Neuroscience. Minds, and Machines, MIT Symposium on Brains (2011)
14. Coecke, B., et al.: Mathematical foundations for a compositional distributional model of meaning. Linguist. Anal. **36**, 345–384 (2010)
15. Davidson, D.: Truth and meaning. Synthese **17**, 304–323 (1967)

16. Domingos, P., Kok, S., Lowd, D., Poon, H., Richardson, M., Singla, P.: Markov logic. In: De Raedt, L., Frasconi, P., Kersting, K., Muggleton, S. (eds.) Probabilistic Inductive Logic Programming. LNCS (LNAI), vol. 4911, pp. 92–117. Springer, Heidelberg (2008). https://doi.org/10.1007/978-3-540-78652-8_4

17. Domingos, P., Lowd, D.: Unifying logical and statistical AI with Markov logic. Commun. ACM **62**, 74–83 (2019)

18. Dummett, M.: The Interpretation of Frege's Philosophy. Duckworth, London (1981)

19. Frege, G.: Compound thoughts. Mind **72**, 1–17 (1963). Originally 1923

20. Galison, P.: Image & Logic: A Material Culture of Microphysics. The University of Chicago Press, Chicago (1997)

21. Garcez, A., Gori, M., Lamb, L., Serafini, L., Spranger, M., Tran, S.: Neural-symbolic computing: an effective methodology for principled integration of machine learning and reasoning. arXiv:1905.06088

22. Gold, K.: Norvig vs. Chomsky and the Fight for the Future of AI. Tor.com, 21 June 2011

23. Grefenstette, E., et al.: Experimental support for a categorical compositional distributional model of meaning. In: Proceedings of EMNLP 2011, pp. 1394–1404 (2011)

24. Janssen, T.: Frege, contextuality and compositionality. J. Log. Lang. Inform. **10**, 87–114 (2001). https://doi.org/10.1023/A:1026542332224

25. Katz, Y.: Noam Chomsky on Where Artificial Intelligence Went Wrong. The Atlantic, 1 November 2012

26. Lederman, G., Rabe, M.N., Lee, E.A., Seshia, S.A.: Learning heuristics for automated reasoning through deep reinforcement learning. arXiv:1807.08058

27. Maruyama, Y.: Fundamental results for pointfree convex geometry. Ann. Pure Appl. Log. **161**, 1486–1501 (2010)

28. Maruyama, Y.: Natural duality, modality, and coalgebra. J. Pure Appl. Algebra **216**, 565–580 (2012)

29. Maruyama, Y.: From operational chu duality to coalgebraic quantum symmetry. In: Heckel, R., Milius, S. (eds.) CALCO 2013. LNCS, vol. 8089, pp. 220–235. Springer, Heidelberg (2013). https://doi.org/10.1007/978-3-642-40206-7_17

30. Maruyama, Y.: Full lambek hyperdoctrine: categorical semantics for first-order substructural logics. In: Libkin, L., Kohlenbach, U., de Queiroz, R. (eds.) WoLLIC 2013. LNCS, vol. 8071, pp. 211–225. Springer, Heidelberg (2013). https://doi.org/10.1007/978-3-642-39992-3_19

31. Maruyama, Y.: Categorical duality theory: with applications to domains, convexity, and the distribution monad. In: International Proceedings in Informatics, vol. 23, pp. 500–520. Leibniz (2013)

32. Maruyama, Y.: AI, quantum information, and external semantic realism: searle's observer-relativity and Chinese room, revisited. In: Müller, V.C. (ed.) Fundamental Issues of Artificial Intelligence. SL, vol. 376, pp. 115–126. Springer, Cham (2016). https://doi.org/10.1007/978-3-319-26485-1_8

33. Maruyama, Y.: Prior's tonk, notions of logic, and levels of inconsistency: vindicating the pluralistic unity of science in the light of categorical logical positivism. Synthese **193**, 3483–3495 (2016). https://doi.org/10.1007/s11229-015-0932-9

34. Maruyama, Y.: Categorical harmony and paradoxes in proof-theoretic semantics. In: Piecha, T., Schroeder-Heister, P. (eds.) Advances in Proof-Theoretic Semantics. TL, vol. 43, pp. 95–114. Springer, Cham (2016). https://doi.org/10.1007/978-3-319-22686-6_6

35. Maruyama, Y.: Meaning and duality: from categorical logic to quantum physics. D.Phil. thesis, University of Oxford (2017)
36. Maruyama, Y.: The dynamics of duality: a fresh look at the philosophy of duality. In: RIMS Kokyuroku (Proceedings of RIMS, Kyoto Univesity), vol. 2050, pp. 77–99 (2017)
37. Maruyama, Y.: The frame problem, Gödelian incompleteness, and the Lucas-Penrose argument: a structural analysis of arguments about limits of AI, and its physical and metaphysical consequences. In: Müller, V.C. (ed.) PT-AI 2017. SAPERE, vol. 44, pp. 194–206. Springer, Cham (2018). https://doi.org/10.1007/978-3-319-96448-5_19
38. Maruyama, Y.: Compositionality and contextuality: the symbolic and statistical theories of meaning. In: Bella, G., Bouquet, P. (eds.) CONTEXT 2019. LNCS (LNAI), vol. 11939, pp. 161–174. Springer, Cham (2019). https://doi.org/10.1007/978-3-030-34974-5_14
39. Maruyama, Y.: Contextuality across the sciences: bell-type theorems in physics and cognitive science. In: Bella, G., Bouquet, P. (eds.) CONTEXT 2019. LNCS (LNAI), vol. 11939, pp. 147–160. Springer, Cham (2019). https://doi.org/10.1007/978-3-030-34974-5_13
40. Maruyama, Y.: Foundations of mathematics: from Hilbert and Wittgenstein to the categorical unity of science. In: Wuppuluri, S., da Costa, N. (eds.) WITTGEN-STEINIAN (adj.). TFC, pp. 245–274. Springer, Cham (2020). https://doi.org/10.1007/978-3-030-27569-3_15
41. Maruyama, Y.: Rationality, cognitive bias, and artificial intelligence: a structural perspective on quantum cognitive science. In: Harris, D., Li, W.-C. (eds.) HCII 2020. LNCS (LNAI), vol. 12187, pp. 172–188. Springer, Cham (2020). https://doi.org/10.1007/978-3-030-49183-3_14
42. Maruyama, Y.: The conditions of artificial general intelligence: logic, autonomy, resilience, integrity, morality, emotion, embodiment, and embeddedness. In: Goertzel, B., Panov, A.I., Potapov, A., Yampolskiy, R. (eds.) AGI 2020. LNCS (LNAI), vol. 12177, pp. 242–251. Springer, Cham (2020). https://doi.org/10.1007/978-3-030-52152-3_25
43. Maruyama, Y.: Topological duality via maximal spectrum functor. Commun. Algebra **48**, 2616–2623 (2020)
44. Maruyama, Y.: Higher-order categorical substructural logic: expanding the horizon of tripos theory. In: Fahrenberg, U., Jipsen, P., Winter, M. (eds.) RAMiCS 2020. LNCS, vol. 12062, pp. 187–203. Springer, Cham (2020). https://doi.org/10.1007/978-3-030-43520-2_12
45. Maruyama, Y.: Universal stone duality via the concept of topological dualizability and its applications to many-valued logic. In: Proceedings of FUZZ-IEEE. IEEE Computer Society (2020)
46. Maxwell, J.C.: The Scientific Letters and Papers of James Clerk Maxwell: 1846–1862. Cambridge University Press, Cambridge (1990)
47. McLear, C., Kant: philosophy of mind. In: Internet Encyclopedia of Philosophy. Accessed 2 Feb 2020
48. Minsky, M.L.: Logical versus analogical or symbolic versus connectionist or neat versus scruffy. AI Mag. **12**, 34–51 (1991)
49. Montague, R.: Universal grammar. Theoria **36**, 373–398 (1970)
50. Norvig, P.: On Chomsky and the two cultures of statistical learning. Berechenbarkeit der Welt?, pp. 61–83. Springer, Wiesbaden (2017). https://doi.org/10.1007/978-3-658-12153-2_3

51. Paliwal, A., Loos, S.M., Rabe, M.N., Bansal, K., Szegedy, C.: Graph representations for higher-order logic and theorem proving. In: Proceedings of AAAI, pp. 2967–2974 (2020)

52. Pelletier, F.J.: Did Frege believe Frege's principle? J. Logic Lang. Inform. **10**, 87–114 (2001). https://doi.org/10.1023/A:1026594023292

53. Thompson, N.C., et al.: The computational limits of deep learning (2020). arXiv:2007.05558

54. Turney, P., Pantel, P.: From frequency to meaning: vector space models of semantics. J. Artif. Intell. Res. **37**, 141–188 (2010)

55. Zhou, W., Li, W.: Safety-aware apprenticeship learning. In: Chockler, H., Weissenbacher, G. (eds.) CAV 2018. LNCS, vol. 10981, pp. 662–680. Springer, Cham (2018). https://doi.org/10.1007/978-3-319-96145-3_38

An Interdisciplinary Model for Graphical Representation

G. Antonio Pierro[1,2(✉)], Alexandre Bergel[3], Roberto Tonelli[2],
and Stéphane Ducasse[1]

[1] Université de Lille, Inria, CNRS, Centrale Lille, UMR 9189 – CRIStAL,
Lille, France
giuseppe.pierro@inria.fr
[2] Università degli Studi di Cagliari, Cagliari, Italy
[3] DCC Universidad de Chile, Santiago, Chile

Abstract. The paper questions whether data-driven and problem-driven models are sufficient for a software to automatically represent a meaningful graphical representation of scientific findings. The paper presents descriptive and prescriptive case studies to understand the benefits and the shortcomings of existing models that aim to provide graphical representations of data-sets. First, the paper considers data-sets coming from the field of software metrics and shows that existing models can provide the expected outcomes for descriptive scientific studies. Second, the paper presents data-sets coming from the field of human mobility and sustainable development, and shows that a more comprehensive model is needed in the case of prescriptive scientific fields requiring interdisciplinary research. Finally, an interdisciplinary problem-driven model is proposed to guide the software users, and specifically scientists, to produce meaningful graphical representation of research findings. The proposal is indeed based not only on a data-driven and/or problem-driven model but also on the different knowledge domains and scientific aims of the experts, who can provide the information needed for a higher-order structure of the data, supporting the graphical representation output.

Keywords: Data visualization · Interdisciplinary model · Data-driven model · Problem-driven model

1 Introduction

Graphical representations of data are fundamental for the understanding of scientific knowledge, as readers often rely on what the experts visually represent in their publications to understand the underlying data-set and interpret their potential scientific meaning [1]. Figures and diagrams not only show the relevant data that support key research findings, but also provide visual information on the interactions among different operations required in scientific reasoning [2,3]. Being able to adequately and precisely visualize data is also a pillar on which decisions can be made, as proposed by different dashboards in the market.

© Springer Nature Switzerland AG 2021
L. Cleophas and M. Massink (Eds.): SEFM 2020 Workshops, LNCS 12524, pp. 147–158, 2021.
https://doi.org/10.1007/978-3-030-67220-1_12

Data visualization has various purposes, such as to make abstract thinking on data series or sets more concrete and (mentally) manipulable, to help readers identify and evaluate some features of the data, to let users see the possible underlying trends, patterns, processes, mechanisms, etc. of the phenomena considered and studied [4]. The way data are visualized can therefore have important epistemic implications for scientific knowledge, as data visualization is not an "interpretation-free" practice, i.e. a neutral process of data presentation in terms of scientific understanding. There are indeed several ways to transform data into a visual format, each of them entailing different possibilities for data interpretation.

Nowadays data visualization plays a significant role in the large adoption of data-driven and machine learning approaches and techniques. In this frame, the definition of what a visualization is can be object of debate. A visualization could be defined as a reusable component, which is achieved through a dedicated software library. For instance, some software for data visualization are MATLAB and Mathematica. Despite the large amount of tools offered by these software, surprisingly, it is left to the practitioner to actually manipulate the data to achieve a ready-to-be-used graphical representation. Previous research proposed data-driven models that exploit existing software libraries or adopt a framework-agnostic approach (D. A. Keim, 2002 [5]) based on data types to be visualized.

The paper aims at designing a framework for a software, named Miró, which instead allows the users to produce meaningful graphical representation in an automatic way without the need to manually transform the data. First of all, we aim to verify the benefits and the shortcomings of existing data-driven and problem-driven models, by presenting some case studies. The case studies focus on the problem of visually representing specific data-sets collected in different scientific domains for different (descriptive vs. prescriptive) scientific aims. The case studies suggest that data-driven models can actually provide a visualization that fits the domain knowledge and scientific aims of the experts in the case of descriptive sciences, but present some limitations in the case of prescriptive sciences. Finally, the paper draws some conclusion from the case studies, presenting an alternative interdisciplinary perspective for data visualization. A comprehensive model for graphical representation is then presented, which integrates a data-driven approach with an approach that guides the experts on a specific domain field to achieve the intended visualization, based on their aims, knowledge and hypotheses. Miró adopts this interdisciplinary perspective and is based on a visualization engine developed in Pharo and named Roassal [6].

2   Data-Driven and Problem-Driven Models

In the field of data visualization computing, researchers proposed different approaches to a comprehensive data-model, i.e. a model able to provide a meaningful graphical representation of a data-set for some scientific aims. Some authors advocated graphical representation techniques or visualization frameworks [7] based on data-driven models. The data-driven model approach is based on the idea that a comprehensive data-model is based on a prior data classification that can guide the automatic creation of a meaningful graphical representation. In general, the data-driven model describes the data characteristics of the data-set, such as the size (the number of rows), the data type (string, number, boolean) and the dimension (the number of the variables to represent), to categorize the data. Keim [5] proposed a data-driven visualization model based on the data types to be visualized, the visualization technique and the technique of visual interaction with data, ranging from standard and projection to distortion and "link&brush".

Other authors, especially in the context of big data visualization, proposed graphical representation techniques based on a problem-driven model [8]. The problem-driven model provides the researchers with the possibility to perform specific tasks on specific variables of the data-set, such as visualizing a variable distribution, performing a linear regression between two variables to see an eventual relationship via a scatter plot, comparing their composition via a pie chart, etc.

On the one hand, adopting a problem-driven model does not necessarily mean abandoning data-driven models. The problem-driven model may be tightly linked to the data-driven model, because the data-driven model imposes constraints on the graphical representation of data which might conditioning how the problem can be solved. For instance, in the case of time series, there are graphs that are less appropriate than others or that are simply wrong depending on the data classification: the time data-type is indeed a constraint given or inferred from the data-driven model. On the other hand, a graphical representation that is guided only by a data-driven model would not allow the users to further act on data to have their final intended graphical representation. In the software where a problem-driven model is also envisaged, the user can interfere with the final graphical representation of the data. The user can indeed act on and guide the graphical representation to be produced.

The main disadvantage of the problem-driven model is that it might be negatively influenced by the users' previous hypotheses or scientific aims. On the contrary, a data-driven model is neutral under this respect: of course it is based on a prior classification, but the users might not know it. Without the users' interference, the final graphical output of a data-driven model might indeed have the advantage of questioning the researchers' prior goals and solicit a belief revision. Especially when a graphical output is unexpected and not corresponding to previous scientific goals, it might bring about further research or action.

Both the models assume that the data-set contains the information useful to produce a meaningful graphic representation. This may not always be the case. Scientific studies based on data-sets make use of graphical representations to better interpret their results. Among these studies, it is possible to find descriptive as well as prescriptive studies. The former aim to describe phenomena as they are, observing, recording, classifying, and comparing them [9]. The latter aim to provide the conditions for how phenomena should be, thus supporting inferences for data interpretation and decision and/or action to perform on data. Of course, a scientific study could be both descriptive and prescriptive, also depending on the scientific goals. The development of new decision-aiding technology should be tailored for both [10], also in the case of graphical representation [11]. The paper is therefore driven by the question on how a model should be to provide a meaningful graphical representation of a data-set to support the inferences and/or the decision a researcher wants to draw, in both the case of descriptive and prescriptive scientific studies.

In the paper we propose a general distinction between a model for descriptive studies and a model for prescriptive studies. Within these two models, it is possible to specify sub-models, specific for scientific domain and particular data types involved in the study [12]. Both the models can be used whenever a study has both descriptive and prescriptive scientific aims, as it is often the case.

3   Research Questions and Hypotheses

The paper aims to discuss the strengths and limitations of existing models for data visualization, by considering and discussing some case studies coming from publications of different scientific domains and having different scientific aims.

The research addresses the following questions: Q1) Are data-driven models sufficient for a software to help the researchers to automatically create the intended visual form for a data-set? Q2) In the case the data-driven models are not sufficient, what could be the best way to overcome their limitations? Q3) Can the existing libraries or programs fit a data-driven model perspective and at the same time overcome their shortcomings?

To answer the research questions, we advanced the following hypotheses: H1) The data-driven models might support the creation of meaningful graphical representation only for some specific scientific aims, such as the researchers' aims to provide a descriptive data analysis. H2) For scientific aims going beyond descriptive analysis, the existing data-driven models might not be sufficient. The data-driven models might need to be integrated into a more comprehensive and interdisciplinary data-model to overcome their eventual limitations. H3) Existing software libraries are data-driven and might not be sufficient to help researchers to find the intended visual form for prescriptive scientific aims. They might need further implementation to allow the users to perform different manipulation on data, such as transformation, accommodation and integration with complementary data, to achieve the intended graphical output.

Several different real-world scenarios and case studies support the hypotheses mentioned above [12,13], a couple of which are discussed in the following Sect. 4.

4   Case Studies Evaluation

We analyzed data-sets which are representative of two different scientific approaches: 1) descriptive and 2) prescriptive studies. In particular we provide a detailed analysis of some case studies, coming from 1) the domain of software metrics, in the wider field of AI, and 2) the field of human mobility and sustainable development. The analysis can be extended to further case studies in different scientific domains.

4.1 Descriptive Case Studies

As to descriptive scientific studies, we considered first of all the case of a study on the performance evaluation of different frameworks in AI [14]. The case study proposes a set of meaningful visual representations of a benchmark data-set for the performance evaluation of different Deep Learning (DL) models and frameworks. The Authors calculated the accuracy and the throughput of five classification problems for the DL models and frameworks. The output data-set was made of a series of two categorical data (the name of the framework and the DL model) and two physical data.

We selected this study for three reasons: 1) The work aims to provide a significant graphical representation of the performance metrics of different frameworks; 2) The work also aims to extend the graphical representation to other frameworks, to be applied to other works and thus be generalized. 3) The study's data-set presents a number of variables and categories, which are not trivial to represent as a whole to obtain a meaningful graphical representation [15].

When analyzing the study case, we found that there is a data-driven model, specifically Keim's data-model, that provides us with a significant representation of the data-set, without any accommodation and/or transformation of the data and, more importantly, without any addition of further information by the user. Indeed, by applying Keim's data-model, the data-set is well within multi-dimensional category and so the meaningful graphical representation technique should be a "heat-map graph", where the colour is represented by the categorical data and the two physical data (accuracy and throughput) are represented in a 2D coordinate system. Therefore, as to what concerns Q1, "Do data-driven models support the creation of meaningful graphical representation", the answer is positive. As the Keim's data-model is sufficient to have a proper graphical representation, we do not need to cope with Q2 on how to improve it for this specific case study. As to what concern Q3, the existing libraries for producing data visualizations alone cannot give that expected output, even though based on a data-driven model. However, throughout a data-driven model such as the Keim's model and some accommodation of the data, the existing libraries could provide the expected automatic visual representation, starting from the raw data-set.

Other descriptive case studies concern, for instance, static programming analysis and focus on the correlation between numerical variables, such as the number of lines of code, cohesion, coupling or cyclomatic complexity [16] and categorical variables, such as the name of the package included in the analyzed software. This type of studies' authors often choose to represent their data-sets via a bar graph where the bar length represents the numerical value and the categorical variable is represented by the different color of the bar or by a label. Also in these cases, the graphical output can thus be provided by a data-driven model such as Keim's model. The analysis can be extended to other descriptive case studies in different disciplines (e.g. biology [17], and sociology [18]), where Keim's data-model is sufficient to provide the categorization for descriptive scientific aims.

4.2 Prescriptive Case Studies

In the case of prescriptive scientific studies we first considered an interdisciplinary study on human mobility [19]. The Authors collected the data using smartphones and smartwatches worn by several participants over 2 weeks. Through these devices, they collected three kinds of data: 1) motion sensor data, 2) physiological data, 3) environmental data. For the purposes of this case study, we are interested in the second data-set collecting information about electrocardiographic (ECG) data, such as heart beat and blood pressure. The data-set has the following characteristics: 1) data are multidimensional, as each row of the data set contains both spatial coordinates (longitude and latitude) and physiological data (heart rate, in beats per minute), provided by the optical heart rate sensor of the smartwatch; 2) the row data series consists of over 1 millions of data.

One of the purposes of the research paper was to use physiological data to infer the user's stress and emotion level to identify places within a University campus area that are perceived as dangerous by the majority of participants. We selected this research for the following reasons:

- The research covers different domains: mobile computing, sensing systems, human mobility profiling and cardiology.
- As in the previous case study, the data-set presents a number of variables and categories, which are not trivial to represent in an overall meaningful graphic representation.

If we apply the Keim's model to the data-set, the graphic representation output is a "heat-map chart", where the position is represented in a 2D-coordinate system and the heart rate beat is represented by color hue. This type of representation may not be enough meaningful for the aims of the study, when based only on the data-set collected by the devices. Indeed, the data-set is not per se sufficient to have a meaningful representation: the danger zones' classification needs other, additional data, such as the normal resting heart rate range and the dangerous heart rate range, to be properly represented.

Figure 1 shows the graphical representation produced considering the additional data, the normal and dangerous heart rate ranges. These additional data are used to represent the different zones on the map with colors having different opacity (color with opacity 1 for the dangerous zones and transparent color for the zones considered safe).

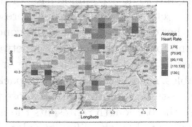

Fig. 1. Places that are perceived as dangerous by the majority of users through the use of colors with different shades.

Therefore, as to what concerns Q1, the answer is that the data-driven model is not sufficient to give the intended graphic representation. Indeed the authors considered complementary data that are not merely added to the existing categories considered by the data-driven model, but rather organize in a higher-order structure and provide the cues to interpret the data-set to have a meaningful representation of the zones considered dangerous. The complementary data do shape the authors' interpretation of the data-set as they provide some intervals (the heartbeat rates intervals), as conditions to classify dangerous vs. safety zones. Indeed, the graphical representation Fig. 1 can be prescriptively used by experts in urban development for strategic planning to improve safety in public places.

As to Q2, the solution to overcome the limitations of the data-driven model could be the possibility of inserting further data types into the data-set, relating the average heartbeat rates stored in the original data-set with the heartbeat rates intervals considered normal and dangerous. Furthermore, the data must be re-sampled taking into account the new knowledge, the normal resting heart rate range, coming from a different domain, the cardiology. However, this solution requires specific knowledge from the cardiology domain which may be different from the researchers' knowledge performing the data analysis.

Finally, regarding Q3, data visualization libraries alone cannot help to obtain the expected output. Indeed, different tasks should be foreseen to achieve the intended outcome through a software, including the data visualization libraries:

- the program should make use of a data-driven model, such as the Keim's model.
- the program should give the user the possibility to add other data type. In the prescriptive case study, the data-type are intervals (conditioning the interpretation of the other data), also coming from a different scientific domain, i.e. cardiology.
- the program should give the researchers the possibility to further categorize the data-set via the additional knowledge. The program must provide the data-set with an higher-order structure to achieve the graphic representation meaningfully corresponding to the authors' scientific aims.
- Once adopting this workflow, the program might use the data visualization library to generate the intended graphic representation.

Another example of prescriptive studies concern the correlation between air pollution and respiratory illnesses [20]. The research findings come from data belonging to different domains such as 1) prescriptive data conditions in health information systems, 2) the air quality index (AQI) data provided by the World Health Organization (WHO), and 3) the descriptive data coming from particular air pollution electrical sensors. The descriptive data alone, in particular the concentra-

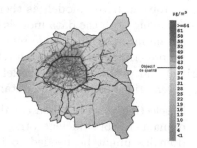

Fig. 2. Areas affected by air pollution.

tion of microscopic particles with a diameter of 2.5 μm or less, are not sufficient to produce a graphical representation apt to meet the prescriptive aims of the study (see Fig. 2), i.e the sustainable development program in urban and rural areas affected by air pollution.

5   An Interdisciplinary Model

In the field of graphical representation, interdisciplinary models have been proposed to cope with the limitations of both previous data-driven and problem-driven models. For instance, Hall et al. [21] proposed a trans-disciplinary model which allow the experts in a particular domain to be supported by visualization experts. Their work is very interesting as the interaction between experts with skills in different domains could greatly influence the production of meaningful graphical representations to display cues for scientific findings.

However, the prescriptive case study examined in this paper cannot be solved through this trans-disciplinary approach. Of course a competence in visualization is welcome, but cannot per se highlight the conditions of meaningfulness, which come from another scientific domain in the prescriptive case studies. Therefore an interdisciplinary model is needed which integrates knowledge and practice coming from different scientific domains in the process of visualization. Figure 3 proposes the main elements of the interdisciplinary model.

- The source domain/s is/are the domain/s from which the data are collected.
- The complementary domain/s is/are the domain/s from where to collect the data required to interpret the source domain/s data.
- The blended domain [22] is given by the intersection between the source domain/s and the complementary domain/s, where some new insight could emerge.
- The data model is the model driving the software in the process of data categorization and visualization.

As a solution to the prescriptive case studies examined, we propose an interdisciplinary problem-driven approach for the visualization of data coming from different domains. For the aims of descriptive studies, the source domain and the data-driven model are usually sufficient to have meaningful graphical representations. The prescriptive case studies instead show the limits of both data-driven and problem-driven model, as there are scientific aims for which it is not sufficient having both the data models and the data coming from a scientific domain to obtain meaningful graphic reports for the research findings.

In prescriptive studies, two further processes - not envisaged in previous data-driven and problem-driven models - are needed to have meaningful graphical representations of the source data:

- A selection process: when the data collected by the researchers in the source domain are not sufficient, other specific data selected from a different scientific domains might be needed to interpret the source data. These data might indeed be the condition of meaningfulness for data interpretation, and thus for the visual output of the software.

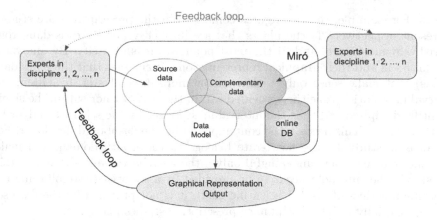

Fig. 3. Interdisciplinary model

- A transformation process: specific tasks might be needed for the reinterpretation of the data in light of the selected complementary data and
  the scientific aims of the study. For instance, the source data might need to
  be re-sampled considering the complementary knowledge.

The scientist's insight needs, therefore, to be entered as complementary data
in any software's visual framework, which in turn should make it possible to
enter them, interacting with the scientist. In the prescriptive case studies, the
interdisciplinary approach is driven by the interaction among experts in different
domains (mobile computing and cardiology) and guides the production of graphical representations, meaningfully representing the areas perceived as dangerous
(see Fig. 1). The insertion of the relevant complementary data might come not
only from experts of another domain, but also from online interactions among
experts in different domains and/or online web-based crowd-sourcing selected
by the expert users themselves.

This interdisciplinary model might then overcome the limitations of both
the data-driven and the problem-drive models, especially when it automatically
proposes the complementary data based on the scientific aims of the expert
and the relative missing expertise, which could come from an expert in another
domain. This approach is the framework for Miró, a software intended to be
a guide to build meaningful graphical representations for both descriptive and
prescriptive studies, based on a data-set coming from the source domain/s and on
a data model eventually able to provide complementary online data. Differently
from softwares based on previous data-driven and/or problem-driven models,
the Miró's interdisciplinary model allows the user to insert data or select data
coming from complementary domain/s, and transform the source data-set to
have the intended graphical representation.

In the case study requiring data from both human mobility and cardiology,
when the participants to the experiment are considered as a group, their information provides other meaningful cues to identify critical geographical or temporal

points. For example, the two figures coming from the prescriptive case studies represent respectively 1) the places that are implicitly perceived as dangerous or risky by most users and 2) the most polluted areas of a city. By analyzing the data-sets and their graphical representations, it emerges that there are data (fields) that make sense only within one or more interval/s [a, b]. Often, the interval information is neither provided within the data-set nor within the single scientific discipline and thus the interval must be set by the scientist and/or by another expert. This needs to be contemplated by the dashboard developer. For instance, in study 1), the heart-rate belongs to the health domain and make the place dangerousness meaningful only if the average value is above a certain threshold. The threshold needs to be provided by a scientist (also following the scientific practices of his/her scientific field), it is not provided by the data-set per sé, especially in interdisciplinary prescriptive studies like 1).

Some data actually come from the data-sets, some other data come from the scientist's interpretation of the data in light of the scientific hypotheses in her/his study. The latter should be provided by the scientist and a dashboard should make it possible to enter them. Prescriptive scientific studies are more likely to need interval information as a condition of meaningfulness to make sense of the data-sets when compared to descriptive scientific studies, which can instead provide meaningful graphical representations based on traditional models. Of course, scientific studies can be both descriptive and prescriptive: Miró can provide a meaningful graphical representation also for these studies as it does not abandon traditional models, but it instead proposes further functionalities.

6   Conclusion and Future Works

The paper shows how important might be an interdisciplinary data model, especially in prescriptive studies, to have a software able to provide meaningful graphical representations of data. In the case of descriptive studies, existing models - data-driven models and/or problem-driven models - might be sufficient to produce meaningful graphical representations when providing the data coming from the source domain/s. In the case of prescriptive studies, the existing models might fail to produce meaningful graphical representations when just the collected data coming from the source domain/s are provided. The paper proposed an interdisciplinary approach to overcome the limitations of the existing models via a software-expert interaction. In this framework, the software allows the users to reinterpret and transform the collected source data in the light of the scientific knowledge coming from (online) interaction with other experts or data-sets coming from complementary domain/s. The graphical representation is made meaningful in the blended domain, thus providing a visual support for new findings.

References

1. Mahling, A., Herczeg, J., Herczeg, M., Böcker, H.-D.: Beyond visualization: knowing and understanding. In: Gorny, P., Tauber, M.J. (eds.) IPsy 1988. LNCS, vol. 439, pp. 16–26. Springer, Heidelberg (1990). https://doi.org/10.1007/3-540-52698-6_2

2. Bechtel, W., Abrahamsen, A.: Explanation: a mechanist alternative. Stud. Hist. Philos. Biol. Biomed. Sci. **36**(2), 421–441 (2005)

3. Eklund, P., Haemmerlé, O. (eds.): Conceptual Structures: Knowledge Visualization and Reasoning. Springer, Heidelberg (2008). https://doi.org/10.1007/978-3-540-70596-3

4. Zacks, J., Tversky, B.: Bars and lines: a study of graphic communication. Mem. Cogn. **27**(6), 1073–1079 (1999). https://doi.org/10.3758/BF03201236

5. Keim, D.A.: Information visualization and visual data mining. IEEE Trans. Vis. Comput. Graph. **8**(1), 1–8 (2002)

6. Bergel, A., et al.: A domain-specific language for visualizing software dependencies as a graph. In: 2014 Second IEEE Working Conference on Software Visualization, pp. 45–49 (2014)

7. Zhu, J., et al.: A data-driven approach to interactive visualization of power systems. IEEE Trans. Power Syst. **26**(4), 2539–2546 (2011)

8. Marai, G.E.: Activity-centered domain characterization for problem-driven scientific visualization. IEEE Trans. Vis. Comput. Graph. **24**(1), 913–922 (2018)

9. Grimaldi, D.A., Engel, M.S.: Why descriptive science still matters. Bioscience **57**(8), 646–647 (2007)

10. Brown, R.V., Vári, A.: Towards a research agenda for prescriptive decision science: the normative tempered by the descriptive. Acta Psychol. **1–3**, 33–48 (1992)

11. Kim, I., Cho, G., Hwang, J., Li, J., Han, S.: Visualization of neutral model of ship pipe system using X3D. In: Luo, Y. (ed.) CDVE 2010. LNCS, vol. 6240, pp. 218–228. Springer, Heidelberg (2010). https://doi.org/10.1007/978-3-642-16066-0_33

12. Kerren, A., et al.: Information Visualization: Human-Centered Issues and Perspectives. LNCS. Springer, Heidelberg (2008). https://doi.org/10.1007/978-3-540-70956-5

13. Hansen, C.: Scientific Visualization: Uncertainty, Multifield, Biomedical, and Scalable Visualization. Springer, London (2014). https://doi.org/10.1007/978-1-4471-6497-5

14. Velasco-Montero, D., et al.: Optimum selection of DNN model and framework for edge inference. IEEE Access **6**, 51680–51692 (2018)

15. Godfrey, P., et al.: Interactive visualization of large data sets. IEEE Trans. Knowl. Data Eng. **28**(8), 2142–2157 (2016)

16. Chidamber, S.R., Kemerer, C.F.: A metrics suite for object oriented design. IEEE Trans. Softw. Eng. **20**(6), 476–493 (1994)

17. Roux, O., Bourdon, J. (eds.): Computational Methods in Systems Biology. Springer, Cham (2015). https://doi.org/10.1007/978-3-319-23401-4

18. Salerno, J., Yang, S.J., Nau, D., Chai, S.-K. (eds.): Social Computing, Behavioral-Cultural Modeling and Prediction. Springer, Heidelberg (2011). https://doi.org/10.1007/978-3-642-19656-0

19. Faye, S., et al.: Characterizing user mobility using mobile sensing systems. Int. J. Distrib. Sens. Netw. **13**(8) (2017). https://doi.org/10.1177/1550147717726310

20. Forkan, A., et al.: AqVision: a tool for air quality data visualisation and pollution-free route tracking for smart city. In: 2019 23rd InfoVis, pp. 47–51 (2019)
21. Hall, K.W., et al.: Design by immersion: a transdisciplinary approach to problem-driven visualizations. IEEE Trans. Vis. Comput. Graph. **26**(1), 109–118 (2020)
22. Turner, M., Fauconnier, G.: A mechanism of creativity. Poetics Today **20**(3), 397–418 (1999)

Information Retrieval from Semantic Memory: BRDL-Based Knowledge Representation and Maude-Based Computer Emulation

Antonio Cerone[✉][iD] and Diana Murzagaliyeva

Department of Computer Science, Nazarbayev University, Nur-Sultan, Kazakhstan
{antonio.cerone,diana.murzagaliyeva}@nu.edu.kz

Abstract. This paper presents a formal model for the representation of relational information in semantic memory and for its retrieval as a reaction to triggering questions which are normally used in experimental psychology. Information is represented using the Behaviour and Reasoning Description Language (BRDL), while the engine for its retrieval is given by the real-time extension of the Maude rewrite language. Maude's capability of specifying complex data structures as many sorted algebras and the time features of Real-Time Maude are essential in providing a means for formalising alternative human memory models. Furthermore, using Maude's object-oriented modelling style, aspects of such alternative memory models may be implemented in separate, interchangeable modules, thus providing a way for their comparison through in silico experiments. Finally, the results of in silico experiments may be contrasted with the data produced through lab experiments and natural observations to yield, on the one hand, a calibration of the emulation engine underlying BRDL and, on the other hand, important insights into alternative theories of cognition.

Keywords: Cognitive science · Behaviour and Reasoning Description Language (BRDL) · Formal methods · Rewriting logic · Real-Time Maude

1 Introduction

Since the end of the 1960 s, experimental psychology has shown that semantic memory has a complex network-like structure and that information is retrieved by 'navigating' such a network while following relationships between the stored representations of concepts. Collins and Quillian's experiments [10] have shown that the time to retrieve information is proportional to how far we need to

Work partly funded by Project SEDS2020004 "Analysis of cognitive properties of interactive systems using model checking", Nazarbayev University, Kazakhstan (Award number: 240919FD3916).

L. Cleophas and M. Massink (Eds.): SEFM 2020 Workshops, LNCS 12524, pp. 159–175, 2021.
https://doi.org/10.1007/978-3-030-67220-1_13

navigate the network to find the requested information and support a hierarchically organised memory model. Such a hierarchical model allows us to explain human understanding of simple propositions about class membership and property statement as well as the different retrieval times needed to understand a given proposition or to answer a given question.

After Qullian's hierarchical network model [21], further work was carried out in the 1970s by replacing the hierarchical structure assumption with a general network representing concept relatedness [9], for dealing with more formal propositions involving universal quantification [11,17] and by focusing on the role of connotative relationships between conceptual components of propositions [22]. The last approach resulted in set-theoretical models, such as the one developed by Smith *et al.* [23]. Although some researchers argued that these two different classes of models, network-based and set-theoretical, are formally isomorphic [13], experiments conducted in the 1980s on the semantic distance in the retrieval of conceptual relationships showed that both classes of models may be fallacious in some contexts that discriminate between them [3].

The strong emphasis on models for the representation of information in semantic memories continued throughout the 1990s and 2000s with little success in producing convincing computational models of the retrieval of the represented relations. Even more recent distributed models such as LISA [12], cognitive architectures such as ATC-R [1] and connectionist models [15] provide only limited retrieval mechanism, and mostly as part of inferential or analogical reasoning engines. Holyoak uses the term 'retrieval gap' to denote this limitation of the current computational models and observes that there is no generally accepted model yet [14].

The purpose of our work is the development of an approach in which different models of semantic memory can be formally described, executed to perform in silico experiments and formally analysed, with the objective of comparing different alternative models of semantic memory. In previous work, we have developed modelling languages for this purpose. The Human Behaviour Description Language (HBDL) [5,6] aims at the modelling of automatic and deliberate human behaviour while interacting with an environment consisting of heterogenous physical components. The Behaviour and Reasoning Description Language (BRDL) [7] originates from and extends HBDL with the linguistic constructs to specify reasoning goals, inference rules and unsolved problems.

Recently we have developed a cognitive engine using the Maude rewrite system [16,19] and its real-time extension, Real-Time Maude [18,20], to execute HBDL models of human behaviour [5,6] and BRDL models that emulate human reasoning [8]. All these implementations are based on direct access to the information in semantic memory. Human behaviour is modelled in terms of basic activities whose representation is stored in semantic memory and which are directly triggered by the content of short-term memory (STM) and the perceptions available in the environment. Human reasoning is modelled in terms of inference rules whose representation is stored in semantic memory and which are directly triggered by the presence of its premises in STM.

A similar approach to ours was developed by Broccia *et al.* [4], who, driven by the specific objective of modelling human multitasking, used Real-time Maude to extend our initial untimed framework [5]. In their work, however, time is used to model non-cognitive aspects, such as the duration of the task, which is an interface-dependent outcome of the interaction process, and external aspect, such as the delay due to the switching from one task to another. In contrast to Broccia *et al.* we focus on the human component and model the duration of the mental process, which is an important aspect of human cognition.

In this paper we consider the subset of BRDL that models, on the one hand, propositions that express facts of the real world and questions related to such facts and, on the other hand, the representation of such facts in semantic memory. We develop a modular implementation of such a subset of BRDL using Real-Time Maude, thus providing an emulation of the process of information retrieval from semantic memory to be used to carry out in silico experiments.

The rest of this paper is organised as follows. Section 1.1 provides a brief highlight of Real-Time Maude and refers to the sections of the paper where the different aspects of the language are illustrated. Section 2 introduces the BRDL syntax for facts and questions and shows how they are modelled in Real-Time Maude. Section 3 presents the implementation of a variant of Quillian's model, which is used to illustrate our approach. Section 4 illustrates how to plan and carry out experiments. Section 5 discusses the timed evolution of the system and the information retrieval process implementation. Section 6 concludes the paper.

1.1 Real-Time Maude

Real-Time Maude [18,20] is a formal modeling language and high-performance simulation and model checking tool for distributed real-time systems. It is based on Full Maude, the object-oriented extension of Core Maude, which is the basic version of Maude. Real-Time Maude makes use of

- algebraic equational specifications in a functional programming style to define data types;
- labeled rewrite rules to define local transitions;
- tick rewrite rules to advance time in the entire system state.

The definition of data types is illustrated in Sect. 2.3. The Full Maude syntax is illustrated in Sect. 3.3 for classes and in Sect. 4.1 for messages. Labelled rewrite rules and tick rewrite rules are illustrated in Sect. 4.3 and Sect. 5, respectively.

2 Natural Language Constructs: Facts and Questions

BRDL has a concise, functional-like syntax, which is presented elsewhere [7]. Its conciseness is thought to provide an essential description of the model able to present the bigger picture of the modelled system, and its functional flavour makes it suitable for direct mathematical manipulation without computer support. In this paper we mostly use an alternative, somehow verbose syntax, which is more similar to the English natural language, although ungrammatical in some details, and can be directly manipulated by the Maude system.

2.1 Facts

Humans, throughout their lives, acquire knowledge of the *facts of the real world* and are able to refer to them and reason about them using *declarative propositions*. Since a declarative proposition is just a natural language description of a fact, we will often use the word 'fact' also to denote the declarative proposition that describes it.

A *fact* is modelled using the functional-like BRDL syntax [7] as

$$type(category, attribute)$$

where *category* is an object of our knowledge and *attribute* is an attribute of type *type* associated with that category. For example, the fact that 'a dog is an animal' is represented using the functional-like BRDL syntax as

$$is\_a(dog, animal).$$

The equivalent English-like syntax manipulated by the Maude system is

```
a "dog" is a "animal".
```

The article 'a' is used for any noun, although this is ungrammatical when the noun starts with a vowel as in the case of 'animal'. Other examples of facts are:

```
a "dog" has "four legs"
a "animal" can "breathe"
a "dog" can "move"
a "dog" can "bark"
a "cat" cannot "bark"
a "cat" is not a "dog"
a "bird" has not "four legs".
```

In such examples `"animal"`, `"dog"`, `"cat"` and `"bird"` are categories, `"four legs"`, `"breathe"`, `"move"`, `"bark"` are attributes and `is a`, `has`, `can`, `is not a`, `has not` and `cannot` are types to be applied to attributes. Categories may also be used as attributes as in `a "dog" is a "animal"`. The application of a type to an attribute, such as `is a "animal"` or `has "four legs"` is called *typed attribute*. The reason why categories and attribute are between double apices will be explained in Sect. 2.3.

Finally, we also consider declarative propositions that describe the absence of knowledge about facts, such as

```
I do not know if a "animal" "bark"
```

We could also model declarative propositions that describe our knowledge about facts, but this would not be more interesting that the fact itself. Actually, if I state that `a "dog" has "four legs"`, it is obvious that I know that a `"dog"` has `"four legs"`.

2.2   Questions

Questions can be of different kinds. Examples are:

```
what can a "animal" do?
can a "dog" "breathe" ?
is a "dog" a "animal" ?
has a "cat" "four legs"
```

We can note here that the first question may be answered by a set of declarative propositions, whereas the other three questions, which all have the same structure, will be answered by one declarative proposition, either the one stating the fact, negatively or positively, or the one describing the absence of knowledge about the fact. For example, can a "dog" "breathe" ? will be answered by one of the following three declarative propositions: a "dog" can "breathe", a "dog" cannot "breathe" or I do not know if a "dog" can "breathe".

2.3   Modelling Facts and Question in Real-Time Maude

Maude *equational logic* supports declaration of *sorts*, with keyword sort for one sort, or sorts for many. A sort A may be specified as a subsort of a sort B by subsort A < B. Operators are introduced with the op (for a single definition) and ops (for multiple definitions) keywords:

$$\text{op } f : s_1 \ldots s_n \text{ -> } s.$$
$$\text{ops } f_1 \ f_2 : s_1 \ldots s_n \text{ -> } s.$$

Operators can have user-defined syntax, with underbars '_' marking the argument positions and '' to denote a space. Some operators can have *equational attributes*, such as assoc, comm, and id, stating that the operator is associative, commutative and has a certain identity element, respectively. Such attributes are used by the Maude engine to match terms *modulo* the declared axioms. An operator can also be declared to be a constructor (ctor) that defines the carrier of a sort. Axioms are introduced as equations using the eq keyword or, if they can be applied only under a certain condition, using the ceq keyword, with the condition introduced by the if keyword. Variables used in equations are placeholders in a mathematical sense and cannot be assigned values. They must be declared with the keyword var for one variable, or vars for many. The use of the owise (or otherwise) equational attributes in an equation denotes that the axiom is used for all cases that are not matched by the previous equations. All Maude statements are ended by a dot.

The English-like syntax of facts and questions is defined in the Real-Time Maude module INFORMATION given in Figure 1. The module imports the predefined modules NAT and STRING, which support the specification of natural numbers and strings, respectively. Sorts Fact and Question model facts and questions, respectively, and are subsorts of BasicItem, which is in turn subsort of Item. Only the BasicItem subsort of Item is relevant for the subset of the

```
(tomod INFORMATION is
  protecting NAT .
  protecting STRING .

  sorts Fact Question BasicItem Item BasicItemSet ItemSet EmptyItemSet .
  subsorts Fact Question < BasicItem < Item .
  subsort BasicItem < BasicItemSet .
  subsorts EmptyItemSet < BasicItemSet < ItemSet .
  subsort Item < ItemSet .
  op none : -> EmptyItemSet [ctor] .
  op _;_ : BasicItemSet BasicItemSet -> BasicItemSet
                        [ctor assoc comm id: none format (b o n b)] .
  op _;_ : ItemSet ItemSet -> ItemSet [ctor ditto] .
  op _;_ : EmptyItemSet EmptyItemSet -> EmptyItemSet [ctor ditto] .

  sorts Category Attribute TypedAttribute .
  subsort String < Category < Attribute .
  ops can_ is'a_ has_ : Attribute -> TypedAttribute [ctor] .
  ops cannot_ is'not'a_ has'not_ : Attribute -> TypedAttribute [ctor] .
  op what'can'a_do? : Category -> Question [ctor] .
  ops can'a__? is'a_a_? has'a__? : Category Attribute -> Question [ctor] .
  op a__ : Category TypedAttribute -> Fact [ctor] .
  ops I'dont'know'if'a_can_
      I'dont'know'if'a_is_ : Category Attribute -> Fact [ctor] .
  op I'dont'know'what'a_can'do : Category -> Fact [ctor] .
  op _is'negative'of_ : TypedAttribute TypedAttribute -> Bool .

  var A : Attribute .    vars TA1 TA2 : TypedAttribute .
  eq (cannot A) is negative of (can A) = true .
  eq (is not a A) is negative of (is a A) = true .
  eq (has not A) is negative of (has A) = true .
  eq TA1 is negative of TA2 = false [owise] .

  op isItemIn : Item ItemSet -> Bool .
  vars I1 I2 : Item .    var IS : ItemSet .
  eq isItemIn(I1, I2 ; IS) = if I1 == I2 then true
                                          else isItemIn(I1, IS) fi .
  eq isItemIn(I1, none) = false .
endtom)
```

Fig. 1. Module INFORMATION.

BRDL implementation considered in this paper. Other subsorts, which are not introduced here, are used in other parts of the BRDL implementation [6,7].

Both BasicItem and Item are organised into sets by defining the two sorts BasicItemSet and ItemSet using the ; user-defined infix operator, which is given the appropriate equational attributes for the properties that characterise sets. The ditto equational attribute is a short form for all attributes of the previous sort declaration. The format equational attribute is used to format the output with spaces, colours and newlines in order to make it more readable. By declaring `BasicItem` as a subsort of `BasicItemSet` and `Item` as a subsort of `ItemSet` we implicitly defined singletons of sorts `BasicItemSet` and `ItemSet`. However, the `none` empty set needs to be explicitly introduced as the only element of sort `EmptyItemSet`, which is subsort of `BasicItemSet`, in turn subsort of `ItemSet`.

The sorts `Category` and `Attribute` include Maude-predefined sort `String` as a subsort. This allows us to freely use any string, which is enclosed by double quotes in Maude syntax, as a category or attribute, while leaving open the option to use other representations in possible extensions of the module. The elements of sorts `TypedAttribute`, `Fact` and `Question` are instead defined using constructors, since they have special relationships between each other and need to be manipulated in special, distinct ways by the Maude engine.

One of these special relationships is the negation: `cannot`, `is not` and `has not` are the negations of `can`, `is` and `has`, respectively. Negation is expressed as an infix boolean operator `is negative of` characterised by three axioms for the three pairs of attribute types above which return `true`. The last axiom has the `owise` equational attribute, thus it is applied to all other cases and returns `false`.

Finally, the boolean operator `isItemIn` returns `true` if an element of sort `Item` belongs to an element of sort `ItemSet`.

3   Human Memory Model

BRDL is based on Atkinson and Shiffrin's *multistore model* of human memory [2]. This model is characterised by three stores between which various forms of information flow: *sensory memory*, where information perceived through the senses persists for a very short time, *short-term memory (STM)*, which has a limited capacity and where the information that is needed for processing activities is temporarily stored with rapid access and rapid decay, and *long-term memory (LTM)*, which has a virtually unlimited capacity and where information is organised in structured ways, with slow access but little or no decay. We consider a further decomposition of LTM: *semantic memory*, which refers to our *knowledge* of the world and consists of the *facts* that can be *consciously* recalled, and *procedural memory*, which refers to our *skills* and consists of *rules* and *procedures* that we *unconsciously* use to carry out tasks, particularly at the motor level.

This paper focuses on STM and on the part of semantic memory devoted to the storage of *fact representations*, that is, the representations of our knowledge of the facts of the real world in the form of a *hierarchical network*, often called a *semantic network*, as first introduced by Collins and Quillian [10,21]. In Sect. 2.1 we illustrated how to model facts in a natural-language-like fashion.

In Sect. 3.1 we present how to model fact representations and their hierarchical organisation in semantic memory. The hierarchy among categories is expressed using the is a type applied to the more general category. For example, the type attribute is a "animal" denotes a *generalisation* to category "animal". The more specific category inherits all attributes of the more generic category unless the attribute is redefined at the more specific category level.

3.1 Fact Representation in Semantic Memory

A fact representation in semantic memory is modelled using the functional-like BRDL [7] syntax as

$$domain : category \mid \xrightarrow{delay} \mid type(attribute)$$

where *delay* is the mental processing time needed to retrieve the association between category *category* and type attribute *type(attribute)* within the given knowledge domain *domain*. The knowledge domain is used to set a boundary under which the mental processing is retrieving information from the semantic memory and manipulating it. The role of such a boundary is clarified in Sect. 5.

As an example, the fact that 'a dog is an animal' is represented within the semantic domain *dogs* as

$$dogs : dog \mid \xrightarrow{1} \mid is\_a(animal)$$

and this generalisation can be retrieved from semantic memory in 1 time unit. The more specific category of a generalisation inherits all typed attributes of the more generic category unless the attribute is redefined at the more specific category level. Therefore,

$$animals : animal \mid \xrightarrow{1} \mid can(move)$$

which is an association of a category with a typed attribute rather than a generalisation, specifies that an animal can move and, since an animal is a generalisation of a dog, such a typed attribute is inherited by the category *dog*.

The equivalent English-like syntax manipulated by the Maude system of the fact representations above is

```
"dogs" : "dog" |- 1 ->| is a "animal"
"animals" : "animal" |- 1 ->| can "move"
```

Such an English-like syntax of fact representations is defined in the Real-Time Maude module SEMANTIC-MEMORY shown in Figure 2. This module imports the module INFORMATION and the predefined module NAT-TIME-DOMAIN, which defines the sort Time to model discrete time. Fact representations are defined as element of the sort FactRepresentetion. The semantic memory is modelled by the sort SemanticMemory, which is defined as a set of fact representations.

```
(tomod SEMANTIC-MEMORY is
  protecting NAT-TIME-DOMAIN .
  including INFORMATION .

  sorts Domain FactRepresentation SemanticMemory .
  subsort String < Domain .
  subsort FactRepresentation < SemanticMemory .
  op emptySemantic : -> SemanticMemory [ctor] .
  op _:_|-_->|_ : Domain Category Time TypedAttribute ->
                        FactRepresentation [ctor format (!r o b o r o b o)] .
  op __ : SemanticMemory SemanticMemory -> SemanticMemory
                        [ctor assoc comm id: emptySemantic format (o n o)] .
  op _is'negated'in_ : Fact SemanticMemory -> Bool .

  var M : Time .    var D : Domain .
  var C : Category .    var A : Attribute .
  vars TA1 TA2 : TypedAttribute .    var S : SemanticMemory .
  eq ( a C TA1 ) is negated in ( ( D : C |- M ->| TA2 ) S ) =
                                TA2 is negative of TA1 .
  eq ( a C TA1 ) is negated in S = false [owise] .
endtom)
```

Fig. 2. Module SEMANTIC-MEMORY.

The constructor __ denotes that sets of fact representations are created by justap-position, with no written operator. The constructor emptySemantic denotes an empty semantic memory.

Given the "dogs" : "dog" |- 1 ->| can "bark" fact representation, let us consider the following downward extension of the animal–dog hierarchy:

```
"dogs" : "hund" |- 1 ->| is a "dog"
"dogs" : "basenji" |- 1 ->| is a "hund"
"dogs" : "basenji" |- 1 ->| cannot "bark"
```

The category "basenji" has the cannot "bark" typed attribute, which rede-fines the can "bark" typed attribute of the "dog" category. In fact, a basenji is an exceptional dog breed that cannot bark.

In order to make sure that the category "basenji" does not inherit the can "bark" typed attribute from the more general "dog" category, we introduce the is negated in user-defined infix operator, which returns true only if the fact passed as the left argument is negated by some fact representation in the seman-tic memory passed as the right argument. As shown in Sect. 5, this operator is used by the Maude engine to prevent the retrieval of general information that is negated for the more specific category we are considering. In this way, given the fact representations above, (a "basenji" can "bark") is negated in ("dog" : "basenji" |- 1 ->| cannot "bark") equals true, thus the ques-tion can a "basenji" "bark" ? does not trigger the retrieval of the typed attribute can "bark".

```
(tomod TIMED-INFORMATION is
  protecting INFORMATION .
  protecting NAT-TIME-DOMAIN-WITH-INF .

  sorts TimedItem TimedItemSet TimedBasicItem FutureBasicItem .
  subsort TimedItem < TimedItemSet .

  op _<'decay_> : Item TimeInf -> TimedItem [ctor] .
  op _for_ : BasicItem TimeInf -> TimedBasicItem [ctor] .
  op _in_ : TimedBasicItem Time -> FutureBasicItem [ctor] .
  op emptyTIS : -> TimedItemSet [ctor] .
  op _;_ : TimedItemSet TimedItemSet -> TimedItemSet
                      [ctor assoc comm id: emptyTIS format (b o n b)] .
  eq ITEM:Item < decay 0 > = emptyTIS .

  op removeTime : TimedItemSet -> ItemSet .
  var I : Item .
  var TIS : TimedItemSet .
  var T : TimeInf .
  eq removeTime(emptyTIS) = none .
  eq removeTime((I < decay T >) ; TIS) =  I ; removeTime(TIS) .

  ops DECAY-TIME MAX-RETRIEVAL-TIME : -> TimeInf .
endtom)
```

Fig. 3. Module TIMED-INFORMATION.

3.2 Short-Term Memory (STM) Model

STM is normally used as a buffer where the information that is needed for processing activities is temporarily stored. In our model the kind of information stored in the STM belongs to sort Item. However, the limited capacity of STM requires the presence of a mechanism to empty it when the stored information is no longer needed. In fact, information in STM decays very quickly, normally in less than one minute. To implement STM decay, we need to associate a time with the elements of sort Item.

Module TIMED-INFORMATION in Figure 3 imports the module INFORMATION and the predefined module NAT-TIME-DOMAIN-WITH-INF, which defines a new sort TimeInf by extending the sort Time with value INF to model an infinite time. Sort TimedItem associates a time with the elements of sort Item using the constructor _<'decay_> and sort TimedItemSet define sets of elements of TimedItem using the constructor ;. The equation on the constructor _<'decay_> ensures that if the time to decay has reached zero, the term is removed. Therefore, STM is modelled as an element of sort TimedItem. The operator removeTime removes the time from an element of sort TimedItem and returns the corresponding element of sort TimedItem. Finally, undefined operation DECAY-TIME is declared to be used as a constant holding the user-defined STM decay time as we will see in Sect. 4.3.

3.3   Human Memory as a Maude Class

Full Maude supports the definition of classes. Class objects are elements of the pre-defined sort `Configuration`, which defines the state of the overall system.

We model the structure of human memory using the following Real-Time Maude class.

```
class HumanMemory | shortTermMem : TimedItemSet,
                    semanticMem : SemanticMemory .
```

STM is modelled by the field `shortTermMem`. Semantic memory is modelled by the field `SemanticMem`.

4   Experimental Environment and Its Evolution

We model an environment consisting of perceptions in the form of questions to which a human subject has to answer. Once they appear in the environment, questions persist for a certain time before disappearing. In a typical experiment, questions will be shown to the human subject for a few seconds.

4.1   Modelling Perceptions

We use Full Maude *messages* to model perceptions. A message is an element of the pre-defined sort `Msg` and has the same syntax as an operation but, in addition, is also an element of the pre-defined sort `Configuration`.

The persistence of a perception is modelled by the user-defined infix operator `for` in module `TIMED-INFORMATION` given in Fig. 3. Note that the time may be infinite to denote that the perception persists forever.

Therefore, a perception is modelled by operation `perc` as follows.

```
sorts Perception .
subsort Perception < Msg .
op perc : TimedBasicItem -> Perception [ctor] .
var BI : BasicItem .
eq perc(BI for 0) = none .
```

The equation ensures that if the persistence time has reached zero, the perception is removed from the configuration (`none` is the empty configuration).

4.2   Planning Experiments

We call *planned experiment* any single question presented to a human subject together with the time that must pass before the question is actually presented. Such a time is modelled by the user-defined infix operator `in` in module `TIMED-INFORMATION` given in Fig. 3.

Therefore, a planned experiment is modelled by operation `exp` as follows.

```
sorts Experiment .
subsorts Experiment < Msg .
op exp : FutureBasicItem -> Experiment [ctor] .
```

For example, we can plan an experiment similar to the one by Collins and Quillian [10] to find how *response time* (RT) may depend on the hierarchical structure of semantic memory. The subjects are presented with questions on a screen, each question for 30 s. Within these 30 s the subject has to answer the question. The experimental setting is as follows.

```
op init : -> Configuration .
op human : -> Oid .
op initSemanticMem : -> SemanticMemory .
op aHuman : -> Object .
eq aHuman = < human : Human | shortTermMem : emptyTIS,
                              semanticMem : initSemanticMem > .

eq init = (exp(((can a "dog" "breathe" ?) for 30000) in 0))
          (exp(((can a "animal" "move" ?) for 30000) in 30000))
          (exp(((has a "dog" "four legs" ?) for 30000) in 60000))
          (exp(((can a "hound" "track" ?) for 30000) in 90000))
          (exp(((can a "basenji" "bark" ?) for 30000) in 120000))
          (exp(((is a "armadillo" a "mammal" ?) for 30000) in 150000))
          (exp(((can a "giraffe" "bark" ?) for 30000) in 180000))
          (exp(((is a "swallow" a "bird" ?) for 30000) in 210000))
          aHuman .
```

The time is expressed in milliseconds.

The operator `initSemanticMem` must be defined with an equation whose right part comprises the content of the semantic memory. The purpose of the experiment above could be the validation of the following research hypotheses:

1. Higher RTs may be due to the fact that the typed attribute is associated with a more general category than the one mentioned in the question;
2. In some cases, the RT for a negative answer may be considerably smaller than the average RT for a positive answer to a similar question;
3. The retrieval of less known or seldom used categories or associations may result in a higher RT.

Hypothesis 1 can be validated using questions like `can a "dog" "breathe" ?` and `can a "animal" "move" ?`: the RT for the former question is higher if the typed attribute `can "breathe"` is associated with the more general category `"animal"`. Hypothesis 2 can be validated using questions like `can a "basenji" "bark" ?` and `can a "giraffe" "bark" ?`: the RT for the former question is lower if the typed attribute `cannot "bark"` is directly associated with category `"basenji"`. Hypothesis 3 can be validated using questions like `is a "armadillo" a "mammal" ?` and `is a "swallow" a "bird" ?`: the RT for the former question is higher if the fact representation stating that the less known category `"armadillo"` is a `"mammal"` has a higher mental processing time than the fact representation stating that the well-known category `"swallow"` is a `"bird"`.

Initially, in silico experiments can be conducted with no aim at validating research hypotheses and contrasted to experiments with human subjects to calibrate the Maude emulation engine. Once the calibration process is concluded, in silico experiments and experiments with human subjects aiming at the validation of research hypothesis can be conducted and contrasted. These kinds of experiment may be used to compare alternative models corresponding to different models of semantic memory.

As a final remark, we would like to clarify that the experiment above has been defined for a purely illustrative purpose. In a real experimental context, a separate set of planned experiments should be considered for each of the research questions.

4.3   Environment Evolution

Maude models system evolution using rewrite logic. Labeled rewrite rules

$$\texttt{rl } [l]: t \texttt{ => } t' \quad \text{or} \quad \texttt{crl } [l]: t \texttt{ => } t' \texttt{ if } cond$$

define local transitions from state t to state t'.

The following labelled rewrite rule transforms a planned experiments into a perception when the scheduled time has been reached.

```
rl [activate-perception] :
   (exp((BI for T) in 0))
   REST
=>
   (perc (BI for T))
   REST .
```

The following labelled rewrite rule make a 'copy' of the perception and stores it in the STM at any time when the perception is available in the environment and associates the value of the constant operator `DECAY-TIME` with such a copy, provided that the untimed version of the STM (`IS := removeTime(TIS)`) does not contain the perception yet (`not isItemIn(BI, IS)`).

```
crl [perceive] :
   (perc(BI for T))
   < H : Human | shortTermMem : TIS >
   REST
=>
   < H : Human | shortTermMem : (BI < decay DECAY-TIME >) ; TIS >
   (perc(BI for T))
   REST
   if IS := removeTime(TIS) /\ not isItemIn(BI, IS) .
```

5   Tick Rewrite Rules for Information Retrieval

Information retrieval from semantic memory and time passing are modelled using tick rewrite rules. Tick rewrite rules

$$\text{rl } [l]: \{t\} => \{t'\} \text{ in time } \Delta$$
$$\text{crl } [l]: \{t\} => \{t'\} \text{ in time } \Delta \text{ if } cond$$

advance time in the *global* state t by Δ time units.

We use tick rewrite rules to model the navigation through semantic memory to retrieve the information needed to answer the question in STM. The RT is added to the current time and the retrieved information is used to build the answer and store it in STM together with the associated decay time DECAY-TIME.

The following idle function is used to update all time-related components of the system:

```
op idle : Configuration Time -> Configuration [frozen (1)] .
op idle : TimedItemSet Time -> TimedItemSet .
op idle : TimedBasicItem Time -> TimedBasicItem .
op idle : FutureBasicItem Time -> FutureBasicItem .

eq idle(none, T) = none .
eq idle(< H : Human | > REST, T) = < H : Human | > idle(REST, T) .
eq idle(perc(TBI) REST, T) = perc(idle(TBI, T)) idle(REST, T) .
eq idle(exp(FBI) REST, T) = exp(idle(FBI, T)) idle(REST, T) .

eq idle(emptyTIS, T) = emptyTIS .
eq idle((I < decay T1 >) ; TIS , T) =
                          (I < decay (T1 monus T) >) ; idle(TIS, T) .
eq idle(BI for T1, T) = BI for (T1 monus T) .
eq idle(TBI in FT , T) = TBI in (FT monus T) .
```

The module TIMED-EVOLUTION contains the tick rewrite rules that implement a variant of Qullian's hierarchical network model [21]. As an illustrative example, we present the conditional tick rule for answering a can-question, such as can a "dog" "breathe" ?.

```
crl [can] :
  < H : Human | cognitiveLoad : N,
     shortTermMem : TIS ; ((can a C A ?) < decay T1 >),
     semanticMem : S >
   REST
=>
  < H : Human | cognitiveLoad : N,
     shortTermMem : ((a C can A) < decay DECAY-TIME >) ; idle(TIS, T),
     semanticMem : S >
   idle(REST,T)
in time T
if IS := removeTime(TIS) /\
   not isItemIn(a C can A, IS) /\
   T := canRetrievalTime(C, A, S) /\ T <= MAX-RETRIEVAL-TIME /\
   not ( ( a C can A ) is negated in S ) .
```

The retrieval time is calculated using the canRetrievalTime operator, which searches in the semantic memory S for a fact representation with category C1 and typed attribute can A, where either C1 = C or C1 is a generalisation of C.

However, an additional condition for the application of the rule is that the RT T is less than or equal to the `MAX-RETRIEVAL-TIME` constant value. When the RT is greater than the `MAX-RETRIEVAL-TIME`, instead, the application of another conditional tick rule with complementary condition results in the answer `I dont know if C can A`. The use of such a constant as an RT upper bound supports the modelling of situations in which, although the information is in semantic memory, the semantic distance is actually too high for retrieval. Two more rules model the two possible situations in which the answer is `C cannot A`. The most explicit situation is when there is a fact representation with category `C` and typed attribute `cannot A` in the semantic memory. An implicit situation occurs when `C` is the generalisation of another category `C1`, such that the fact representation with category `C1` and typed attribute `can A` is in the semantic memory. As an example of such an implicit situation, the question `can a "animal" "bark" ?` results in the fact `a "animal" cannot "bark" ?`.

Further rules are used for covering the situations in which we answer `is a` and `has a` questions. The `what can` question results in the retrieval of a number of facts. This is achieved by repeated applications of the rule. Moreover, the answer to this question depends on the knowledge domain on which the human subject focuses. For example, the question `what can a "dog" do?` returns only `a "dog" can "bark"` if the focus is on the knowledge domain `dogs`, whereas it returns also `a "dog" can "breathe"` and `a "dog" can "move"` if the focus is on the knowledge domain `animals`. This requires the use of two distinct rules, one for the presence and one for the absence of focus. However, the technical details concerning these rules are beyond the illustrative purpose of this paper.

Finally, the following conditional tick rule models the time passing.

```
crl [time-passing] :
   < H : Human | shortTermMem : TIS >
     REST
 =>
   < H : Human | shortTermMem : idle(TIS,1) >
     idle(REST,1)
 in time 1
 if not questionFound(TIS) /\ noExperimentStart(REST)
                           /\ noPerceptionAvailable(REST) .
```

The condition ensure that time is increased by this rule only when there is no question in STM (`not questionFound(TIS)`), there is no planned experiment at the current time (`noExperimentStart(REST)`) and there is no perception available in the environment (`noPerceptionAvailable(REST)`). In this way the `time-passing` rule may be applied only if no other rule can be applied.

6   Conclusion and Future Work

We have developed an approach for formally modelling fact representations in semantic memory and carrying out the in silico emulation of experiments aiming at the comparison of different models of semantic memory. This comparison is an essential part of our future work. The code illustrated in this paper can be downloaded at

http://github.com/AntonioCerone/Pubblications/tree/master/2020/CIFMA

Although the code refers to the Quillian's hierarchical network model [21] and Collins and Quillian's experiments [10], the module TIMED-EVOLUTION can be replaced with a module that implements another semantic memory model.

This code is also simplified for illustrative purposes. It does not include the mechanisms that allow to output the outcome of the in silico experiments in a form suitable to carry out the comparison of alternative semantic memory models.

Acknowledgments. We would like to thank Graham Pluck for reading the manuscript and providing important comments. The first author would also like to thank Peter Csaba Ölveczky for his kind hospitality in Oslo in May 2019 and for many insightful discussions about Maude, which strongly influenced the way Maude was used in this paper.

References

1. Anderson, J.R., Lebiere, C.J.: The Atomic Components of Thought. Lawrence Erlbaum, Hillsdale (1998)
2. Atkinson, R.C., Shiffrin, R.M.: Human memory: a proposed system and its control processes. In: Spense, K.W. (ed.) The Psychology of Learning and Motivation: Advances in Research and Theory II, pp. 89–195. Academic Press (1968)
3. Berry, C., Grove, C.: Semantic distance in memory structure: the retrieval of conceptual relationships. Q. J. Exp. Psychol. **35A**, 553–570 (1983)
4. Broccia, G., Milazzo, P., Ölveczky, P.C.: Formal modeling and analysis of safety-critical human multitasking. Innovations Syst. Softw. Eng. **15**(3–4), 169–190 (2019)
5. Cerone, A.: A cognitive framework based on rewriting logic for the analysis of interactive systems. In: De Nicola, R., Kühn, E. (eds.) SEFM 2016. LNCS, vol. 9763, pp. 287–303. Springer, Cham (2016). https://doi.org/10.1007/978-3-319-41591-8_20
6. Cerone, A.: Towards a cognitive architecture for the formal analysis of human behaviour and learning. In: Mazzara, M., Ober, I., Salaün, G. (eds.) STAF 2018. LNCS, vol. 11176, pp. 216–232. Springer, Cham (2018). https://doi.org/10.1007/978-3-030-04771-9_17
7. Cerone, A.: Behaviour and reasoning description language (BRDL). In: Camara, J., Steffen, M. (eds.) SEFM 2019. LNCS, vol. 12226, pp. 137–153. Springer, Cham (2020). https://doi.org/10.1007/978-3-030-57506-9_11
8. Cerone, A., Ölveczky, P.C.: Modelling human reasoning in practical behavioural contexts using real-time maude. In: Sekerinski, E. (ed.) FM 2019. LNCS, vol. 12232, pp. 424–442. Springer, Cham (2020). https://doi.org/10.1007/978-3-030-54994-7_32
9. Collins, A.M., Loftus, E.F.: A spreading-activation theory of semantic processing. Psychol. Rev. **82**, 407–428 (1975)
10. Collins, A.M., Quillian, M.R.: Retrieval time from semantic memory. J. Verbal Learn. Verbal Behav. **8**, 240–247 (1969)

11. Glass, A.L., Holyoak, K.T.: Alternative onceptions of semantic memory. J. Verbal Learn. Verbal Behav. **5**, 598–606 (1975)
12. Hammel, J.E., Hollok, K.J.: Distributed representations of structures: a theory of analogical access and mapping. Psychol. Rev. **104**, 427–466 (1997)
13. Hollan, J.D.: Feature and semantic model: seth theoretic or network model? Psychol. Rev. **82**, 154–155 (1975)
14. Holyoak, K.T.: Analogy and relational reasoning. In: Holyoak, K.T., Morrison, R.G., (eds.) The Oxford handbook of thinking and reasoning, pp. 234–259. Oxford University Press (2012)
15. Leech, R., Mareschal, D., Cooper, R.P.: Analogy as relational priming: a developmental and computational perspective on the origin of a complex cognitive skill. Behav. Brain Sci. **31**, 357–378 (2008)
16. Martí-Oliet, N., Meseguer, J.: Rewriting logic: roadmap and bibliography. Theoret. Comput. Sci. **285**(2), 121–154 (2002)
17. Meyer, D.E.: On the representation and retrieval of stored sematic information. Cogn. Psychol. **1**, 242–300 (1970)
18. Ölveczky, P.C.: Real-time Maude and its applications. In: Escobar, S. (ed.) WRLA 2014. LNCS, vol. 8663, pp. 42–79. Springer, Cham (2014). https://doi.org/10.1007/978-3-319-12904-4_3
19. Ölveczky, P.C.: Designing Reliable Distributed Systems. Undergraduate Topics in Computer Science. Springer (2017)
20. Ölveczky, P.C., Meseguer, J.: Semantics and pragmatics of real-time-Maude. Higher-order symbolic Comput. **20**(1–2), 161–196 (2007)
21. Quillian, M.R.: The teachable language comprehender: a simulation program and theory of language. Commun. ACM **12**, 459–476 (1969)
22. Rips, L.J., Shoben, E.J., Smith, E.E.: Semantic distance and the verification of semantic relations. J. Verbal Learn. Verbal Behav. **12**, 1–20 (1973)
23. Smith, E.E., Shoben, E.J., Rips, L.J.: Comparison processes in semantic memory. Psychol. Rev. **81**, 214–241 (1974)

A Multi-Agent Depth Bounded Boolean Logic

Giorgio Cignarale[1]($\boxtimes$) and Giuseppe Primiero[2]

[1] Embedded Computing Systems, TU Wien, Vienna, Austria
giorgio.cignarale@tuwien.ac.at
[2] Department of Philosophy, University of Milan, Milan, Italy
giuseppe.primiero@unimi.it

Abstract. Recent developments in the formalization of reasoning, especially in computational settings, have aimed at defining cognitive and resource bounds to express limited inferential abilities. This feature is emphasized by Depth Bounded Boolean Logics, an informational logic that models epistemic agents with inferential abilities bounded by the amount of information they can use. However, such logics do not model the ability of agents to make use of information shared by other sources. The present paper provides a first account of a Multi-Agent Depth Bounded Boolean Logic, defining agents whose limited inferential abilities can be increased through a dynamic operation of becoming informed by other data sources.

Keywords: Logic of information · Resource bounded reasoning · Information transmission

1 Introduction

Knowledge, belief and information are notions that have received increasing attention in the formal representation of cognition since the 1960s for their relevance to AI, distributed and multi-agent systems. Epistemic Logic (EL) [9, 15], with its interpretation of modal operators, and its Dynamic counterpart (DEL) [2,10] have modelled knowledge from an agent-based perspective and its changes through the use of private and public announcement operations. Informational logics aim at a similar task by offering formal interpretations for the epistemic states of being informed [1,11,12]; of holding the information [6–8]; and of becoming informed [16,17]. The former still interprets information states explicitly through modal operators, while the logics that refer to states of holding the information and becoming informed interpret propositional contents

This research was funded by the Department of Philosophy "Piero Martinetti" of the University of Milan under the project "Departments of Excellence 2018–2022" awarded by the Ministry of Education, University and Research (MIUR); and supported by the Austrian Science Fund (FWF) project ByzDEL (P33600). The authors wish to thank Marcello D'Agostino and Pere Pardo for comments on previous versions of this work.

L. Cleophas and M. Massink (Eds.): SEFM 2020 Workshops, LNCS 12524, pp. 176–191, 2021.
https://doi.org/10.1007/978-3-030-67220-1_14

epistemically, in the vein of intuitionistic logic. In particular, Depth-Bounded Boolean Logic (DBBL) [4,5,7,8] captures single-agent reasoning on the basis of informational states qualified as follows, see [14, p.177]:

1. **Informational truth** (T): agent j holds the information that ϕ is true;
2. **Informational falsehood** (F): agent j holds the information that ϕ is false;
3. **Informational indeterminacy** (*): agent j does not hold either the information that ϕ is true, or that it is false.

If bound to information actually held, an agent is said to reason at 0-depth. To resolve those cases in which the agent has informational indeterminacy, i.e. she does not hold either the information that ϕ is true, or that it is false, she may use information not actually held. If the agent is able to derive ϕ by supposing another formula ψ to be true on the one hand, and false on the other, then ϕ must hold irrespective of the information state concerning ψ. In this way, agents are able to simulate informational states richer than the ones they actually are in, so as to infer further contents. For k nested instances of such process, the agent is said to reason at k-depth. This informal concept is formalized both semantically and proof-theoretically. Each k-consequence and derivability relation characterizes a tractable logic, with classical logic obtained as the limit of their sequence.

DBBL models so far only the case in which virtual information is simulated by an agent's inner inferential abilities. But in many practical applications information is obtained through communication, as modelled for example by the epistemic operation of becoming informed, see [17]. When in a context of informational indeterminacy, the cognitive process of an agent may be aided by becoming informed by another agent with stronger cognitive or computational capabilities. For k distinct instances of such communication operation, the system may be said to implement multi-agent reasoning at k-depth. In the following, we propose MA-DBBL, a multi-agent version of DBBL, extending the latter with an operator of information sharing between agents, and one expressing information held by every agent in the system. This opens to multiple considerations on how to attribute such degrees of cognitive abilities to different agents, be they human or artificial. In a human-machine interaction setting, for example, cognitively bounded human agents may be aided in their processes by mechanical ones. Then, a measure of their inferential depth should express the ability to access information. Such measure can be expressed in terms of information quality (IQ, [13]), which provides qualitative parameters for evaluating data available for the inferential process. MA-DBBL uses such account of information access to qualify agents with different inferential abilities, and models bounds over sets of agents through communication operations.

The remaining of this paper is structured as follows. In Sect. 2 we introduce our logic and motivate the order on the agents by data quality criteria that justify inferential abilities. In Sect. 3 we provide the semantics, and in Sect. 4 the proof-theory for MA-DBBL. In Sect. 5 we provide soundness and completeness results. We conclude in Sect. 6 highlighting limitations and further possible extensions of the present system.

2   Multi-Agent DBBL

In Depth-Bounded Boolean logic, agents are characterised by two aspects: first, the information they currently hold as true, false or indeterminate; second, their inferential abilities. The latter aspect is typical of DBBL, considered especially in order to distinguish tractable fragments of classical logic. We consider a situation in which different epistemic agents are ordered by such characteristics. While current information held is a static property, the inferential abilities are usually linked to cognitive or computational capacities which allow to extend the information base of the agent. In the following, the inferential ability of an agent corresponds to a semantic interpretation of the data available to her. The result is a hierarchy of agents in which at each level the agent holds more information and also has better inferential abilities than the agent below.

Definition 1 (Syntax of MA-DBBL).

$$\mathcal{S} := \{i \preceq j \preceq k \preceq \cdots \preceq z\}$$

$$\mathcal{P} := \{\phi_i, \psi_i, \rho_i, \ldots, \}$$

$$\mathcal{D} := \{Ac, Un, Us\}$$

$$\mathcal{C} := \{\wedge, \vee, \rightarrow, \neg\}$$

$$\mathcal{K} := \{BI_i, I_i\}$$

$\mathcal{S}$ is a finite set of agents where each agent is identified with the information she holds. A total preorder $\preceq$ is imposed on $\mathcal{S}$, such that $i \preceq j$ means that agent i holds at least as much information as agent j. This is a preference relation, i.e. it is reflexive ($\forall i \in \mathcal{S}, i \preceq i$); connexive ($\forall i, j \in \mathcal{S}, i \preceq j$ or $j \preceq i$); and transitive ($\forall h, i, j \in \mathcal{S}$ if $h \preceq i$ and $i \preceq j$, then $h \preceq j$). This guarantees that any two agents are comparable, and when considered equivalent in the hierarchy they have the same information. $\mathcal{P}$ is an enumerable set of propositional variables denoting data available to agents in $\mathcal{S}$, although not every agent has a determinate truth value for every such data. Propositional variables are indexed by agents, thereby expressing a source for that data; and are closed under a set of propositional connectives $\mathcal{C}$ and a set of indexed epistemic operators $\mathcal{K}$. In order to define the preorder $\preceq$ on agents and give a precise interpretation of the relation underlying it, we use elements of the set $\mathcal{D}$ over elements of the set $\mathcal{P}$. Elements in $\mathcal{D}$ are data quality functions, namely expressing for each agent i which data is accessible, which is understandable and which is usable. Note that it is essential to have $\mathcal{P}$ accessible in principle to all agents in order for them to be ordered with respect to one another. Formally, these are functions on formulas or sets of formulas with the following meaning:[1]

---

[1] We use here a two-value valuation function v on formulas as a mapping to truth and falsity. Its formal definition, extended to a three-valued function, is postponed to Sect. 3.

$Ac(\phi_i) \leftrightarrow v(\phi_i) \in \{1, 0\}$
$Un(\phi_i) \leftrightarrow Ac(\psi_i)$, for some ψ_i such that ψ_i implies ϕ_i
$Us(\phi_i) \leftrightarrow Ac(\psi_i), \forall \psi_i$ implied by ϕ_i

The first function expresses the condition that information is accessible if the agent holds it true or false. Accessibility is a monotonic function, i.e. if the cardinality of the set Γ_i of formulas accessible to agent i is greater than the cardinality of the set Γ_j of formulas accessible to agent j, then there is at least one formula ϕ_i which is accessible to i, but which is not accessible to j. Formally: $\mid Ac(\Gamma_i) \mid > \mid Ac(\Gamma_j) \mid \leftrightarrow \exists \phi_i \in Ac(\Gamma_i) \wedge \phi \notin Ac(\Gamma_j)$. In other words, all agents are characterized by a common database represented by $\mathcal{P}$, but each agent can have access to different elements of $\mathcal{P}$, depending on which truth values she has for those elements. We shall see in Sect. 3 that the semantics of DBBL is three-valued and non-deterministic, hence accessible information is a subset of all information available to agents. The second function expresses the condition that information is understandable if the agent can access it from other accessible information. This, informally, means that understandable information is such when it is recognised as a consequence of accessible information. Finally, the third function expresses the condition that information is usable if it allows to access other information. This, informally, means that information is qualified as usable when it allows to perform inferences that make other information accessible. Accessibility is a monotonic function, hence agents higher in the hierarchy have access to an increasing set of formulas; so are therefore also the set of formulas that can be understood and used by them. We shall see in Sect. 4 how usability is the criterion which underlies the inferential depth of agents. Let us for now consider an informal example.[2]

Example 1. Suppose that agent j doesn't hold information about whether the sentence ϕ: "C is father of both A and B" is true or false, nor is she able to derive it by herself. Then ϕ is not accessible to agent j i.e. $\phi \notin Ac(\phi_j)$. Suppose also that agent i has not access to ϕ, that is $\phi \notin Ac(\phi_i)$, but she holds true the information ψ: "C is not father of A" and true the information χ: "C is not father of B". Then agent i has access to both ψ and χ i.e. $Ac(\psi_i, \chi_i)$. Moreover, agent i holds true the information ξ: "C is not father of A and C is not father of B". Then we obtain that ψ and χ are understandable to agent i, written $Un(\psi_i, \chi_i)$, for i is able to use those propositional contents in order to access new information. Finally, if agent i is able to establish that ϕ is false, then we conclude that $Us(\psi_i, \chi_i)$ and $Ac(\phi_i)$.

Information is usable if understood and accessible. Given a shared information set, not every agent might have the same accessible, understandable or usable information. Information which is not accessible, understandable or usable denotes contents for which a truth value cannot be determined, or whose consequences cannot be inferred, or which cannot be inferred from other information. We link the inferential ability of an agent i to her accessible, understandable and

---

[2] This example is based on an unpublished one formulated by Marcello D'Agostino.

usable information. We define therefore our total preorder $(\mathcal{S}, \preceq)$ on agents as follows:

Definition 2 (Source Order by Information Usability).

$$i \preceq j \text{ iff } \mid Us(\Gamma_i) \mid \geq \mid Us(\Gamma_j) \mid$$

Where Γ_i and Γ_j are set of formulas, and $\mid Us(\Gamma_i) \mid$ represents the cardinality of the set of formulas usable to agent i. Hence $i \preceq j$ holds if agent i has more information than j with a determined truth-value, from which she can infer new information and which she can infer from determined information.

Example 2. Take $\mathcal{S} = \{h, i, j\}$, $Ac(p_i \wedge q_i)$, $Un(p_i)$, $Us(p_i)$ so that $Ac(p_i \vee r_i)$. Now suppose $Ac(\{\}_j)$, it also holds $Un(\{\}_j)$ and $Us(\{\}_j)$. Hence $i \preceq j$. Moving to agent h, $Ac(r_h \rightarrow t_h)$, $Un(p_h)$ $Un(t_h)$, $Us(p_h)$ so that $Ac(p_h \vee r_h)$ and $Us(t_h)$ so that $Ac(t_h \wedge s_h)$. Now since $\mid Us(p_h, t_h) \mid \leq \mid Us(p_i) \mid$, $h \preceq i$ is the case, and given transitivity of $\preceq$ the order among agents is established.

Our goal is now to model reasoning in a context in which agents higher in the hierarchy can communicate information which is not directly accessible to those below. From now on, we say that information is held by an agent always meaning that it is usable by that agent in the sense defined above.

Definition 3 (Language).

$$\mathcal{L}^{\mathcal{CK}} ::= p_j | \phi_j \wedge \phi_j | \phi_j \vee \phi_j | \phi_j \rightarrow \phi_j | \neg \phi_j \mid BI_j\phi_i \mid I_j\phi_i$$

Given p_j denoting an indexed atomic variable, a metavariable for a formula ϕ_j denotes information ϕ held – i.e. usable – by an agent j. The set of information held by agent j is denoted as Γ_j for a set of formulae $\{\phi_j, \ldots, \psi_j\}$. Formulas are closed under standard connectives. We use brackets to enclose composed formulas and use the index after the closing bracket, while avoiding brackets in the case of negated formulas for aiding readability. We refer to the non-epistemic fragment of our language as $\mathcal{L}^{\mathcal{C}}$, and the corresponding logic as MA-DBBL$^{\mathcal{C}}$. We refer to the epistemic fragment of our language as $\mathcal{L}^{\mathcal{K}}$, and the corresponding logic as MA-DBBL$^{\mathcal{K}}$. The latter is obtained from the former by adding two epistemic formulae with the indexed operators:

– $BI_j\phi_i$ says that "agent j becomes informed that ϕ by an agent $i \preceq j$";
– $I_j\phi_i$ says that "agent j is informed that ϕ by all agents $i \preceq j$".

$BI_j\phi_i$ simulates a private announcement of ϕ received by agent j from agent i. $I_j\phi_i$ expresses consensus on ϕ, similar in meaning to the epistemic operator "everybody knows".[3]

---

[3] The two operators reflect the distinction between an alethically neutral and a veridical conception of information, see [17] and [11] respectively. Technically, it is possible to reformulate the present monotonic version of MA-DBBL without the I operator without loss of expressiveness. We keep it both in the language to preserve the mentioned conceptual distinction, and because it offers the basis for a planned extension of the present system with contradictory information updates.

Table 1. Non-deterministic 3-valued truth tables for agent i

$\wedge$	1_i	0_i	$*_i$	$\vee$	1_i	0_i	$*_i$	$\rightarrow$	1_i	0_i	$*_i$	$\neg$	
1_i	1_i	0_i	$*_i$	1_i	1_i	1_i	1_i	1_i	1_i	0_i	$*_i$	1_i	0_i
0_i	0_i	0_i	0_i	0_i	1_i	0_i	$*_i$	0_i	1_i	1_i	1_i	0_i	1_i
$*_i$	$*_i$	0_i	$0_i,*_i$	$*_i$	1_i	$*_i$	$1_i,*_i$	$*_i$	1_i	$*_i$	$1_i,*_i$	$*_i$	$*_i$

3 Semantics

MA-DBBL$^{\mathcal{C}}$ has a three-valued semantics which formalizes reasoning by an agent based on her actually held information, captured as evaluation of $\mathcal{L}^{\mathcal{C}}$ formulas. The matrices exposed in Table 1 express the informational meaning of the logical operators for this fragment, see [6]. We include values for information that agent i holds as true (1_i), information that agent i holds as false (0_i), and information which agent i cannot establish whether it is true or false ($*_i$). For formulas with determined values (i.e. where $*$ is not present), the tables are the classical ones. When only one element is undetermined, those tables work exactly as their classical counterparts: for example, the truth value of a conjunction is undetermined only when the element that is not undetermined is true, and false otherwise. But when considering, e.g. the conjunction of two undetermined elements, the resemblance with the classical tables falls apart. In such instances, the conjunction of two undetermined elements can have two different outputs: it could be false or undetermined. Which is the case, does not depend merely on the truth values of the components of the conjunction, but it also depends on the background information possessed by the agent. Example 1 from the previous section has already shown a case for this unusual behaviour.

Formula valuation for $\mathcal{L}^{\mathcal{C}}$ is defined as follows:

Definition 4 (Valuation of $\mathcal{L}^{\mathcal{C}}$ formulas). *A 3ND-valuation is a mapping* $v : \Gamma_i \rightarrow \{1, 0, *\}^{\mathcal{L}^{\mathcal{C}}}$, *satisfying the following conditions for all* $\phi_i, \psi_i \in \Gamma_i$:

1. $v(\neg\phi_i) = \mathcal{F}_{\neg}(v(\phi_i))$
2. $v(\phi_i \circ \psi_i) \in \mathcal{F}_{\circ}(v(\phi_i), v(\psi_i))$

where

(i) $\mathcal{F}_{\neg}$ is the deterministic truth-function defined by 3-valued table for $\neg$, and
(ii) $\mathcal{F}_{\circ}$ is the non-deterministic truth-function defined by the 3-valued table for $\circ = \{\wedge, \vee, \rightarrow\}$.

This valuation defines therefore the way in which an agent establishes truth, falsity and indeterminacy for the information she holds, without any aid offered by other sources. The consequence relation for the fragment $\mathcal{L}^{\mathcal{C}}$ is dubbed 0-depth and it is defined as follows:

Definition 5 (0-depth consequence relation). *For every set of formulas Γ_i and formula ϕ_i of $\mathcal{L}^{\mathcal{C}}$, we say that ϕ_i is a 0-depth consequence of Γ_i, denoted by*

$\Gamma_i \vDash_0 \phi_i$, if $v(\phi_i) = 1$ for every v as of Definition 4 such that $v(\psi_i) = 1, \forall \psi_i \in \Gamma_i$.

Accordingly, a notion of inconsistency for $\mathcal{L}^{\mathcal{C}}$ holds as follows:

Definition 6 (0-depth inconsistency). *A set of formulas Γ_i of $\mathcal{L}^{\mathcal{C}}$ is inconsistent, denoted by $\Gamma_i \vDash_0 \bot$, if no valuation exists such that all formulas $\phi_i \in \Gamma_i$ are satisfied.*

When adding the BI and I operators, we move to the fragment $\mathcal{L}^{\mathcal{K}}$ of the language, with formulas indexed by multiple agents.

Definition 7 (Valuation of $\mathcal{L}^{\mathcal{K}}$ formulas).

A 3ND-valuation is a mapping $v : \Gamma_j \to \{1, 0, *\}^{\mathcal{L}^{\mathcal{K}}}$, satisfying the following conditions for all $\phi_j \in \Gamma_j$, and $\circ = \{1, 0, *\}$:

$$- v(BI_j\phi_i) = \circ \text{ iff } v(\phi_i) = \circ$$
$$- v(I_j\phi_i) = \circ \text{ iff } v(BI_j\phi_i) = \circ \text{ for all } i \preceq j.$$

Example 3. Suppose that agent j doesn't hold information about whether the sentence ϕ: "C is father of both A and B" is true or false, nor is she able to derive it by herself. Agent i, despite not holding a truth value for ϕ, is able to use the information that "C is not father of A" and that "C is not father of B", thus concluding that "C is father of both A and B" is false. Hence, agent j becomes informed that "C is father of both A and B" is false from agent i.

Example 4. Suppose $h \preceq j$ and $i \preceq j$. Suppose that agent j doesn't hold information about whether ϕ: "C is father of both A and B" is true or false, nor is she able to derive it by herself. Agent i, despite not holding a truth value for ϕ, is able to use the information that "C is not father of A" and that "C is not father of B", thus concluding that "C is father of both A and B" is false. Assume moreover that agent h is able to infer that "C is father of both A and B" is false, because she knows that "C is father of D" and "D is sibling of A and not sibling of B" are both true. Hence, agent j becomes informed that "C is father of both A and B" is false from agent i and from agent h separately. Then we say that j is informed that "C is father of both A and B" is false.

The notion of consequence relation for MA-DBBL expresses the principle that given $i \preceq j$, an agent j either establishes the truth of some formula ϕ by using the information she holds – i.e. consequence relation at 0-depth; or else, she becomes informed of a formula ϕ from agent i who is able to infer it on her own by using some additional information, or what are called virtual assumptions: the latter is a consequence relation at $k > 0$-depth. In order to bound the number of formulas that are allowed to be used as virtual assumptions up to a certain fixed depth, we introduce the notion of "virtual space", see [8, p.84]. Given any set Δ_i of formulae of $\mathcal{L}^{\mathcal{CK}}$; the function $Sub(\Delta_i)$ that returns the set of subformulae of Δ_i; and the function $At(\Delta_i)$ that returns the set of all its atomic subformulae; consider all operations f such that:

1. for all Δ_i, $At(\Delta_i) \subseteq f(\Delta_i)$
2. $Sub(f(\Delta_i)) = f(\Delta_i)$
3. $| f(\Delta_i) | \leq p(| \Delta_i |)$, for some fixed polynomial p and $| \Delta_i |$ denoting the size of Δ_i, i.e. the number of occurrences of symbols in that set.

The first criterion ensures that for every set of formulas Δ_i the set of its atomic subformulae is contained in the set resulting from applying f to Δ_i; the second states that $f(\Delta_i)$ is closed under subformulae; the third ensures that any function f is at most polynomial in some size. While $Sub(\Delta_i)$ and $At(\Delta_i)$ are examples of such operations, in fact all operations that are built from them and have bounded logical complexity satisfy the above criteria. Our virtual space can now be defined as a function f of the set $\Gamma_i \cup \{\phi_i\}$ made of premises Γ_i and conclusion ϕ_i of the given inference for agent i.

Definition 8 (Consequence relation for MA-DBBL).
   For every set of formulas Γ_j, formula ϕ_i of $\mathcal{L}^C$, formulas $BI_j\phi_i$ and $I_j\phi$ of $\mathcal{L}^K$, all operations f satisfying points 1–3 above and $i \preceq j$:

- $\Gamma_j \models_0^f \phi_i$ *iff* $\Gamma_j \models_0 \phi_i$;
- $\Gamma_j \models_{k+1}^f BI_j\phi_i$ *iff* $\Gamma_i, \psi_i \models_k^f \phi_i$ *and* $\Gamma_i, \neg\psi_i \models_k^f \phi_i$ *for some* $\psi_i \in f(\Gamma_i \cup \{\phi_i\})$ *and some* $i \preceq j$;
- $\Gamma_j \models_{k+1}^f I_j\phi$ *iff* $\Gamma_i, \psi_i \models_k^f \phi_i$ *and* $\Gamma_i, \neg\psi_i \models_k^f \phi_i$ *for some* $\psi_i \in f(\Gamma_i \cup \{\phi_i\})$ *and all* $i \preceq j$.

This definition follows from Definitions 5 and 7, together with the function f defined above. The first case holds trivially, as f is irrelevant. The second case is shown to hold by reasoning by induction on the depth of k. Suppose that $k = 0$ and that $\Gamma_j \models_{k+1}^f BI_j\phi_i$ is valid, but that $\Gamma_i \cup \psi_i \not\models_k^f \phi_i$ and $\Gamma_i \cup \neg\psi_i \not\models_k^f \phi_i$ for some $\psi_i \in f(\Gamma_i \cup \{\phi_i\})$. Take $\Gamma_i \cup \psi_i \not\models_k^f \phi_i$. As said, with $k = 0$ we have $\Gamma_i \cup \psi_i \not\models_0^f \phi_i$. Hence, by Definition 4, $v(\phi_i) = 0$ and $v(\Gamma_i \cup \psi_i) = 1$. Now by Definition 7, if $v(\phi_i) = 0$ then $v(BI_j\phi_i) = 0$, which (by monotonicity of Us_i) means that $\Gamma_j \not\models_{k+1}^f BI_j\phi_i$, against the assumption. The case for $k > 0$ is straightforward and the case for I formulas only generalises to all agents.

Definition 9 (Inconsistency for MA-DBBL). *A set of formulas Γ_i of $\mathcal{L}^{CK}$ is k-depth inconsistent if and only if $\Gamma_i \cup \{\psi_i\}$ and $\Gamma_i \cup \{\neg\psi_i\}$ are both $(k-1)$-depth inconsistent for some $\psi_i \in f(\Gamma_i)$.*

Since $\models_0$ is monotonic, $\models_k^f \subseteq \models_{k+n}^f$, and the increase in depth corresponds to the use of additional virtual information by more agents, restricted by the space defined by f. Note also that we can define a partial order among operations in f, such that $f_1 \trianglelefteq f_2$ iff $f_1(\Delta_i) \subseteq f_2(\Delta_i)$, for every Δ_i. Hence $\models_k^{f_1} \subseteq \models_k^{f_2}$ whenever $f_1 \trianglelefteq f_2$. Then (see [8, p.85]):

Proposition 1. *The relation $\models_\infty^f = \bigcup_{k \in N} \models_k^f$ is the consequence relation of classical propositional logic*

$$\frac{\phi_i}{(\phi \lor \psi)_i} \lor - intro1 \qquad \frac{\psi_i}{(\phi \lor \psi)_i} \lor - intro2 \qquad \frac{\neg\phi_i \quad \neg\psi_i}{\neg(\phi \lor \psi)_i} \lor - intro3$$

$$\frac{\phi_i \quad \psi_i}{(\phi \land \psi)_i} \land - intro1 \qquad \frac{\neg\phi_i}{\neg(\phi \land \psi)_i} \land - intro2 \qquad \frac{\neg\psi_i}{\neg(\phi \land \psi)_i} \land - intro3$$

$$\frac{\neg\phi_i}{(\phi \to \psi)_i} \to - intro1 \qquad \frac{\psi_i}{(\phi \to \psi)_i} \to - intro2 \qquad \frac{\phi_i \quad \neg\psi_i}{\neg(\phi \to \psi)_i} \to - intro3$$

$$\frac{(\phi \land \psi)_i}{\phi_i} \land - elim1 \qquad \frac{(\phi \land \psi)_i}{\psi_i} \land - elim2 \qquad \frac{\neg(\phi \land \psi)_i \quad \phi_i}{\neg\psi_i} \land - elim3$$

$$\frac{\neg(\phi \land \psi)_i \quad \psi_i}{\neg\phi_i} \land - elim4$$

$$\frac{(\phi \lor \psi)_i \quad \neg\phi_i}{\psi_i} \lor - elim1 \qquad \frac{(\phi \lor \psi)_i \quad \neg\psi_i}{\phi_i} \lor - elim2$$

$$\frac{\neg(\phi \lor \psi)_i}{\neg\psi_i \quad \neg\phi_i} \lor - elim3$$

$$\frac{(\phi \to \psi)_i \quad \phi_i}{\psi_i} \to - elim1 \qquad \frac{(\phi \to \psi)_i \quad \neg\psi_i}{\neg\phi_i} \to - elim2$$

$$\frac{\neg(\phi \to \psi)_i}{\phi_i} \to - elim3 \qquad \frac{\neg(\phi \to \psi)_i}{\neg\psi_i} \to - elim4$$

Fig. 1. Introduction and elimination rules for MA-DBBL^C

If we constrain one instance of a *BI*-formula for each distinct agent, higher depth expresses nesting of information requests from distinct sources, which means we obtain the full classical propositional logic only with an infinite number of agents.

4   Proof-Theory

To illustrate the proof-theory of MA-DBBL, we start again by separating the inferential abilities of the agent reasoning only on the basis of her actually held information, from her inferential abilities supported by information received from agents with a larger set of usable information. Introduction and elimination rules for MA-DBBL^C capture the idea of manipulating actual information, expressing inferential ability at 0-depth, see Fig. 1.

Definition 10 (Derivability for MA-DBBL^C). *Given a set of formulas Γ_i of $\mathcal{L}^C$:*

- *An intelim sequence for Γ_i is a sequence $\phi_i, ..., \psi_i$ such that*
    - *each formula ϕ_i in the sequence is a member of Γ_i, or*
    - *is the conclusion of the application of a MA-DBBL^C rule;*
- *An intelim proof of ϕ_i from Γ_i is such that*
    - *either it is a closed sequence, i.e. it contains both ϕ_i and $\neg\phi_i$;*
    - *or ϕ_i is the last formula in the sequence;*
- *A formula ϕ_i is 0-depth derivable from Γ_i, written $\Gamma_i \vdash_0 \phi_i$, if and only if there is an intelim proof of ϕ_i from Γ_i according to MA-DBBL^C rules.*

Definition 11 (Refutability for MA-DBBL^C). *Given a set of formulas Γ_i of $\mathcal{L}^C$:*

- *An intelim refutation of Γ_i is a closed intelim sequence for Γ_i;*
- *Γ_i is intelim-refutable if there is a closed intelim sequence for Γ_i.*

Reasoning at 0-depth proves to be a weighty limit for an agent, for she cannot make any supposition that exceeds the information she actually holds, bounding her inferential abilities. The next step consists in finding a way to manipulate virtual information: when an agent is not able to infer the truth value of a formula ϕ by the information she holds, she can be informed about ϕ by an agent with higher inferential abilities. The rules of Fig. 2 describe operations in which agent j receives information from one or several agents with access to more information. By convention the formula in the conclusion of a rule with multiple premises always carries the index of the highest agent in the hierarchy: this ensures that the most informed source is always referenced. The RB rule (for Rule of Bivalence) corresponds to the introduction rule for the BI operator. When an agent can infer the truth value of a formula ϕ in any state of affairs (i.e. from both ψ and $\neg\psi$), then an agent $j \succeq i$ becomes informed about it. The BI-operator is closed under standard connectives. The I-intro rule infers from an operation of becoming informed from every source strictly higher in the hierarchy a state of being informed about the formula. This rule also plays the role of elimination rule for the BI operator. The I-elim rule makes information shared by all agents into a valid formula for the current agent.

Example 5. Consider the following scenario, with three agents involved in the reasoning process and ordered as $h \preceq i \preceq j$:

$$\frac{\dfrac{(\chi \rightarrow (\psi \vee \phi))_i \qquad (\neg\chi \rightarrow (\psi \vee \phi))_i}{BI_j(\psi \vee \phi)_i} \qquad \dfrac{(\neg\rho \rightarrow \neg\phi)_h \qquad (\rho \rightarrow \neg\phi)_h}{BI_j\neg\phi_h}}{BI_j\psi_h}$$

We assume that j is at k-depth and she does not hold the truth value of the formula ψ. The derivation of such formula occurs at $k + 2$ because in the derivation of $BI_j\psi_h$ two instances of RB involving distinct agents occur, namely the derivation of $(\psi \vee \phi)$ from both χ and $\neg\chi$ for agent i, and the derivation of

$$\frac{(\psi \to \phi)_i \qquad (\neg\psi \to \phi)_i}{BI_j \phi_i} \; RB$$

$$\frac{BI_j \, \psi_h \qquad BI_j \, \phi_i}{BI_j \, (\phi \wedge \psi)_h} \qquad \frac{BI_j \, \neg\phi_i}{BI_j \, \neg(\phi \wedge \psi)_i} \qquad \frac{BI_j \, \neg\psi_i}{BI_j \, \neg(\phi \wedge \psi)_i}$$

$$\frac{BI_j \, \phi_i}{BI_j \, (\phi \vee \psi)_i} \qquad \frac{BI_j \, \psi_i}{BI_j \, (\phi \vee \psi)_i} \qquad \frac{BI_j \, \neg\phi_h \qquad BI_j \, \neg\psi_i}{BI_j \, \neg(\phi \vee \psi)_h}$$

$$\frac{BI_j \, \neg\phi_i}{BI_j \, (\phi \to \psi)_i} \qquad \frac{BI_j \, \psi_i}{BI_j \, (\phi \to \psi)_i} \qquad \frac{BI_j \phi_h \qquad BI_j \neg\psi_i}{BI_j \, \neg(\phi \to \psi)_h}$$

$$\frac{BI_j \, (\phi \wedge \psi)_i}{BI_j \, \phi_i} \qquad \frac{BI_j \, (\phi \wedge \psi)_i}{BI_j \, \psi_i}$$

$$\frac{BI_j \, \neg(\phi \wedge \psi)_h \qquad BI_j \, \phi_i}{BI_j \, \neg\psi_h} \qquad \frac{BI_j \, \neg(\phi \wedge \psi)_h \qquad BI_j \, \psi_i}{BI_j \, \neg\phi_h}$$

$$\frac{BI_j \, (\phi \vee \psi)_h \qquad \neg BI_j \, \phi_i}{BI_j \, \psi_h} \qquad \frac{BI_j \, (\phi \vee \psi)_h \qquad \neg BI_j \, \psi_i}{BI_j \, \phi_h}$$

$$\frac{BI_j \neg(\phi \vee \psi)_i}{BI_j \neg \phi_i} \qquad \frac{BI_j \neg(\phi \vee \psi)_i}{BI_j \neg \psi_i}$$

$$\frac{BI_j \, (\phi \to \psi)_h \qquad BI_j \, \phi_i}{BI_j \, \psi_h} \qquad \frac{BI_j \, (\phi \to \psi)_h \qquad BI_j \, \neg\psi_i}{BI_j \, \neg\phi_h}$$

$$\frac{BI_j \, \neg(\phi \to \psi)_i}{BI_j \, \neg\psi_i} \qquad \frac{BI_j \, \neg(\phi \to \psi)_i}{BI_j \, \phi_i}$$

$$\frac{BI_j \phi_i, \, \forall i \preceq j \in \mathcal{S}}{I_j \phi_i} \; I\text{-intro} \qquad \frac{I_j \phi_i}{\phi_j} \; I\text{-elim}$$

Fig. 2. Introduction and elimination rules for MA-DBBL$^{\mathcal{K}}$ with $h \preceq i \preceq j$

$\neg\phi$ from both ρ and $\neg\rho$ for agent h. Hence, the elements of the set of premises of the derivation of ψ are four formulas: $(\chi \to (\psi \vee \phi))_i, (\neg\chi \to (\psi \vee \phi))_i, (\neg\rho \to \neg\phi)_h, (\rho \to \neg\phi)_h$.[4]

---

[4] We stress here that while in single agent DBBL the depth of the reasoning process is given by *nested* applications of RB by the agent, in MA-DBBL are the instances of RB indexed by *distinct* agents that determine such depth.

Example 6. Consider the following scenario:

$$\frac{(\phi \vee \psi)_j \qquad \neg\psi_j}{(\phi \wedge \gamma)_j} \qquad \frac{\dfrac{\dfrac{(\chi \rightarrow \gamma)_i \qquad (\neg\chi \rightarrow \gamma)_i}{BI_j\gamma_i} \qquad BI_j\gamma_n}{I_j\gamma_n} \; \forall n \preceq j}{\gamma_j}$$

Let's begin with the longer branch. Agent j (assuming is at $k \geq 0$ depth), does not hold a truth value for formula γ. However, agent i can derive the truth value of γ from both χ and $\neg\chi$ at $k+1$-depth. Hence, $BI_j\gamma_i$ can be inferred. Now suppose that not only i, but every agent $n \preceq j$ (i.e. every agent with more information than j) holds γ, and thus agent j becomes informed about it by all agents higher in the hierarchy. As such, by the I-intro rule, it holds that $I_j\gamma_i$. By the I-elim rule, agent j holds that γ, and hence γ_j holds. Moving to the other branch, agent j holds $(\phi \vee \psi)$. In addition, she holds that $\neg\psi$ is the case. Hence, by the $\vee$-elim2 rules of MA-DBBL$^{\mathcal{C}}$, agent j is able to infer that ϕ is the case, hence ϕ_j holds. Using the $\wedge$-intro1 rule, agent j is able to infer that $(\phi \wedge \gamma)$ is the case, hence $(\phi \wedge \gamma)_j$ holds. This example shows how rules of MA-DBBL$^{\mathcal{C}}$ and of MA-DBBL$^{\mathcal{K}}$ operate harmoniously at two different levels in the same tree: the rules of the first allows the formalization of the operations that an agent performs with her own information, while the second formalizes the operations of becoming informed and being informed.

The number of instances of RB in a derivation executed by distinct agents establishes the depth of the reasoning in which they are (collectively) involved, and f is still an operation to establish the depth of formulas. Then we generalize derivability and refutability for the logic MA-DBBL:

Definition 12 (Derivability for MA-DBBL). *Given a set of formulas Γ_j, formula ϕ_i of $\mathcal{L}^{\mathcal{C}}$, formulas $BI_j\phi_i$ and $I_j\phi$ of $\mathcal{L}^{\mathcal{K}}$, all operations f and $i \preceq j$:*

- $\Gamma_j \vdash_0^f \phi$ *iff* $\Gamma_j \vdash_0 \phi$;
- $\Gamma_j \vdash_{k+1}^f BI_j\phi_i$ *iff*
    - *there are $k+1$ distinct instances of RB such that for each application of the rule there is some formula $\psi_i \in f(\Gamma_i \cup \{\phi_i\})$, such that $\Gamma_i, \psi_i \vdash_k^f \phi_i$ and $\Gamma_i, \neg\psi_i \vdash_k^f \phi_i$;*
    - *or $BI_j\phi_i$ is obtained from $k+1$-depth derivable BI-formulae by an intelim rule of MA-DBBL$^{\mathcal{K}}$;*
- $\Gamma_j \vdash_{k+1}^f I_j\phi_i$ *iff*
    - *there is an instance of an I-introduction rule with n premises of the form $\Gamma_j \vdash_{k+1}^f BI_j\phi_i$, for n agents $i \preceq j$;*
    - *or ϕ_i is obtained from a $k+1$-depth derivable I-formula by an I-elim rule of MA-DBBL$^{\mathcal{K}}$.*

The first case holds trivially. For the second case and $k = 0$ suppose that $\Gamma_j \vdash^f_{k+1} BI_j\phi_i$, but that there is no formula ψ_i such that $\Gamma_i, \psi_i \vdash^f_k \phi_i$ nor $\Gamma_i, \neg\psi_i \vdash^f_k \phi_i$ can be satisfied: then the RB rule from Fig. 2 cannot be applied and $\Gamma_j \nvdash^f_{k+1} BI_j\phi_i$ against the hypothesis. For $k > 0$ the previous case requires multiple distinct instances of RB. The second part of the second point follows from the rules of Fig. 2. For the third case, suppose that $\Gamma_j \vdash^f_{k+1} I_j\phi_h$ is valid for $h \preceq i \preceq j$. Suppose also that $\Gamma_j \vdash^f_{k+1} BI_j\phi_i$ holds, but $\Gamma_j \vdash^f_{k+1} BI_j\phi_h$ does not. Hence, the I-introduction rule from Fig. 2 cannot be applied and thus $\Gamma_j \nvdash^f_{k+1} I_j\phi_i$, against the hypothesis. The second part of the third case follows from the definition of the I-elimination rule of Fig. 2.

Definition 13 ($k+1$ Inconsistency). *Γ_i is $k+1$-depth inconsistent if and only if $\Gamma_i \cup \{\psi_i\}$ and $\Gamma_i \cup \{\neg\psi_i\}$ are both k-depth inconsistent for some $\psi_i \in f(\Gamma_i)$.*

Definition 14 (MA-DBBL Refutability). *Given a set of formulas $\Gamma_i \in \mathcal{L}^{C\mathcal{K}}$ and f:*

- *An MA-DBBL derivation is closed when it contains both ϕ_i and $\neg\phi_i$, for some $\phi_i \in f(\Gamma_i)$;*
- *An MA-DBBL refutation of Γ_i is a closed MA-DBBL derivation for Γ_i;*
- *Γ_i is MA-DBBL-refutable if there is a closed MA-DBBL derivation for Γ_i.*

5   Meta-theory

In the present section we provide essential meta-theoretic results.

Theorem 1 (Soundness). *If $\Gamma_i \vdash^f_k \phi_j$ then $\Gamma_i \vDash^f_k \phi_j$.*

Proof. The proof proceeds by induction on the depth of the derivability relation:

- for $k = 0$-depth: consider the case $\phi_j \wedge \psi_j$, with ϕ_j, ψ_j in Γ_j. Then, $\Gamma_j \vdash_0 \phi_j$, $\Gamma_j \vdash_0 \psi_j$ and by $\wedge$-intro1 their conjunction is derivable at $k = 0$. There is a matching case in the semantics such that if $1_j, 1_j$ then the conjunction is in the consequence set $Cn_0(\Gamma_j)$. The same holds for all rules of MA-DBBL^C.
- for $k > 0$-depth:
  - BI formulae: consider $\Gamma_j \vdash^f_{k+1} BI_j\phi_i$, obtained by an instance of RB. Then $\Gamma_i \vdash^f_k \psi_i \to \phi_i$ and $\Gamma_i \vdash^f_k \neg\psi_i \to \phi_i$ for some $\psi_i \in f(\Gamma_i)$, while $\Gamma_j \nvdash^f_k \phi_j$. By the former $\Gamma_i \vdash^f_k \phi_i$, which grants $\Gamma_i \vDash^f_k \phi_i$, i.e. $v(\phi_i) = 1$; the latter means $\Gamma_j \nvDash^f_k \phi_j$, i.e. $v(\phi_j) = 0$ or $v(\phi_j) = *$; in both cases, according to the evaluation function in Definition 7, if 1_i then $BI_j\phi_i = 1$.
  - I formulae: if $\Gamma_j \vdash^f_{k+1} I_j\phi_j$ then $BI_j\phi_i, \forall i \preceq j$. The above argument holds for versions of the RB rules with premises indexed by every $i \preceq j$, with matching case in the truth-table for the I operator.

Theorem 2 (Completeness). *If $\Gamma_j \vDash^f_k \phi_j$ then $\Gamma_j \vdash^f_k \phi_j$.*

Proof. The proof proceeds by induction on the depth of the consequence relation:

- for $k = 0$-depth we follow the proof in [8]: suppose that $\Gamma_j \vDash_0 \phi_j$ and $\Gamma_j \nvDash_0 \phi_j$. Then Γ_j is not 0-depth refutable; otherwise, by definition of 0-depth proof, it should hold that $\Gamma_j \vdash_0 \phi_j$ (as well as $\Gamma_j \vdash_0^f \neg\phi_j$) against the hypothesis. Now, consider the set $\Gamma_j^0 = \{\psi_j \mid \Gamma_j \vdash_0 \psi_j\}$; since Γ_j is not 0-depth refutable, there are no formulas χ_j, and $\neg\chi_j$ such that they are both in Γ_j^0. Then, the function $\mathcal{V}$, defined as follows:

$$\mathcal{V}(\chi_j) = \begin{cases} 1 \text{ if } \chi_j \in \Gamma_j^0 \\ 0 \text{ if } \neg\chi_j \in \Gamma_j^0 \\ * \text{ otherwise} \end{cases} \tag{1}$$

is a three-value valuation that agrees with Table 1. Consider now the following formula case: $\mathcal{V}(\chi_j) = \mathcal{V}(\rho_j) = *$; then $\neg(\chi \vee \rho)_j \notin \Gamma_j^0$. Otherwise, by definition of Γ_j^0 and by the $\vee$-elim3 rule of Fig. 1, $\neg\chi_j$ and $\neg\rho_j$ should also be in Γ_j^0. Therefore, by definition of $\mathcal{V}$, $\mathcal{V}(\chi_j) = \mathcal{V}(\rho_j) = 0$, against our assumption. Hence $\mathcal{V}(\chi \vee \rho)_j \neq 0$. Moreover, $(\chi \vee \rho)_j$, may or may not belong to Γ_j^0, and so either $\mathcal{V}(\chi \wedge \rho)_j = 1$ or $\mathcal{V}(\chi \wedge \rho)_j = *$. Finally:
  i) $\psi_j \in \Gamma_j^0$, for all $\psi_j \in \Gamma_j$ and so, by definition of $\mathcal{V}$, $\mathcal{V}$ satisfies all $\psi_j \in \Gamma_j$;
  ii) by the hypothesis that $\Gamma_j \nvdash_0^f \phi_j$, $\phi_j \notin \Gamma_j^0$ and so $\mathcal{V}$ does not satisfy ϕ_j.
  Hence $\Gamma_j \nvDash_0^f \phi_j$, against the initial assumption.
- for $k > 0$-depth; we denote, as above, with Γ_i^k the theoremhood set of Γ_i and depth k. Functions are defined for the theoremhood sets of BI and I formulas:

$$\mathcal{V}(BI_j\chi_i) = \begin{cases} 1 \text{ if } \chi_i \in \Gamma_i^k, \chi_j \notin \Gamma_j^k \text{ and } \neg\chi_j \notin \Gamma_j^k \\ 0 \text{ if } \neg\chi_i \in \Gamma_i^k, \chi_j \notin \Gamma_j^k \text{ and } \neg\chi_j \notin \Gamma_j^k \\ * \text{ if } \chi_i \notin \Gamma_i^k \text{ and } \neg\chi_i \notin \Gamma_i^k \end{cases} \tag{2}$$

$$\mathcal{V}(I_j\chi) = \begin{cases} 1 \text{ if } \chi_i \in \Gamma_i^k \text{ for all } i \preceq j \in \mathcal{S}, \text{ and } \chi_i \in \Gamma_j^{k+1} \\ 0 \text{ if } \neg\chi_i \in \Gamma_i^k \text{ for all } i \preceq j \in \mathcal{S}, \text{ and } \neg\chi_i \in \Gamma_j^{k+1} \\ * \text{ if } \chi_i \notin \Gamma_i^k \text{ and } \neg\chi_i \notin \Gamma_i^k \text{ for all } i \preceq j \in \mathcal{S}. \end{cases} \tag{3}$$

- For BI formulas: Suppose that $\Gamma_j \vDash_{k+1}^f BI_j\phi_i$ and $\Gamma_j \nvDash_{k+1}^f BI_j\phi_i$. Then Γ_j is not $k+1$-depth refutable; otherwise, by definition of $k+1$-depth proof, it should hold that $\Gamma_i \vdash_k^f \phi_i$ (as well as $\Gamma_i \vdash_k^f \neg\phi_i$), for some $\phi \in f(\Gamma_i)$; hence $\Gamma_j \vdash_{k+1}^f BI_j\phi_i$ is derivable by RB, against the hypothesis. Now, consider the set $\Gamma_j^{k+1} = \{BI_j\phi_i \mid \Gamma_j \vdash_{k+1}^f BI_j\phi_i\}$. Since Γ_j is not $k+1$-depth refutable, there is no formula such that $BI_j\chi_i$ and $BI_j\neg\chi_i$ are both in Γ_j^{k+1}. Then, consider $\mathcal{V}(BI_j\phi_i) = 1$. Then $\phi_i \in \Gamma_i^k$, and both ϕ_j and $\neg\phi_j \notin \Gamma_j^k$. Otherwise $\mathcal{V}(\phi_i) = 0$ and $\mathcal{V}(\phi_j) = 1$ or $\mathcal{V}(\phi_j) = 0$:
  -- if $\mathcal{V}(\phi_i) = 0$, the premises of the relevant RB cannot be satisfied, and $\mathcal{V}(BI_j\phi_i) \neq 1$ against our assumption;

-- if $\mathcal{V}(\phi_j) = 1$, by 0-depth evaluation $\phi_j \in \Gamma_j^k$ which is against the definition of $\mathcal{V}(BI_j\phi_i)$ from the assumption $\Gamma_j \vDash_{k+1}^f BI_j\phi_i$;

-- and if $\mathcal{V}(\phi_j) = 0$, by 0-depth evaluation $\neg\phi_j \in \Gamma_i^k$, as above against the definition of $\mathcal{V}(BI_j\phi_i)$ from the assumption $\Gamma_j \vDash_{k+1}^f BI_j\phi_i$;

- For I-formulas: Consider $\Gamma_j^{k+1} = \{I_j\phi_i \mid \Gamma_j \vdash_{k+1}^f I_j\phi_i\}$ and $\mathcal{V}(I_j\phi_i) = 1$. Then, $\phi_i \in \Gamma_i^k$ for any $i \preceq j \in \mathcal{S}$. Then, by the definition of I, $\phi_j \in \Gamma_j^{k+1}$. Otherwise $\phi_i \notin \Gamma_i^k$ for at least one $i \preceq j \in \mathcal{S}$. But then $\phi_j \notin \Gamma_j^{k+1}$ and hence $\Gamma_j \nvDash_{k+1}^f I_j\phi_i$, against the initial assumption.

6   Conclusions and Future Work

MA-DBBL is an extension of Depth Bound Boolean Logic to express reasoning by agents ordered on their cognitive abilities, in terms of inferential power based on data access. The language includes a one-to-one information transmission operator between source and receiver which, for one instance of its application, expresses an increase of one degree in the depth of the reasoning power of the receiving agent. The language also includes an operator between every source and a given receiver, which expresses a state of the system where everybody shares the same information. Such operations allow agents with poorer cognitive capacities (inferential abilities and data access) to become informed of content otherwise available only to better equipped agents. This feature is especially well fitted for human-machine interaction settings or in distributed systems, where agents with different inferential-bounds work towards a common goal.

Planned extensions of this logic are: an appropriate relational semantics by using three-valued Kripke models, where the BI-operator corresponds to an accessibility relation indexed by the receiving agent, restricting the accessible worlds to some of those accessible by the source; a non-monotonic extension, to allow updates with contradictory information and the manipulation of disjoint sets of information for agents; the definition of a negative trust operation on informational contents held and shared by agents, on the lines of [18]; trustworthiness evaluations on sources and updates on their order, on the lines of [3]; the combination with different resource bounds, e.g. by considering access to information as costly, i.e. leading to a consumption of resources, see e.g. [19].

References

1. Allo, P.: The logic of 'being informed' revisited and revised. Philos. Stud. **153**(3), 417–434 (2011). https://doi.org/10.1007/s11098-010-9516-1
2. van Benthem, J.: Dynamic logic for belief revision. J. Appl. Non-Classical Logics **17**(2), 129–155 (2007). https://doi.org/10.3166/jancl.17.129-155
3. Ceolin, D., Primiero, G.: A granular approach to source trustworthiness for negative trust assessment. In: Meng, W., Cofta, P., Jensen, C.D., Grandison, T. (eds.) IFIPTM 2019. IAICT, vol. 563, pp. 108–121. Springer, Cham (2019). https://doi.org/10.1007/978-3-030-33716-2_9

4. D'Agostino, M.: Depth-bounded logic for realistic agents. Logic Philos. Sci. **11**, 3–57 (2013)
5. D'Agostino, M.: Informational semantics, non-deterministic matrices and feasible deduction. In: Fernández, M., Finger, M. (eds.) Proceedings of the 8th Workshop on Logical and Semantic Frameworks, LSFA 2013, São Paulo, Brazil, 2–3 September 2013, Electronic Notes in Theoretical Computer Science, vol. 305, pp. 35–52. Elsevier (2013). URL https://doi.org/10.1016/j.entcs.2014.06.004
6. D'Agostino, M.: Semantic information and the trivialization of logic: Floridi on the scandal of deduction. Information **4**(1), 33–59 (2013). https://doi.org/10.3390/info4010033
7. D'Agostino, M.: Analytic inference and the informational meaning of the logical operators. Logique Anal. **227**, 407–437 (2014)
8. D'Agostino, M.: An informational view of classical logic. Theor. Comput. Sci. **606**, 79–97 (2015). https://doi.org/10.1016/j.tcs.2015.06.057
9. van Ditmarsch, H., Halpern, J., van der Hoek, W., Kooi, B.: Handbook of Epistemic Logic. College Publications, London (2015)
10. Ditmarsch, H., van der Hoek, W., Kooi, B.: Dynamic Epistemic Logic and Philosophy Of Science. Springer, Berlin (2007)
11. Floridi, L.: The logic of being informed. Logique et Analyse **49**(196), 433–460 (2006). http://www.jstor.org/stable/44085232
12. Floridi, L.: Semantic information and the correctness theory of truth. Erkenntnis **74**(2), 147–175 (2011). https://doi.org/10.1007/s10670-010-9249-8
13. Stegenga, J.: Information quality in clinical research. In: Floridi, L., Illari, P. (eds.) The Philosophy of Information Quality. SL, vol. 358, pp. 163–182. Springer, Cham (2014). https://doi.org/10.1007/978-3-319-07121-3_9
14. Larese, C.: The principle of analyticity of logic, a philosophical and formal perspective. Ph.D. thesis, Scuola Normale Superiore, Classe di Scienze Umane (2019)
15. Meyer, J.J.C., Hoek, W.v.d.: Epistemic Logic for AI and Computer Science. Cambridge Tracts in Theoretical Computer Science. Cambridge University Press (1995). https://doi.org/10.1017/CBO9780511569852
16. Primiero, G.: An epistemic constructive definition of information. Logique et Analyse **50**(200), 391–416 (2007)
17. Primiero, G.: An epistemic logic for becoming informed. Synthese **167**(2), 363–389 (2009). https://doi.org/10.1007/s11229-008-9413-8
18. Primiero, G.: A logic of negative trust. J. Appl. Non Class. Logics **30**(3), 193–222 (2020). https://doi.org/10.1080/11663081.2020.1789404
19. Primiero, G., Raimondi, F., Bottone, M., Tagliabue, J.: Trust and distrust in contradictory information transmission. Appl. Netw. Sci. **2**(1), 1–30 (2017). https://doi.org/10.1007/s41109-017-0029-0

The Intensional Structure of Epistemic Convictions

Reinhard Kahle[1,2]([⊠])

[1] Carl Friedrich von Weizsäcker-Zentrum, Universität Tübingen,
Keplerstr. 2, D-72074 Tübingen, Germany
kahle@mat.uc.pt
[2] FCT, CMA, Universidade Nova de Lisboa, P-2829-516 Caparica, Portugal

Abstract. We discuss an axiomatic setup as an appropriate account to the intensional structure of epistemic convictions. This includes a resolution of the problem of logical omniscience as well as the individual rendering of knowledge by different persons.

In this position paper we present a model for epistemic convictions, which provides the general framework for different applications as belief revision [12,18], modalities [13–15], Frege's *mode of presentation* [16], and counterfactuals [17].

The idea is to model *knowledge* or *belief* of a person in an *axiomatic setup*. The purpose is to make good use of the concept of *derivation*, in fact *performed derivations*, to overcome some of the well-known problems of knowledge representation in formal frameworks[1].

To avoid an intricate discussion of the conflicting terms of "knowledge" and "belief" we prefer to use the neutral designation of *epistemic conviction* for a person's knowledge or belief. In the last section, however, we address how our account may help to explicate, at least in part, the traditional understanding of *knowledge as true and justified belief*.

1 Axiomatic Setup

We presuppose that epistemic convictions of a person can be represented by sentences in a formal(izable) language. We do not specify a particular formal language, but assume that, at least, propositional connectives and quantifiers are available.

---

[1] Our approach reassembles ideas which one also finds in Doyle's *Truth Maintenance Systems* (TMSs), [5]. As TMSs were developed in the context of expert systems in Computer Science, they soon fell victim to complexity issues. For us it is, however, just the *qualitative* setup which matters from a philosophical point of view. The *quantitative* aspect may go out of control when one tries to explain and store every single step of a derivation.

© Springer Nature Switzerland AG 2021
L. Cleophas and M. Massink (Eds.): SEFM 2020 Workshops, LNCS 12524, pp. 192–200, 2021.
https://doi.org/10.1007/978-3-030-67220-1_15

Some *basic convictions* are fixed (for an individual person; they might differ from one person to another[2]). A sentence belonging to these basic convictions is considered as an *axiom*, an axiom which can then be used to derive further epistemic convictions. A sentence is added to the person's convictions, if it is actually *derived* from the basic convictions by a *correct* derivation.

For the discusion which follow one has to distinguish two complementary aspects of the axiomatic setup.

1. The *internal* aspect, which takes the "axioms" of a person—at a certain moment—as fixed and studies the *derived* consequences.
2. The *external* aspect, which considers the choice, justification, and changes of these axioms.

The word "axiom", when we use it for *basic convictions*, is not used with its traditional meaning as *evident truth*, but rather in the modern understanding as distinguished starting point for derivations (see also Remark 1 below).

The *intensional structure*, discussed in this paper, concerns, first of all, only the internal aspect of an axiomatic setup. To keep the presentation clean, we presuppose that only correct reasoning is used by a person. But our account may also be used to study the effects of *incorrect reasoning*. "Correct" does not need to be restricted to purely logical reasoning, but could include *inductive* reasoning in the way it is vindicated by common sense. And one may address the question, to which extent the rules for correct reasoning have to be part of the basic convictions. A detailed discussion of these aspects, however, would go beyond the scope of this paper which should just illustrate the qualitative potential of the intensional structure of an axiomatic setup.

We will touch occasionally on the external aspect of the axiomatic setup, which can be located in the narthex of axiomatics[3]. Here, one would have to discuss how an axiom enters in the individual convictions of a person (*empirical observation* will be one way; another way is by *learning in school*; but one may also consider just *believing*, see Remark 1 below; and a good Platonist may include *anamnesis*). Also, the axioms of a person's convictions are subject to change (one is continuously making new empirical observation and learning new facts; one may also retract old convictions or revise the present ones). The dynamics of the change of convictions is a wide field, and here we like to point out only that our axiomatic setup provides a tool to deal with it (see, in particular, the end of the following section).

---

[2] For a fruitful discussion amoung different people it is, however, desirable that one can agree on the same starting points.

[3] See the reference to "Vorhalle der Geometrie" for work of Moritz Pasch, one of the founding fathers of modern axiomatics, in [24, p. 80].

2   Avoiding Logical Omniscience

Many approaches to knowledge and belief presuppose that they are closed under logical consequences. Although it comes sometimes under the label of *rationality criteria*, as in AGM [2], this presupposition is quite inadequate[4].

Alan Turing formulated this neatly [25, p. 451][5]:

> The view that machines cannot give rise to surprises is due, I believe, to a fallacy to which philosophers and mathematicians are particularly subject. This is the assumption that as soon as a fact is presented to a mind all consequences of that fact spring into the mind simultaneously with it. It is a very useful assumption under many circumstances, but one too easily forgets that it is false.

Under the heading of *logical omniscience*, this problem is acknowledged in the literature[6]. One way out is to turn to a theory of *implicit knowledge*, which is supposed to come from the logical closure of initial knowledge. But it seems to be of little help. Just consider mathematical knowledge: it doesn't seem to be of particular interest to study a person's implicit knowledge of mathematics, which, in some sense, should contain the solutions of all open questions in Mathematics[7].

There a some attempts to address logical omniscience with "non-standard" logical frameworks. Some of them are semantically in nature, drawing essentially on (modified) possible world semantics, as, for instance, *awareness structures* and *local-reasoing structures* (Halpern and Fagin [6]), *impossible worlds* (Hintikka [9]), or *situation semantics* (Barwise and Perry, [3]). On a more proof-theoretic side, one can consider *depth-bounded logics* [4] and *resource-aware logics*. As much as they tame logical omniscience in general, it is not clear how they should cope with examples from Mathematics. And we don't find any counterpart to the specific ingredient of our account: the relation between presupposed axioms and

---

[4] See [18] for a detailed discussion of AGM in our perspective.

[5] Also cited in [1, p. 261].

[6] The paragraph on logical omniscience in the article on Epistemic Logic in the Stanford Encyclopedia of Philosophy exposes here a certain helplessness [22, § 4].

[7] Of course, one may study an idealized notion of *knowability* which should be closed under logical omniscience. But, in our view, this form of idealization goes to far to provide a tool to take up challenges concerned with *knowledge of a person*. Thus, we are explicitly at odds with the first part of the justification of this idealization by Gabbay and Woods [7, p. 158]:

> A logic is an idealization of certain sorts of real-life phenomena. By their very nature, idealizations misdescribe the behavior of actual agents. This is to be tolerated when two conditions are met. One is that the actual behavior of actual agents can defensibly be made out to approximate to the behavior of the ideal agents of the logician's idealization. The other is the idealization's facilitation of the logician's discovery and demonstration of deep laws.

the derived formulas. Conceptionally, this relation is located on the meta-level, i.e., the derivability relation, rather than inside the logical framework[8].

This relation is manifest in Mathematics by a proof; and for mathematical knowledge by a proof "at hand": we consider a theorem only as a part of a person's knowledge, when (s)he is able to provide this proof when asked[9].

This should be the normal case for mathematical knowledge, but we may discuss two special cases, which belong, however, to the external aspect of our axiomatic setup.

First, you may have learned a mathematical theorem from a trustworthy source, normally that means from Mathematicians which did perform the proof of the theorem in question. We would say that most of the Mathematicians *know* that Fermat's Last Theorem is true, without being able to provide Wiles's proof. In this case, it is simply added as an axiom, and the justification of the axiom comes from reference to the reliable source.

Secondly, it is possible to be convinced of the truth of an open conjecture without having a proof yet. This, of course, does not result in knowledge, but is just *belief.* But it can be treated in our framework in the same way as knowledge, and one may *flag* such conjectures when they enter the convictions of a person only as "axioms" with lower credibility. Of course, consequences derived from such *flagged axioms* will also only be "believed theorems", rather than known ones. Here, we will not work out any technical theory of mathematical beliefs, but just pointing to one important aspect, which contrasts our approach to those aiming for logical omniscience (even, if only implicitely): *believing* in a conjecture might not be irrational, even if the conjecture is false. It can be treated as an axiom of the person's convictions, and there is no fundamental problem as long as no contradiction is actually derived.

Remark 1. David Hilbert gave a bold example of this situation [8, p. 160]:

> Nothing prevents us from taking as axioms propositions which are provable, or which we believe are provable. Indeed, as history shows, this procedure is perfectly in order: [...] Riemann's conjecture about the zeroes of $\zeta(s)$, [...]

Only, when a contradiction is actually derived, it should trigger a *belief revision.* Belief revisions might be caused for quite different reasons, but the encounter of a contradiction is probably the most important one. The way such a belief revision can be performed is part of the external aspect of our axiomatic setup. But the concrete derivation of a contradiction is of fundamental help: it singles out those axioms (basic convictions) which are involved in the derivation

---

[8] There are other criticisms of the alternative approaches, like, for example, the "ontological overkill" of possible worlds [14], which we cannot discuss here. All these criticisms, of course, do not mean that those approaches do not have their merits; we just like to point to the conceptional difference with our account.

[9] Ryle's distinction of *knowing how* and *knowing that* [23] points into another direction. But we share with him, at least, the opposition to a raw *knowing that*. We like to complement it by *knowing why*.

and point to the fact that the resolution has to revise at least one of them (while, at the same time, basic conviction not used in this derivation do not need to be considered)[10]. For a more elaborated approach to belief revision in terms of our axiomatic setup, see [12].

3 Individual Structuring of Convictions

The axiomatic setup exhibits the intensional structure of a person's convictions. These are not just the derived sentences, but the concrete derivations come into play in, at least, two ways: first, a derivable sentence has many different proofs in an axiomatic framework. Choosing one or another derivation can influence the trust in the derived sentences as well as the way the conviction can be defended when it comes under scrutiny[11]. Secondly, and more important, the same set of sentences can be derived from different sets of axioms. Thus, extensionally equal sets of convictions can be represented by different sets of axioms.

We would like to illustrate the latter situation by a simplified example from Astronomy.

Person A may observe a good number of positions of the planets of our solar system, with the precision available in the 18th century. The coordinates of these positions are A's axioms, justified by empirical evidence. By intelligent inductive reasoning, A derives from these positions Kepler's laws for the movement of planets.

Person B learned at school Kepler's laws and, without ever looking to the sky at night, may derive the positions of the planets using some given initial data.

By construction, the astronomical knowledge of A and B should be equivalent. However, the difference in the intensional structure should become visible, when both learn about the more exact perihelion precession of Mercury as available in the 19th century. For A these are "just" new empirical data, which question, of course, the derived Kepler's laws, but which do not contradict the original axioms. For B, however, the very axioms are *falsified* and B's knowledge as such is called into question.

In [16] we mentioned as another example two axiomatizations of the natural numbers, the first one as a commutative semigroup, the second by the Peano

---

[10] This sketches the qualitative aspect of our account only; in the presence of a plausibility order of beliefs, for instance, one may revise first the basic convictions with lowest plausibility.

[11] The question how different derivations should compared with each other was rised by Hilbert in his *24th problem*. This problem was not included in his famous *problems lecture* at the International Congress of Mathematicians in 1900 in Paris but remained unpublished in his notebook before it was discovered in 2000. Since then, it triggered a lot of a research, including the question of identity of proofs [10]. Our approch is not intended to contribute to a solution of Hilbert's 24th problem, but rather the other way around: a satisfactory concept for identity of proofs may allow to abstract from the concrete derivations we are relying on.

Axioms. While in the former one, commutativity of addition is "built in", in the latter one this property requires a proof by induction. Thus, one may introduce a notion of *sense*—in Fregean terms—such that the sense of the two sum terms $t+s$ and $s+t$ is equal in the former but different in the latter axiomatic presentation.

In general, whenever one has two different axiomatizations which result in the same set of derived formulas, the very difference of the axiomatization can be considered as an intensional difference—a difference which could, at best, only be encoded artifically in a semantic setup.

Due to the axiomatic nature of mathematical theories, they will provide many more examples to illustrate our point; you may think, for instance, of Geometry given in terms of points, straight lines, etc., following Euclid and Hilbert, or in terms of reflections, following Bachmann. In Physics, for instance, the Heisenberg picture [19] can be contrasted by the Schrödinger picture [20]. We expect that it should be possible to find examples even from "every day" concepts which may be represented extensionally equivalently, but intensionally differently.

One may ask whether our approach reduces the knowledge of a person to a "blob of derived sentences" which is somehow arbitrary. In some sense, this is the case, as the knowledge of different people is usually quite different. But it is arbitrary only in the way that it cannot be determined "from the outside" or "in advance" which knowledge a particular person is deriving for himself or herself. *Ex post* the particular person is, however, supposed to provide the justification for the derived knowledge (see also the next section) and, thus, one should not characterizes the knowledge as "arbitrary". But we turn this argument around, when we ask whether the knowledge of the person is *computable*. Assuming that the life span of every person is finite, and that, in finite time, you may perform only finitely many derivations, it is clear that every person's knowledge is finite. Thus, it appears to be *computable*, as we should have no problems to compute finite sets. But this works only *ex post* (and with a protocol of all knowledge of the person at hand). Given a certain set of axioms, we may go different routes to derive new consequences; as these choice are not determined in advance, it is not to see how to program a machine which could anticipate which formulas are actually derived by a person; thus, in this way the knowledge of person remains uncomputable[12].

4   The Question "Why?"

The axiomatic setup is also the adequate model to study the way a person answers the question: "Why do you believe this?" or "How do you know this?" The given answer should reveal the argument the person used in the derivation of the sentence in question. Turning the perspective around, we can use why-questions to uncover the axiomatic structure of a person's epistemic convictions.

---

[12] More moderately expressed: unpredictable. The situation is not too different from the question whether you can compute the roulette results of the Monte Carlo casino. Of course, you can do it for those outcomes which already took place in a finite time period; but you will not be able to compute it in advance.

A by-product of this analysis is that it gives support for the classical characterization of knowledge as *true and justified belief*. While the justification comes from derivations[13], the mentioned analogy to mathematical proof shows that, of course, the sentences used in the derivation need to be hereditary knowledge, i.e., being themselves all true. Leaving aside the notorious question how truth should be established, it rules out, at least, flawed justifications.

Let us draw on a known example of Mathematics. When, in the last decades of the 19th century, the proof attempts of Kempe and Tait were considered by the mathematical community as correct proofs of the four colour problem, of course, this community did not have *knowledge* of the four colour theorem. Even if we know today, by the proofs of the second half of the 20th century, that the four colour theorem is, indeed, true, the alleged proofs of the 19th century could not serve as justification, for the simple reason that they were flawed.

The problem confusing a correct justification with a raw justifiability, not respecting the heritability of truth was discussed a lot in context of the notorious Gettier examples, and the flawed knowledge based on it could well be called *Gettier knowledge*[14].

The question: "Why do you believe this 'axiom'?" makes part of the external aspect of our axiomatic setup. In contrast to the internal aspect, here we do not expect a derivation as answer[15], but some other type of justification. For example, the famous equation $E = mc^2$ can be derived in the approriate formal framework of relativity theory. However, many people will "know" it without having ever studied relativity theory. If one asks why, an answer like "because I learned it in school" provides a proper justification; an answer of the sort "because I read in my horoscope" would not serve as justification.

The question "Why?" also poses a challange for Artificial Intelligence, in particular to modern AI based on statistics. As mentioned in Footnote 1, Truth Maintenance Systems [5] were an attempt to store information which could be used ot answer Why-questions in expert systems; this turned out to be unfeasible due to complexity constraints. Still, in SAT solving we may localize clauses "responsible" for a contradiction which may provide answers to a Why-question (see [21]). But modern AI cannot provide any justification for its results—and this is *by design* due to the used *black box* technologies. In fact, recovering justifications for the answers of such AI software would probably amount to *open the black box*, a task *explainable AI* is aiming for. As long as the black box

---

[13] For knowledge, we tactically assume here logical reasoning, only. The case of inductive reasoning is—in the case of knowledge—more delicate as it questions the status of knowledge in general.

[14] When we subscribe the *No False Lemmas* condition, it goes without saying that also the "axioms" presupposed for the knowledge need to be true; this is not the case for the example discussed in [11, § 4].

[15] Although inductive reasoning, used to justify a universal statement which is supposed to be taken as an axiom, might enter here, rather than as part of the internal aspect of knowledge.

stays closed, we would even predict that AI software will fail the Turing Test when asked to answer a Why-question[16].

Acknowledgement. Research supported by the Udo Keller Foundation and by the Portuguese Science Foundation, FCT, through the project UIDB/00297/2020 (Centro de Matemática e Aplicações).

References

1. Aaronson, S.: Why philosophers should care about computational complexity. In: Copeland, B.J., Posy, C.J., Shagrir, O. (eds.) Computability, pp. 261–327. MIT Press (2013)
2. Alchourrón, C., Gärdenfors, P., Makinson, D.: On the logic of theory change: partial meet functions for contraction and revision. J. Symbolic Logic **50**(2), 510–530 (1985)
3. Barwise, J., Perry, J.: Situations and Attitudes. MIT Press, Bronx (1983)
4. D'Agostino, M.: Tractable depth-bounded logics and the problem of logical omniscience. In: Hosni, H., Montagna, F., (eds.) Probability, Uncertainty and Rationality, pp. 245–275. Edizioni Scuola Normale Superiore, Springer (2010)
5. Doyle, J.: A truth maintenance system. Artif. Intell. **12**, 231–272 (1979)
6. Fagin, R., Halpern, J.Y.: Belief, awareness, and limited reasoning. Artif. Intell. **34**, 39–76 (1988)
7. Gabbay, D.M., Woods, J.: The new logic. Logic J. IGPL **9**(2), 141–174 (2001)
8. Hilbert, D.: Neubegründung der Mathematik. Abhandlungen aus dem Mathematischen Seminar der Hamburgischen Universität **1**, 157–177 (1922)
9. Hintikka, J.: Impossible possible worlds vindicated. J. Philos. Logic **4**(4), 475–484 (1975)
10. Hipolito, I., Kahle, R.: Theme issue on "The notion of simple proof - Hilbert's 24th problem". Philosophical Transactions of the Royal Society A, 377 (2019)
11. Ichikawa, J.J., Steup, M.: The analysis of knowledge. In: Zalta, E.N. (eds.), The Stanford Encyclopedia of Philosophy. Metaphysics Research Lab, Stanford University, summer 2018 edition (2018)
12. Kahle, R.: Structured belief bases. Logic Logical Philos. **10**, 49–62 (2002)
13. Kahle, R.: A proof-theoretic view of necessity. Synthese **148**(3), 659–673 (2006)
14. Kahle, R.: Against possible worlds. In: Degremont, C., Keiff, L., Rückert, H. (eds.), Dialogues, Logics and Other Strange Things. Essays in Honour of Shahid Rahman, vol. 7 of Tributes, pp. 235–253. College Publications (2008)
15. Kahle, R.: Modalities without worlds. In: Rahman, S., Primiero, G., Marion, M. (eds.), The Realism-Antirealism Debate in the Age of Alternative Logics, vol. 23 of Logic, Epistemology and the Unity of Science, pp. 101–118. Springer, Dordrecht (2012) https://doi.org/10.1007/978-94-007-1923-1_6
16. Kahle, R.: Towards a proof-theoretic semantics of equalities. In: Piecha, T., Schroeder-Heister, P. (eds.) Advances in Proof-Theoretic Semantics. TL, vol. 43, pp. 153–160. Springer, Cham (2016). https://doi.org/10.1007/978-3-319-22686-6_9

---

[16] Here, we like to express serious doubts that justifications, in terms of derivations, can be "learned" just statistically, in the same way, as it is unlikely that AI could statistically generate a C++ compiler.

17. Kahle, R.: The Logical Cone. If CoLog J. Logics their Appl. **4**(4), 1087–1101 (2017). Special Issue Dedicated to the Memory of Grigori Mints. Dov Gabbay and Oleg Prosorov (Guest Editors)
18. Kahle, R.: Belief Revision Revisited. In: Pombo, O., Pato, A., Redmond, J. (eds.), Epistemologia, Lógica e Linguagem, vol. 11 of Colecção Documenta. Centro de Filosofia das Ciências da Universidade de Lisboa (2019)
19. Kukin, V D.: Heisenberg representation. Encyclopedia of Mathematics. http://encyclopediaofmath.org/index.php?title=Heisenberg_representation&oldid=44704
20. Kukin, V.D.: Schrödinger representation. Encyclopedia of Mathematics. http://encyclopediaofmath.org/index.php?title=Schr
21. Küchlin, W.: Symbolische KI für die Produktkonfiguration in der Automobilindustrie. In: Mainzer, K. (eds.), Philosophisches Handbuch Künstliche Intelligenz. Springer, New York (2021) https://doi.org/10.1007/978-3-658-23715-8_53-1
22. Rendsvig, R., Symons, J.: Epistemic logic. In: Zalta, E.N. (ed.), The Stanford Encyclopedia of Philosophy. Metaphysics Research Lab, Stanford University, summer 2019 edition (2019)
23. Ryle, G.: Konwing how and knowing that. Proc. Aristotelian Soc. **46**, 1–16 (1945–46)
24. Tamari, D.: Moritz Pasch (1843–1930). Shaker Verlag (2007)
25. Turing, A.: Computing machinery and intelligence. Mind **59**, 433–460 (1950)

Short-Circuiting the Definition
of Mathematical Knowledge
for an Artificial General Intelligence

Samuel Allen Alexander[(✉)]

The U.S. Securities and Exchange Commission, Washington D.C., USA
samuelallenalexander@gmail.com
https://philpeople.org/profiles/samuel-alexander/publications

Abstract. We propose that, for the purpose of studying theoretical properties of the knowledge of an agent with Artificial General Intelligence (that is, the knowledge of an AGI), a pragmatic way to define such an agent's knowledge (restricted to the language of Epistemic Arithmetic, or EA) is as follows. We declare an AGI to know an EA-statement ϕ if and only if that AGI would include ϕ in the resulting enumeration if that AGI were commanded: "Enumerate all the EA-sentences which you know." This definition is non-circular because an AGI, being capable of practical English communication, is capable of understanding the everyday English word "know" independently of how any philosopher formally defines knowledge; we elaborate further on the non-circularity of this circular-looking definition. This elegantly solves the problem that different AGIs may have different internal knowledge definitions and yet we want to study knowledge of AGIs in general, without having to study different AGIs separately just because they have separate internal knowledge definitions. Finally, we suggest how this definition of AGI knowledge can be used as a bridge which could allow the AGI research community to import certain abstract results about mechanical knowing agents from mathematical logic.

Keywords: AGI · Machine knowledge · Quantified modal logic

1   Introduction

It is difficult to define knowledge, or what it means to know something. In Plato's dialogues, again and again Socrates asks people to define knowledge[1], and no-one ever succeeds. Neither have philosophers reached consensus even in our own era [15].

At the same time, the problem is often brushed aside as something only philosophers care about: pragmatists rarely spend time on this sort of debate. One exception is in the study of agents with Artificial General Intelligence (*AGIs*,

---

[1] Perhaps the best example being in the *Theaetetus* [18].

© Springer Nature Switzerland AG 2021
L. Cleophas and M. Massink (Eds.): SEFM 2020 Workshops, LNCS 12524, pp. 201–213, 2021.
https://doi.org/10.1007/978-3-030-67220-1_16

or *Type II AIs* in the terminology of [7]), where even the staunchest pragmatists admit the importance of the question.

In this paper, we narrow down the question "what is knowledge" and offer a simple answer within that narrow context: we propose a definition of what it means for a suitably idealized AGI to know a mathematical sentence[2] in the language of Epistemic Arithmetic [21] (hereafter EA). EA is the language of Peano Arithmetic along with an additional modal operator K for knowledge. To be precise:

- EA-terms are built up from variables $x, y, \ldots$ and a constant symbol for 0, the unary function symbol S for the successor function, and binary function symbols for addition and multiplication.
- EA-formulas are built up from atomic EA-formulas (which are of the form $s = t$ where s and t are EA-terms), propositional connectives $\rightarrow, \neg$, universal quantifiers $\forall x, \forall y, \ldots$, existential quantifiers $\exists x, \exists y, \ldots$, and modal operator K.

The EA-sentence $K(1 + 1 = 2)$ might be read "I know $1 + 1 = 2$" or "the knower knows $1 + 1 = 2$". Our proposed definition is parsimonious (at the price of appearing deceptively circular). We say that an AGI knows an EA-sentence ϕ if and only if ϕ would be among the sentences which that AGI would enumerate if that AGI were commanded:

"Enumerate all the EA-sentences which you know."

This is non-circular because an AGI, being capable of practical English communication, is therefore capable of understanding the everyday English word "know" in the above command, independently of how any philosopher formally defines knowledge. We discuss this further in Subsect. 3.1.

A primary motivation for this paper was the author's experience in the AGI research community where applications of mathematical logic are hindered by questions like "What does it mean for an AGI to know something?" For example, philosophers have long known that a suitably idealized mechanical knowing agent cannot know its own code and its own truthfulness[3]. But in informal conversations, we find AGI researchers struggle with this assertion, and we can hardly blame them, since, without agreeing what it means for the AGI to know something, of course the question arises, "What does it mean for an AGI to know something?" Likewise, we have proposed [6] an AGI intelligence measure based on the AGI's knowledge, and this, too, often provokes the response "What does it mean for an AGI to know something?" In Sect. 5 we will consider these examples, and related examples from the same area, using our proposed definition to translate them into a more concrete form, not in terms of what the AGI knows, but in terms of the AGI's stimulus-responses.

---

[2] By a *sentence*, we mean a formula with no free variables. Thus, $x^2 > 0$ is not a sentence, but $\forall x(x^2 > 0)$ is.

[3] Often phrased more like "cannot know its own code", with knowledge-of-own-truthfulness taken for granted.

The structure of this paper is as follows.

- In Sect. 2 we discuss the AGIs whose knowledge we are attempting to define.
- In Sect. 3 we propose a knowledge definition for AGIs for EA-sentences.
- In Sect. 4 we extend our knowledge definition to formulas with free variables.
- In Sect. 5 we use this knowledge definition as a bridge to translate some ideas from mathematical logic into the field of AGI.
- In Sect. 6 we summarize and make concluding remarks.

2   Idealized AGIs

In this paper, we approach AGI using what Goertzel [13] calls the Universalist Approach: we adopt "...an idealized case of AGI, similar to assumptions like the frictionless plane in physics", hoping that by understanding this "simplified special case, we can use the understanding we've gained to address more realistic cases." At the same time, an AGI might serve as a kind of hyper-idealized proxy for human cognition, and we hope that the development of the logic of AGI may serve as a step toward development of "new forms of logic as the basis of cognitive and substrate-independent studies of intelligent interaction" [11].

We do not have a formal definition for what an AGI is, but whatever it is, we assume an AGI is a deterministic machine which repeatedly reads sensory input from its environment and outputs English words based on what sensory inputs it has received so far[4]. When we say that this AGI is a "deterministic machine", we mean that said outputs (considered as a function of said inputs) could be computed by a Turing machine. We further assume the AGI can understand English commands and is capable of practical English communication. Thus, if we were to command the AGI in English, "Tell us the value of $1 + 1$", the AGI would respond in English and reply "2", or "$1 + 1 = 2$", or something along those lines[5].

We assume an AGI is capable of everyday English discussions which would cause no difficulty to a casual English speaker, even if these discussions involve topics, such as "knowledge", which might be philosophically tricky. A casual English speaker does not get stuck in philosophical questions about the nature of knowledge just in order to answer a question like "Do you know that $1+1 = 2$?", and therefore neither should our AGI.

We also assume an AGI is better than a casual human English speaker in certain ways. We assume an AGI would have no objections to performing tedious

---

[4] We should note that, with the AGI research field being so young, there is little consensus even on basic things. Some researchers would consider some things to be AGI which have no communication ability (applying the term to entities who have certain adaptation abilities or pattern-matching abilities, for example, even if those entities have no means of communicating), however, we believe that to be a minority opinion.

[5] We assume the AGI explicitly follows commands (that it is "under explicit control", to use Yampolskiy's terminology [24]).

tasks indefinitely, if so commanded. If we asked a casual human English speaker to begin computing and reciting all the prime numbers until further notice, and then we waited silently forever listening to the results, said human would eventually get tired of the endless tedium and would disobey our command (and would probably make arithmetic errors along the way). We assume an AGI has no such limitations and would happily compute and recite prime numbers for all eternity, if so commanded (without arithmetic mistakes). Of course, in reality the AGI would eventually run out of memory, terminate when the world ends, etc., but we are speaking of idealized AGI here and we intentionally ignore such possibilities, in the same way a Turing machine is assumed to have infinite tape and infinite time to run.

3   An Elegant Definition of Mathematical Knowledge

The following definition may initially look circular, but we will argue it is not.

Definition 1. *Let X be an AGI. For any EA-sentence ϕ, we say that X knows ϕ if and only if X would eventually include ϕ in the resulting enumeration if X were commanded:*

"Enumerate all the EA-sentences which you know."

Definition 1 is non-circular because the AGI is capable (see Sect. 2) of practical English communication, including that involving everyday English words such as the word "know", independently of how any philosophers formally define things[6]. More on this in Subsect. 3.1.

One of the strengths of Definition 1 is that it is uniform across different AGIs: many different AGIs might internally operate based on different definitions of knowledge, but Definition 1 works equally well for all these different AGIs regardless of those different internal knowledge definitions[7]. We can contrast this with difficulties that could arise from a more experimental approach. Scientists could carefully examine one particular AGI and eventually discover how that AGI's knowledge works, and attempt to define AGI knowledge accordingly, for example, they might define knowledge in terms of the contents of Memory Bank 35, which exists in that particular AGI. But then another AGI might come along which functions completely differently than the first AGI, and does not even have said Memory Bank 35 at all. Definition 1 is not tied to the particular form of the AGI, just so long as the AGI obediently follows English commands.

Remark 1. In Definition 1 when we speak of what the AGI would do if given such a command, implicitly we intend this to be understood as what the AGI would do if given such a command and then allowed to respond to the command

---

[6] This is reminiscent of Williamson's contextualism [23].

[7] This is reminiscent of Elton's proposal that instead of trying to interpret an AI's outputs by focusing on specific low-level details of a neural network, we should instead let the AI explain itself [12].

in isolation, without outside distractions. An AGI could potentially update its knowledge based on observations of the world, and so its knowledge might change from one time to the next: its knowledge at a given instant is defined by Definition 1 to consist of what the AGI would enumerate if the AGI were so commanded at that particular instant (and immediately secluded from further distracting observations).

Although Definition 1 may differ significantly from a particular AGI X's own internal definition of knowledge, the following theorem states that materially the two definitions have the same result.

Theorem 1. *Suppose X is an AGI. For any EA-sentence ϕ, the following are equivalent:*

1. X is considered to know ϕ (based on Definition 1).
2. X knows ϕ (based on X's own internal understanding of knowledge).

Proof. By Definition 1, (1) is equivalent to the statement that X would include ϕ in the list which X would output if X were commanded:

> "Enumerate all the EA-sentences which you know."

Since we have assumed (in Sect. 2) that X is obedient, X would output ϕ in the resulting list if and only if (2). □

Theorem 2. *Let X be an AGI. The set of EA-sentences ϕ such that X knows ϕ (based on Definition 1) is computably enumerable.*

Proof. This follows from our assumption (in Sect. 2) that X is a deterministic machine. □

3.1 Non-circularity of Definition 1

> 'What is said by a speaker (what she meant to say, her "meaning-intention") is understood or misunderstood by a hearer ("an interpreter").'
> —Albrecht Wellmer [22]

Definition 1 is non-circular because an AGI's response to an English command only depends on how the AGI understands the words in that command, not on how *we* (the speakers) understand those words. Recall from Sect. 2 that we are assuming an AGI is a deterministic machine which outputs English words based on sensory inputs from its environment. Those outputs depend *only* on those environmental inputs, and not on any decisions made by philosophers.

If the reader wants to further convince themselves of the non-circularity of Definition 1, we need only point out that the apparent circularity would disappear if we changed Definition 1 to define what it means for X to "grok" sentence ϕ, rather than to "know" sentence ϕ (without changing the command itself). In other words, we could define that X "groks" ϕ if and only if X would include ϕ in the list of sentences that would result if X were commanded,

"Enumerate all the EA-sentences which you know."

This would make the non-circularity clearer, because the word "grok" does not appear anywhere in the command.

We will further illustrate the non-circularity of Definition 1 with two examples.

- (The color blurple) Bob could (without Alice's awareness) define "blurple" to be the color of the card which Alice would choose if Bob were to run up to Alice, present her a red card and a blue card, and demand: "Quick, choose the blurple card! Do it now, no time for questions!" There is nothing circular about this, because Alice's choice cannot depend on a definition which Alice is unaware of.
- (Zero to the zero) If asked to compute 0^0, some calculators output 1, and some output an error message or say the result is undefined[8]. For any calculator X, it would be perfectly non-circular to define "the 0^0 of X" to be the output which X outputs when asked to compute 0^0. Said output is pre-programmed into the calculator; the calculator does not read the user's mind in order to base its answer on any definitions that exist there.

These considerations hinge on the AGI being separate from the reader. The human reader can apply Definition 1 to AGIs which she creates, but not to herself. An AGI X could apply Definition 1 to child AGIs that X created, but X could not apply the definition to X's own knowledge[9].

3.2   Sentences Using the Knowledge Operator

Definition 1 is particularly interesting when ϕ itself makes use of EA's K operator for knowledge.

Example 1. Applying Definition 1, we consider an AGI X to know $K(1+1=2)$ if and only if X would output $K(1+1=2)$ when commanded to enumerate all the EA-sentences he knows. X would (when so commanded) output $K(1+1=2)$ if and only if X knows (in his own internal sense of the word "know") that he knows (in his own internal sense of the word "know") $1+1=2$.

3.3   A Simpler Definition, and Why It Does Not Work

"It is difficult to be aware of whether one knows or not. For it is difficult to be aware of whether we know from the principles of a thing or not— and that is what knowing is. (...) Let that demonstration be better which,

---

[8] Which is incorrect—see [16].
[9] This is reminiscent of a recent argument [17] that humans maintain superiority over the AIs they create, as, for example, today's latest and greatest chess-playing AI is better at tactically playing individual games of chess, but is incapable of designing its own replacement (tomorrow's latest and greatest chess-playing AI), which will instead be designed by humans (making humans still better at chess in a higher-level sense).

other things being equal, depends on fewer postulates or suppositions. For if they are equally familiar, knowing will come about more quickly in this way; and that is preferable." —Aristotle [8]

The reader might wonder why we would not further simplify Definition 1 and declare that X knows ϕ if and only if X would respond "yes" if X were asked: "Do you know ϕ? (Yes or no)". We will argue that this would be a poor candidate for an idealized knowledge definition.

Definition 2. *If X is an AGI and ϕ is an EA-sentence, say that X quick-knows ϕ if and only if X would respond "yes" if X were asked, "Do you know ϕ? (Yes or no)".*

The following should be contrasted with Theorem 2.

Theorem 3. *Let X be an AGI. The set of EA-sentences ϕ such that X quick-knows ϕ is computable.*

Proof. This follows from our assumption (in Sect. 2) that X is a deterministic machine.   □

By Theorem 3, it seems that if we used Definition 2 as a knowledge definition, it would contradict Aristotle's claim that "it is difficult to be aware of whether one knows or not". It is more plausible that knowledge be *computably enumerable* (as in Theorem 2) than that knowledge be *computable*. A prototypical example of a set which is computably enumerable but not computable is: the consequences of Peano arithmetic[10] (hereafter PA). Said consequences cannot be computable, lest they could be used to solve the Halting Problem (because a Turing machine halts if and only if PA proves that it halts).

Theorem 4. *Let X be an AGI and assume X does not quick-know any falsehoods. At least one of the following is true:*

1. *There is an axiom of PA which X does not quick-know.*
2. *There exist PA-sentences ϕ and ψ such that X quick-knows ψ and X quick-knows $\psi \to \phi$, but X does not quick-know ϕ.*

Proof. It is well-known that a sentence ϕ is provable from PA if and only if there is a sequence $\phi_1, \ldots, \phi_n$ such that:

1. ϕ_n is ϕ.
2. For every i, either ϕ_i is an axiom of PA, or else there are $j, k < i$ such that ϕ_k is $\phi_j \to \phi_i$.

(Loosely speaking: proofs from PA can be carried out using no rules of inference besides Modus Ponens.) For any formula ϕ which PA proves, let $|\phi|$ be the smallest n such that there is a sequence $\phi_1, \ldots, \phi_n$ as above.

---

[10] We assume Peano arithmetic is true.

Call a PA-sentence ϕ *elusive* if PA proves ϕ but X does not quick-know ϕ. By Theorem 3, the fact that X does not quick-know any falsehoods, and the unsolvability of the Halting Problem, it follows that some elusive ϕ exists—otherwise, to computably determine whether or not a given Turing machine M halts, we could simply ask X, "Do you know Turing machine M halts? (Yes or no)".

Since some elusive ϕ exists, there exists an elusive ϕ such that $|\phi|$ is as small as possible—that is, such that $|\phi| \leq |\psi|$ for every elusive ψ. Fix such a ϕ.

Case 1: ϕ is an axiom of PA. Then condition (1) of the theorem is satisfied, as desired.

Case 2: ϕ is not an axiom of PA. Let $\phi_1, \ldots, \phi_{|\phi|}$ be as in the first paragraph of this proof (so $\phi_{|\phi|}$ is ϕ). Then since ϕ is not an axiom of PA, there must be $j, k < |\phi|$ such that ϕ_k is $\phi_j \to \phi_{|\phi|}$. Now, the sequence $\phi_1, \ldots, \phi_k$ witnesses that PA proves ϕ_k and $|\phi_k| \leq k < |\phi|$; and the sequence $\phi_1, \ldots, \phi_j$ witnesses that PA proves ϕ_j and $|\phi_j| \leq j < |\phi|$. Thus, since ϕ was chosen to be elusive with $|\phi|$ as small as possible, it follows that ϕ_k and ϕ_j are not elusive. Thus, X quick-knows ϕ_j, and X quick-knows ϕ_k, but ϕ_k is $\phi_j \to \phi$. Thus condition (2) of the theorem is satisfied, as desired. $\square$

Theorem 4 shows that Definition 2 makes a poor notion of idealized knowledge. An AGI should certainly know the axioms of PA, and should certainly be capable of the minimal logical reasoning needed to conclude ϕ from ψ and $\psi \to \phi$. And the way we have established the unsuitability of Definition 2 is nicely anticipated by the words of Aristotle quoted at the beginning of this subsection.

4   Quantified Modal Logic

Definition 1 only addresses sentences with no free variables. In this section, we will extend Definition 1 to formulas which possibly include free variables. We are essentially adapting a trick from Carlson [10].

Definition 3. *We define so-called* numerals, *which are EA-terms, one numeral $\overline{n}$ for each natural number $n \in \mathbb{N}$, by induction: $\overline{0}$ is defined to be 0 (the constant symbol for zero from PA) and for every $n \in \mathbb{N}$, $\overline{n+1}$ is defined to be $S(\overline{n})$ (where S is the successor symbol from PA).*

For example, the numeral $\overline{3}$ is the term $S(S(S(0)))$.

Definition 4. *If ϕ is an EA-formula (with free variables $x_1, \ldots, x_k$), and if s is an assignment mapping variables to natural numbers, then we define ϕ^s to be the sentence*

$$\phi(x_1 | \overline{s(x_1)})(x_2 | \overline{s(x_2)}) \cdots (x_k | \overline{s(x_k)})$$

obtained by substituting for each free variable x_i the numeral $\overline{s(x_i)}$ for x_i's value according to s.

Example 2. Suppose $s(x) = 0$, $s(y) = 1$, and $s(z) = 3$. Then

$$((z > y + x) \land \forall x(K(z > y + x - x)))^s$$

is defined to be

$$((\overline{3} > \overline{1} + \overline{0}) \land \forall x(K(\overline{3} > \overline{1} + x - x)))$$

(note that the numeral is not substituted for the later occurrences of x because these are bound by the $\forall x$ quantifier).

Definition 5. *If ϕ is any $\mathscr{L}$-formula, and s is any assignment mapping variables to $\mathbb{N}$, we say that X knows ϕ (with variables interpreted by s) if and only if X knows ϕ^s according to Definition 1.*

Armed with Definition 5, the Tarskian notion [14] of truth can be extended to EA.

Example 3. Assume an AGI X is clear from context. Suppose ϕ is an EA-formula, of one free variable x, which expresses "the xth Turing machine eventually halts". Suppose we want to assign a truth value to the formula

$$\exists x(\neg K(\phi) \land \neg K(\neg \phi)).$$

We proceed as follows.

- Following Tarski, we should declare $\exists x(\neg K(\phi) \land \neg K(\neg \phi))$ is true if and only if for every assignment s mapping variables to $\mathbb{N}$, $\exists x(\neg K(\phi) \land \neg K(\neg \phi))$ is true (with variables interpreted by s).
- By the semantics of $\exists$, the above is true if and only if for every assignment s, there is some $n \in \mathbb{N}$ such that $\neg K(\phi) \land \neg K(\neg \phi)$ is true (with variables interpreted by $s(x|n)$), where $s(x|n)$ is the assignment that agrees with s except for mapping x to n.
- By Definition 5, this is the case if and only if for every assignment s there is some $n \in \mathbb{N}$ such that X does not know $\phi^{s(x|n)}$ (according to Definition 1) and X does not know $\neg \phi^{s(x|n)}$ (according to Definition 1).
- By Definition 4 and the fact that x is the only free variable in ϕ, the above is the case if and only if there is some $n \in \mathbb{N}$ such that X does not know $\phi(x|\overline{n})$ (according to Definition 1) and X does not know $\neg \phi(x|\overline{n})$ (according to Definition 1).

So ultimately, we consider $\exists x(\neg K(\phi) \land \neg K(\neg \phi))$ to be true if and only if there is some $n \in \mathbb{N}$ such that, in response to the command "Enumerate all the EA-sentences which you know", X would not include $\phi(x|\overline{n})$ nor $\neg \phi(x|\overline{n})$ in the resulting enumeration.

5   Translating Knowledge Formulas

In this section, we will look at some formulas about knowledge and translate them into statements about AGI stimulus-response, using Definitions 1 and 5. First, we will start by translating some simple axioms of knowledge, to give the reader a feel for how this translation works. Then, we will advance to the examples we mentioned in the Introduction, and closely related examples.

Although the statements in the following example may seem plausible, our purpose is not to claim that every AGI must satisfy them. Rather, they serve to classify AGIs: for each axiom schema, one can speak of AGIs who satisfy that axiom schema, and of AGIs who do not satisfy it.

Example 4. (Basic axioms of knowledge) The following axiom schemas, in the language of EA, are taken from Carlson [10] (we restrict them to sentences for purposes of simplicity).

- (E1) $K(\phi)$ whenever ϕ is valid (i.e., true in every model). Translated for an AGI X using Definition 1, this becomes: "If commanded to enumerate his knowledge in EA, X will include all valid EA-sentences in the resulting list."
- (E2) $K(\phi \to \psi) \to K(\phi) \to K(\psi)$. This becomes: "If commanded to enumerate his knowledge in EA, if X would include $\phi \to \psi$ and if X would also include ϕ, then X would also include ψ."
- (E3) $K(\phi) \to \phi$. This becomes: "If commanded to enumerate his knowledge in EA, the resulting statements X enumerates would be true."
- (E4) $K(\phi) \to K(K(\phi))$. This becomes: "If commanded to enumerate his knowledge in EA, if X would list ϕ, then X would also list $K(\phi)$."

Our purpose in Example 4 is not to declare that an AGI must satisfy E1–E4. Rather, our goal is to translate these modal logical axioms into AGI language— note that the translations in quotation marks in Example 4 do not directly depend on the AGI's knowledge, but only on the AGI's stimulus-response. When studying AGI in broadest generality, even E3, the factivity of knowledge, might be questioned (certain AGIs might satisfy it and other AGIs might not). By translating E3 into a concrete statement about the AGI's stimulus-response, we can talk about "AGIs who satisfy E3" or "AGIs who fail E3," without getting stuck on hard questions like "What does it mean to know something?"

Example 5. (Reinhardt's strong mechanistic thesis [10,19,20]) Reinhardt suggested the EA-schema

$$\exists e \forall x (K(\phi) \leftrightarrow x \in W_e)$$

as a formalization of the mechanicalness of the knower. Here, W_e is the eth computably enumerable set of natural numbers[11] (W_e can also be thought of as the set of naturals enumerated by the eth Turing machine). For simplicity, consider the case where x is the lone free variable of ϕ. Then in terms of

---

[11] It can be shown that W_e is definable in the language of Peano arithmetic, therefore we can use expressions like "$x \in W_e$" in EA-formulas as shorthand.

Definition 5, the schema becomes: "If X were commanded to enumerate his knowledge in the language of EA, then the set of $n \in \mathbb{N}$ such that X would include $\phi(x|\overline{n})$ in the resulting list, would be computably enumerable." If Φ is the universal closure[12] of the above EA-schema, then the schema $K(\Phi)$ is *Reinhardt's strong mechanistic thesis*. Reinhardt conjectured that his strong mechanistic thesis is consistent with basic axioms about knowledge (i.e., that it is possible for a knowing machine to know that it is a machine). This conjecture was proved by Carlson [10] using sophisticated structural results about the ordinals [9]. See [4] for an elementary proof of a weaker version of the conjecture.

Example 6. (Reinhardt's absolute version of Gödel's incompleteness theorem) If we vary the formula from Example 5 by requiring that the knower know the value of e, we obtain:

$$\exists e K(\forall x (K(\phi) \leftrightarrow x \in W_e)).$$

Carlson [10] glosses this schema in English as: "I am a Turing machine, and I know which one." Reinhardt showed that this schema is *not* consistent with basic axioms about knowledge. Following Carlson's gloss, this shows that it is impossible for a suitably idealized AGI to know its own code[13].

Remark 2. As far as I know, AGI has not yet received much attention in the mathematical logical literature. Instead, mathematical logicians tend to concern themselves with *knowing agents* or *knowing machines*. Presumably, every suitably idealized AGI is a knowing agent and a knowing machine, but certainly not every knowing agent (or knowing machine) is an AGI. Thus, in general, inconsistency results about knowing agents or knowing machines carry directly over to AGIs (if no knowing agent, or no knowing machine, can satisfy some property, then in particular no suitably idealized AGI can either). Consistency results do not generally carry over to AGIs (it may be possible for a knowing agent or a knowing machine to satisfy some property, but it might be that none of the knowing agents or knowing machines which satisfy that property are AGIs). Nevertheless, a consistency result about knowing agents or knowing machines should at least count as evidence in favor of the corresponding consistency result for AGIs, at least if there is no clear reason otherwise. In the examples above:

- Reinhardt's strong mechanistic thesis (Example 5) was proven to be consistent with basic knowledge axioms, so it is possible for a knowing machine to know that it is a machine (without necessarily knowing which machine). Since not every knowing machine is an AGI, it might still be impossible for an AGI to know it is a machine. But the consistency of Reinhardt's strong mechanistic thesis at least suggests evidence that an AGI can know it is a machine.

---

[12] A *universal closure* of a formula ϕ is a sentence $\forall x_1 \cdots \forall x_k \phi$, and the *universal closure* of a schema of formulas is the schema of universal closures of those formulas.

[13] We have pointed out elsewhere [3] that (i) Reinhardt implicitly assumes that the knower knows its own truthfulness; and (ii) it is possible for a knowing machine to know its own code if it is allowed to be ignorant of its own truthfulness, despite still being truthful. See [1] and [2] for some additional discussion.

– Reinhardt's absolute version of the incompleteness theorem (Example 6) is an inconsistency result. As such, it transfers over directly to AGI, proving that no suitably idealized AGI can know its own code[14].

Example 7. (Intuitive Ordinal Intelligence) In [5] we defined an intelligence measure for idealized mechanical knowing agents (who are aware of the computable ordinals) as follows. If A is such a knowing agent, we define the intelligence of A to be the supremum of the set of ordinals α such that α has some code c such that A knows that c is a code of a computable ordinal. In [6] we specialized this to AGIs, and called it *Intuitive Ordinal Intelligence*. Let $\mathscr{L}$ be a language like EA but including an additional predicate symbol O for the set of codes of computable ordinals. Modifying Definition 1 accordingly, we can systematically perform said specialization to AGIs, and it becomes: "The Intuitive Ordinal Intelligence of an AGI X is the supremum of the set of ordinals α such that α has some code c such that X would include $O(\bar{c})$ in the resulting enumeration if we asked X to enumerate all the $\mathscr{L}$-sentences that he knows."

6   Conclusion

What does it mean to know something? This is a difficult question and there probably is no one true answer. In the field of AGI, how can we systematically investigate the theoretical properties of knowledge, when different AGIs might not even agree about what knowledge really means? So motivated, we have proposed an elegant way to brush these philosophical questions aside. In Definition 1, we declare that an AGI knows an EA-sentence if and only if that AGI would enumerate that sentence if commanded:

"Enumerate all the EA-sentences which you know"

(this definition might look circular at first glance but we have argued that it is not; see Subsect. 3.1). In Definition 5 we extended this to formulas with free variables, not just sentences.

This universal knowledge definition sets the study of AGI knowledge on a firmer theoretical footing. In Sect. 5 we give examples of how our definition can serve as a bridge to translate knowledge-related formulas from mathematical logic into the realm of AGI.

Acknowledgments. We gratefully acknowledge Alessandro Aldini, Phil Maguire, Brendon Miller-Boldt, Philippe Moser, and the reviewers for comments and feedback.

References

1. Aldini, A., Fano, V., Graziani, P.: Do the self-knowing machines dream of knowing their factivity? In: AIC, pp. 125–132 (2015)

---

[14] Or rather, its own code and its own truthfulness—we have pointed out [3] that Reinhardt implicitly assumes the knower knows its own truthfulness.

2. Aldini, A., Fano, V., Graziani, P.: Theory of knowing machines: revisiting Gödel and the mechanistic thesis. In: Gadducci, F., Tavosanis, M. (eds.) HaPoC 2015. IAICT, vol. 487, pp. 57–70. Springer, Cham (2016). https://doi.org/10.1007/978-3-319-47286-7_4

3. Alexander, S.A.: A machine that knows its own code. Stud. Logica. **102**(3), 567–576 (2014)

4. Alexander, S.A.: Fast-collapsing theories. Stud. Logica. **103**(1), 53–73 (2015)

5. Alexander, S.A.: Measuring the intelligence of an idealized mechanical knowing agent. In: CIFMA (2019)

6. Alexander, S.A.: AGI and the Knight-Darwin law: why idealized AGI reproduction requires collaboration. In: Goertzel, B., Panov, A.I., Potapov, A., Yampolskiy, R. (eds.) AGI 2020. LNCS (LNAI), vol. 12177, pp. 1–11. Springer, Cham (2020). https://doi.org/10.1007/978-3-030-52152-3_1

7. Aliman, N.-M., et al.: Error-correction for AI safety. In: Goertzel, B., Panov, A.I., Potapov, A., Yampolskiy, R. (eds.) AGI 2020. LNCS (LNAI), vol. 12177, pp. 12–22. Springer, Cham (2020). https://doi.org/10.1007/978-3-030-52152-3_2

8. Aristotle: Posterior analytics. In: Barnes, J., et al. (eds.) The Complete Works of Aristotle. Princeton University Press (1984)

9. Carlson, T.J.: Ordinal arithmetic and Σ_1-elementarity. Arch. Math. Logic **38**(7), 449–460 (1999)

10. Carlson, T.J.: Knowledge, machines, and the consistency of Reinhardt's strong mechanistic thesis. Ann. Pure Appl. Logic **105**(1–3), 51–82 (2000)

11. Cerone, A., Fazli, S., Malone, K., Pietarinen, A.V.: Interdisciplinary aspects of cognition. In: CIFMA (2019)

12. Elton, D.: Self-explaining AI as an alternative to interpretable AI. In: International Conference on Artificial General Intelligence (2020)

13. Goertzel, B.: Artificial general intelligence: concept, state of the art, and future prospects. J. Artif. General Intell. **5**, 1–48 (2014)

14. Hodges, W.: Tarski's truth definitions. In: Zalta, E.N. (ed.) The Stanford Encyclopedia of Philosophy. Metaphysics Research Lab, Stanford University, fall 2018 edn. (2018)

15. Ichikawa, J.J., Steup, M.: The analysis of knowledge. In: Zalta, E.N. (ed.) The Stanford Encyclopedia of Philosophy. Metaphysics Research Lab, Stanford University, summer 2018 edn. (2018)

16. Knuth, D.E.: Two notes on notation. Am. Math. Monthly **99**(5), 403–422 (1992)

17. Maguire, P., Moser, P., Maguire, R.: Are people smarter than machines? Croatian J. Philosop. **20**(1), 103–123 (2020)

18. Plato: Theaetetus. In: Cooper, J.M., Hutchinson, D.S., et al. (eds.) Plato: complete works. Hackett Publishing (1997)

19. Reinhardt, W.N.: Absolute versions of incompleteness theorems. Nous **19**, 317–346 (1985)

20. Reinhardt, W.N.: Epistemic theories and the interpretation of Gödel's incompleteness theorems. J. Philosoph. Logic **15**(4), 427–474 (1986)

21. Shapiro, S.: Epistemic and intuitionistic arithmetic. In: Studies in Logic and the Foundations of Mathematics, vol. 113, pp. 11–46. Elsevier (1985)

22. Wellmer, A.: Skepticism in interpretation. In: Conant, J.F., Kern, A. (eds.) Varieties of Skepticism: Essays after Kant, Wittgenstein, and Cavell. Walter de Gruyter (2014)

23. Williamson, T.: Knowledge, context, and the agent's point of view. In: Preyer, G., Peter, G. (eds.) Contextualism in Philosophy: Knowledge, Meaning, and Truth, pp. 91–114. Oxford University Press (2005)

24. Yampolskiy, R.: On controllability of artificial intelligence. Technical report (2020)

Reasoning About Ignorance and Beliefs

Alessandro Aldini[✉], Pierluigi Graziani, and Mirko Tagliaferri

University of Urbino Carlo Bo, Urbino, Italy
`alessandro.aldini@uniurb.it`

Abstract. When building artificial agents that have to make decisions, understanding what follows from what they know or believe is mandatory, but it is also important to understand what happens when those agents ignore some facts, where ignoring a fact is interpreted to stand for not knowing/not being aware of something. This becomes especially relevant when such agents ignore their ignorance, since this hinders their ability of seeking the information they are missing. Given this fact, it might prove useful to clarify in which circumstances ignorance is present and what might cause an agent to ignore that he/she is ignoring. This paper is an attempt at exploring those facts. In the paper, the relationship between ignorance and beliefs is analysed. In particular, three doxastic effects are discussed, showing that they can be seen as a cause of ignorance. The effects are formalized in a bi-modal formal language for knowledge and belief and it is shown how ignorance follows directly from those effects. Moreover, it is shown that negative introspection is the culprit of the passage between simply ignoring a fact and ignoring someone's ignorance about that fact. Those results could prove useful when artificial agents are designed, since modellers would be aware of which conditions are mandatory to avoid deep forms of ignorance; this means that those artificial agents would be able to infer which information they are ignoring and they could employ this fact to seek it and fill the gaps in their knowledge/belief base.

Keywords: Ignorance · Beliefs · Modal logics

1 Introduction

A sub-field of artificial intelligence is the one that concentrates on building *expert systems*. An expert system (ES) is a computer system that tries to emulate the decision-making abilities of human beings. ESs often rely on knowledge bases, which can be employed by the systems to infer new information and, thus, allow for better decisions [23]. In this respect, modal logic provides an invaluable contribution to the modelling of such systems, since formal systems of epistemic logic can satisfactorily represent knowledge and inferences based on knowledge [3,17]. Those modal languages are even more impactful when finer grained cognitive phenomena are taken into consideration. For instance, BDI (Belief-Desire-Intention) intelligent systems can decide which plans are better to perform and

© Springer Nature Switzerland AG 2021
L. Cleophas and M. Massink (Eds.): SEFM 2020 Workshops, LNCS 12524, pp. 214–230, 2021.
https://doi.org/10.1007/978-3-030-67220-1_17

then perform them, thus increasing their efficiency compared to expert systems that rely only on knowledge bases [8,16,20,24,25]. This shall not come as a surprise; the more aspects of human cognitive infrastructures systems can emulate, the more the actions and plans of those systems will resemble those of humans.[1] While contemporary systems have become extremely good at emulating positive cognitive elements of human decision-making, one thing which is neglected is the impact of ignorance.[2] Seldom systems make explicit reference to ignorance, despite the abundance of formal work on the notion[7,9,12,13,19]. This paper is an attempt at showing that even when ignorance is not modelled directly into intelligent systems, if those systems have representations of beliefs and certain doxastic effects take place, then, it can be claimed that the systems are ignoring. In order to achieve this goal, a thorough investigation of ignorance within a logical framework is provided. Having a clear idea of how ignorance might be modelled using modal logic and which are the relationships with other cognitive phenomena such as knowledge and belief can greatly enhance the deductive powers of intelligent systems employing those improved formal languages to make their inferences. Moreover, once the formal framework is clear, it is possible to reason about higher-orders of ignorance, to allow intelligent systems to understand which dangerous cognitive stance they should avoid in order to not fall within the black-hole of ignoring to ignore. All of this will be obtained assuming a straightforward definition of ignorance, that can be assumed to be present even when not explicitly modelled into intelligent systems. The aim of the paper is thus to show how doxastic[3] phenomena relate to ignorance and, furthermore, to show what principles must be implemented into an intelligent system to avoid higher-order instances of ignorance.

The structure of the paper will be the following: in Sect. 2, an introduction to the logic of ignorance and beliefs is provided. In Sect. 3, the relation between believing and ignoring is explored and explained, providing insights and novel results into possible ways ignorance can emerge. In Sect. 4, previously proven results [7] on the relationship between higher-orders of ignorance are discussed.[4] In Sect. 5, it will be shown how the lack of negative introspection can be seen as a major cause of second-order ignorance (i.e., ignorance of ignorance), providing novel insights into the relationship between first-order ignorance and second-order ignorance. Finally, concluding remarks and possible ventures for the future will follow.

2   Logic for Ignorance and Beliefs

The origin of the formal discussion on ignorance can be attributed to Jaakko Hintikka's seminal work *Knowledge and Belief: An Introduction to the Logic of*

---

[1] See [1,4,18] for some recent and interesting applications of BDI systems.

[2] It is necessary to clarify that the term *ignorance* employed in this paper is given a specific meaning, i.e., to *not know/not be aware of* something.

[3] In this paper, the term *doxastic* will always refer to the act of believing.

[4] In the paper, ignorance will always be indicated with a specific order, which indicates the depth of the ignoring phenomenon. First-order ignorance means that a given fact is ignored; second-order ignorance means that it is ignored that a given fact is ignored, and so forth.

the Two Notions [11]. In his book, Hintikka provides a propositional axiomatization of the two notions of knowledge and belief, providing insights also on other cognitive notions such as that of ignorance. This work is important because it is the first attempt to try to axiomatize ignorance employing an axiomatization of knowledge as a starting point. Following this path, Hintikka gave birth to the classical approach of formally defining ignorance in terms of lack of knowledge. Specifically, for Hintikka, ignoring a specific proposition ϕ is equivalent to not knowing whether ϕ is true or false (formally $I(\phi) =_{def} \neg K(\phi) \wedge \neg K(\neg\phi)$). This definition of ignorance, while natural, might be considered stronger than the common notion of ignorance interpreted as not knowing, since the phrase "ignoring ϕ" might simply stand for "not knowing that ϕ" (formally $I(\phi) =_{def} \neg K(\phi)$). While this might be true, when dealing with artificial agents, the cognitive stance of those agents can be interpreted as the stronger one, therefore it makes sense to follow the classical approach and employ Hintikka's original definition.

2.1 Defining the Formal Framework

In this paper, the formal definition of ignorance of Hintikka [11] will be assumed. Moreover, the two notions of knowledge and belief that will be discussed are going to be interpreted in the language of propositional modal logic, using, as a semantic basis, Kripke structures. In particular, a bimodal language $\mathcal{L}$ will be employed. For brevity purposes, the syntactic definition of the language will be provided, but the semantics will not be presented.[5]

Definition 1 (Logical Language for Knowledge and Belief). *Given a countable set At of primitive propositions $p_1, \ldots, p_n$ the bimodal logical language $\mathcal{L}$ is defined by the set of all formulas obtained through the following grammar:*

$$\phi := p_i \mid \neg\phi \mid \phi \wedge \phi \mid K(\phi) \mid B(\phi) \text{ with } p_i \in At$$

All the other Boolean connectives are defined in the standard way. The main modalities of the language will be K and B, where $K(\phi)$ should be read as "ϕ is known" and will be called *knowledge formula*; $B(\phi)$ should be read as "ϕ is believed" and will be called *belief formula*. Finally, ignorance (I) is defined in terms of knowledge in the following way: $I(\phi) =_{def} \neg K(\phi) \wedge \neg K(\neg\phi)$, where $I(\phi)$ will be called an *ignorance formula*.[6]

In the language $\mathcal{L}$ different readings for the ignorance formulas will be employed. It will be said that a formula ϕ is *first-order ignored*, whenever there is only one ignorance operator applied to it, the simplest case being $I(\phi)$. It will be said that a formula ϕ is *second-order ignored*, whenever there are at least, and no more than, two nested ignorance operators applied to ϕ. Again, the simplest case is $I(I(\phi))$. Higher-order instances of ignorance follow a similar path.

---

[5] The interested reader is referred to [17] for a standard presentation of Kripke structures.

[6] Note that an ignorance formula could represent instances of ignorance of any order, depending on how many occurrences of ignorance operators appear in the formula ϕ.

Some particular properties of the two notions of knowledge and beliefs will be assumed. For the notion of knowledge, it will be assumed that knowledge is *factual* (*T*) and *positively introspective* (4). The factuality (or truthfulness) of knowledge is pretty straightforward: this comes mainly from philosophical reflections on the notion of knowledge, which is taken to be a rigorous cognitive phenomenon strongly tied with truth, i.e., only true things might be known. In fact, the strength of this axiom is what distinguishes proper knowledge from simple beliefs. Beliefs might be false, but knowledge never is. The positive introspective axiom might be slightly more problematic and it comes from the assumption that agents have a privileged access to their cognitive states. This might not always make sense for human beings, who can often forget what they know and thus, are unable to keep track of everything they know, but it is a reasonable assumption for expert and intelligent systems, which always explicitly compute what they know and thus have records of all the things they know, without major issues on the memory side of things. For the notion of belief, it will be assumed that *beliefs are consistent* (*D*). Consistency of beliefs means that someone cannot believe that a fact is both true and false at the same time. As with positive introspection, this assumption might not always be valid for ordinary human beings, especially the irrational ones; however, since expert and intelligent systems should resemble the behaviour of rational agents, having such a consistency imposition is mandatory.[7] Finally, it will be assumed that the two notions interact in the following way: knowledge will always imply belief (Int_1) and whenever something is believed, it is known that it is believed (Int_2). The first interaction axiom is commonly derived directly from the analysis of knowledge given in Plato's Theatetus [15]: in such an analysis, knowledge is taken to be justified true belief. Unfortunately, the justification component is often neglected in formal languages, even though some attempts have been made to insert it;[8] the truth component is formalized through axiom *T*, while the belief component is given exactly by the interaction axiom Int_1. Int_2, on the other hand, is justified using arguments similar to the ones employed for positive introspection. In fact, it is assumed that agents not only have privileged access to their knowledge, but also to their beliefs. Again, this makes perfect sense when computational systems are involved, since they often can keep a record of what they know and/or believe. Formally, all those properties are axiomatized through the following formulas:

Definition 2 (Properties of Knowledge and Belief). *The following formulas are assumed to be valid in* $\mathcal{L}$ *(in symbols* $\vdash_{\mathcal{L}}$*) for knowledge:*

- ***K**:* $\vdash_{\mathcal{L}} K(\phi \to \psi) \to (K(\phi) \to K(\psi))$.
- ***T**:* $\vdash_{\mathcal{L}} K(\phi) \to \phi$.
- *4:* $\vdash_{\mathcal{L}} K(\phi) \to K(K(\phi))$.

The following formulas are assumed to be valid in $\mathcal{L}$ *for belief:*

---

[7] Note that, given the semantic framework employed to interpret the two notions, those notions also distribute over implications.

[8] See, e.g., [2].

- **B:** $\vdash_{\mathcal{L}} B(\phi \to \psi) \to (B(\phi) \to B(\psi))$.
- **D:** $\vdash_{\mathcal{L}} \neg(B(\phi) \land B(\neg\phi))$.

The following formulas are assumed to be valid in $\mathcal{L}$ for the interaction between knowledge and beliefs:

- **Int$_1$:** $\vdash_{\mathcal{L}} K(\phi) \to B(\phi)$.
- **Int$_2$:** $\vdash_{\mathcal{L}} B(\phi) \to K(B(\phi))$.

A further axiom which will be employed in later sections of the paper, but will not be assumed in the language is the axiom of *negative introspection*, often known as axiom 5 of epistemic logic. Negative introspection is similar in spirit to positive introspection: both axioms attribute to the agents a form of transparency towards their cognition. As was said above, positive introspection allows an agent to know everything he/she knows; on the other hand, negative introspection says that an agent will always know what he/she does not know, i.e., $\neg K(\phi) \to K(\neg K(\phi))$. This axiom, while often assumed in epistemic languages employed in computer science [10], might be too demanding for artificial agents, since it would imply that those agents are aware of all the facts they do not know. Making the reasonable assumption that there are an infinite amount of unknown facts, this would mean that the artificial agent has an infinite memory to stock all those facts that it knows not to know. It will be shown later that negative introspection alone is sufficient to prevent the occurrence of higher-order instances of ignorance. Moreover, it will be shown that when negative introspection is assumed missing, then first-order ignorance and second-order ignorance are tightly tied together.

Note that in the proofs that are given in this paper, various inference rules will be employed. All those rules are standard rules of modal logic. Since indicating all the rules employed would occupy way too much space, the reader is invited to check [14] and [5] for references on all the rules that will be employed in this paper.[9]

Now that all the formal details have been given, it is possible to move on to the reflections concerning the interplay between beliefs and ignorance.

3 Misbelieving, Being Agnostic or Doubting

Understanding the origin of ignorance is quite complicated. Sometimes, it is easy to recognize if someone is ignorant about something, but it is not clear what brought about and fed this ignorance. The main issue is that ignorance is a *negative fact*, i.e., it is a lack of knowledge, and, therefore, there is no specific

---

[9] The abbreviations that will be employed in the proofs of this paper will all be reported here. *Ass.* will stand for "assumption"; *P. Taut.* will stand for "propositional tautology"; *Elim.* will stand for "elimination rule"; *Intr.* will stand for "introduction rule"; *Contrap.* will stand for "contrapposition"; *MP* will stand for "Modus Ponens"; *DM* will stand for "DeMorgan rules"; *DS* will stand for "disjunctive syllogism"; *Nec.* will stand for "necessitation rule"; finally *Distr.* will stand for "distributivity rule".

moment in time when ignorance is generated; it is there the whole time, until it disappears. Simply put, ignorance is not something that can be gained, but only lost. Not having a specific moment in time during which ignorance originates makes it difficult for researchers to focus on specific acts or behaviours that can aid their understanding of the phenomenon. For this reason, a formal research on the notion of ignorance might help to understand what are the constituents of such notion and thus which other phenomena are responsible for its emergence and/or existence. Once the formal links between doxastic effects and ignorance are understood and recognized, modellers will be able to design artificial agents that are better suited to deal with ignorance and the effects ignorance has in planning and pursuing a specific goal. Specifically, three different, alternative doxastic effects will be explored, showing that those individually imply ignorance and, conversely, they are implied by ignorance, making them equivalent to ignorance. The first of those states will be called *the misbelieving effect*, the second will be called *the agnostic effect*, and, finally, the third one will be called *the doubting effect*.

Intuitively, we say that an agent is subject to the *misbelieving effect* either when the agent believes that a given fact is true, while it is false, or when the agent believes that a given fact is false, while it is true.[10]

Definition 3 (Misbelieving Effect). *The misbelieving effect is represented by the following formula:*

$$(B(\phi) \wedge \neg\phi) \vee (B(\neg\phi) \wedge \phi) \tag{1}$$

The misbelieving effect is quite common. Everybody, even the most conscientious human being, will have some misbeliefs about the world that surrounds him/her. Science is full of examples: researchers constantly discover new facts that contradict what was previously thought to be true, thus highlighting many misbeliefs that were held by those scientists. Per se, the fact that this effect is so extensively spread does not cause many problems, since misbelieving, when taken in isolation, only implies ignorance and it is plausible that most scientists will admit to be ignorant about many things. However, if the misbelieving agent is not open to revise his/her beliefs, the misbelieving effect might cause dangerous issues, since both first-order ignorance and higher-order instances of ignorance will be produced.

Intuitively, we say that an agent is subject to the *agnostic effect* when the agent neither believes that a given fact is true nor believes that the fact is false.

Definition 4 (Agnostic Effect). *The agnostic effect is represented by the following formula:*
$$\neg B(\phi) \wedge \neg B(\neg\phi) \tag{2}$$

Again, also the agnostic effect is quite common. People not having an opinion about a specific matter are the prime candidates of agents which are subject to

---

[10] See [6] for a discussion about different aspects that relate misbelieving and ignoring.

this effect. Since they do not have an opinion about a given fact, they simply do not believe neither in the truth of the fact nor in its falsity. Note that this does not mean that they do not believe that the fact is either true or false (which is indeed a tautology and must be believed due to the necessitation rule and Int_1), but they cannot make up their mind in one direction or the other and, thus, suspend their judgement. It is not surprising that the agnostic effect causes ignorance, since the lack of beliefs is just the first step to the lack of knowledge. Again, this is not a problem if the agnostic effect is due to a suspension of judgement about the truth of a specific fact, since this is just a clear acknowledgement that such fact is ignored. The problem begins when the agnostic effect is coupled with unawareness of the possibility of believing that the fact is either true or false. As with misbelieving, also in this latter case, being agnostic does not just cause first-order ignorance, but also higher-order instances of ignorance.

Intuitively, we say that an agent is subject to the *doubting effect* when the agent believes in something which in fact holds, but he/she does not have the guarantee that such fact actually holds.

Definition 5 (Doubting Effect). *The doubting effect is represented by the following formula:*

$$(B(\phi) \wedge \phi \wedge \neg K(\phi)) \vee (B(\neg\phi) \wedge \neg\phi \wedge \neg K(\neg\phi)) \tag{3}$$

The doubting effect is similar in spirit to the misbelieving effect. Since in both cases agents do not have access to the state of the world, from a first-person perspective, it is impossible, for the agent, to recognize whether he/she is misbelieving or is simply doubtful. The main difference between the two cases is that, in the doubting effect, the lack of knowledge of the agent is explicitly specified. This specification is fundamental, since it is the main culprit of the emergence of ignorance. This does not seem to be a great surprise, since, per se, believing something that actually holds should not cause problems.

Even though it seems quite reasonable that the misbelieving effect, the agnostic effect and the doubting effect imply first-order ignorance, such facts must be proven. The existence of these proofs in standard formal systems can be given both a normative and a descriptive reading. On the normative side, they show that the intuitions about misbelieving, being agnostic and doubting are indeed well-guided and, thus, strengthen the relation between beliefs and knowledge; on the descriptive side, if it is assumed that the intuitions are justified, the proofs provided here show that classical epistemic and doxastic formal systems are well-structured and manage to properly describe real world phenomena. The proofs provided will show that the three effects presented above are individually sufficient for ignorance. Subsequently, it will also be shown that ignorance will always imply at least one of the three effects.

3.1   From Misbelieving to Ignoring

Proposition 1. $\vdash_{\mathcal{L}} ((B(\phi) \wedge \neg\phi) \vee (B(\neg\phi) \wedge \phi)) \rightarrow I(\phi).$

Proposition 1 says that, in $\mathcal{L}$, misbelieving and ignorance are tied together. Interestingly, this connection holds also in weaker languages, since axiom 4 and axiom B of $\mathcal{L}$ are not needed in the proof of the proposition, see appendix A.

3.2   From Being Agnostic to Ignoring

Proposition 2. $\vdash_{\mathcal{L}} (\neg B(\phi) \wedge \neg B(\neg\phi)) \rightarrow I(\phi).$

Proposition 2 says that, in $\mathcal{L}$, being agnostic and ignorance are tied together. Again, this result holds also in weaker languages (in fact, systems even weaker than the ones that satisfy Proposition 1), since only the interaction axiom Int_1 is needed to obtain the proof, see appendix A.

3.3   From Doubting to Ignoring

Proposition 3. $\vdash_{\mathcal{L}} ((B(\phi) \wedge \phi \wedge \neg K(\phi)) \vee (B(\neg\phi) \wedge \neg\phi \wedge \neg K(\neg\phi))) \rightarrow I(\phi).$

Proposition 3 shows that, in $\mathcal{L}$, doubting and ignorance are tied together. Note that this connection also holds in weaker systems, since only axiom T of $\mathcal{L}$ has been used in the proof, see appendix A.

3.4   From Ignoring to the Three Effects

The fact that ignorance must imply one among the three effects will now be proven.

Theorem 1 (From Ignorance to the three effects). *The following formula linking Ignorance and the three doxastic effects is valid in $\mathcal{L}$:*

$$\vdash_{\mathcal{L}} I(\phi) \rightarrow$$

$$
\begin{array}{ll}
B(\phi) \wedge \neg\phi & \vee \\
B(\neg\phi) \wedge \phi & \vee \\
\neg B(\phi) \wedge \neg B(\neg\phi) & \vee \\
B(\phi) \wedge \phi \wedge \neg K(\phi) & \vee \\
B(\neg\phi) \wedge \neg\phi \wedge \neg K(\neg\phi) &
\end{array}
$$

Proof (Theorem 1).
The proof will be given by contradiction. Assume:

$$
\begin{array}{ll}
\neg K(\phi) \wedge \neg K(\neg\phi) & \wedge \\
\neg((B(\phi) \wedge \neg\phi) \vee (B(\neg\phi) \wedge \phi)) & \wedge \\
\neg(\neg B(\phi) \wedge \neg B(\neg\phi)) & \wedge \\
\neg((B(\phi) \wedge \phi \wedge \neg K(\phi)) \vee (B(\neg\phi) \wedge \neg\phi \wedge \neg K(\neg\phi)))
\end{array}
$$

The previous formula can be transformed into *Conjunctive Normal Form* (CNF), i.e., a series of clauses connected by ∧s where each clause only contains ∨s. The first step to do so is to apply DeMorgan to the three clauses (double negations will also be eliminated directly):

$$\neg K(\phi) \wedge \neg K(\neg\phi) \qquad\qquad\qquad\qquad\qquad \wedge$$
$$\neg(B(\phi) \wedge \neg\phi) \wedge \neg(B(\neg\phi) \wedge \phi) \qquad\qquad \wedge$$
$$B(\phi) \vee B(\neg\phi) \qquad\qquad\qquad\qquad\qquad\quad \wedge$$
$$\neg(B(\phi) \wedge \phi \wedge \neg K(\phi)) \wedge \neg(B(\neg\phi) \wedge \neg\phi \wedge \neg K(\neg\phi))$$

A second iteration of DeMorgan is possible (again, double negations will be eliminated directly):

$$\neg K(\phi) \wedge \neg K(\neg\phi) \qquad\qquad\qquad\qquad\qquad \wedge$$
$$(\neg B(\phi) \vee \phi) \wedge (\neg B(\neg\phi) \vee \neg\phi) \qquad\qquad \wedge$$
$$(B(\phi) \vee B(\neg\phi)) \qquad\qquad\qquad\qquad\qquad \wedge$$
$$(\neg B(\phi) \vee \neg\phi \vee K(\phi)) \wedge (\neg B(\neg\phi) \vee \phi \vee K(\neg\phi))$$

Now, a row will be given to each clause of the above CNF formula:

$$
\begin{array}{lll}
\text{(a)} & \neg K(\phi) & \wedge \\
\text{(b)} & \neg K(\neg\phi) & \wedge \\
\text{(c)} & \neg B(\phi) \vee \phi & \wedge \\
\text{(d)} & \neg B(\neg\phi) \vee \neg\phi & \wedge \\
\text{(e)} & B(\phi) \vee B(\neg\phi) & \wedge \\
\text{(f)} & \neg B(\phi) \vee \neg\phi \vee K(\phi) & \wedge \\
\text{(g)} & \neg B(\neg\phi) \vee \phi \vee K(\neg\phi) &
\end{array}
$$

Taking the list above as a reference, it is possible to prove that the set of formulas (a)-(g) leads to a contradiction.

Note first that clause (e) produces two separate cases, i.e., either $B(\phi)$ holds or $B(\neg\phi)$ holds. It will be shown that both cases lead to a contradiction.

$$
\begin{array}{lll}
\text{Case 1:} & & \\
(1) & B(\phi) & \text{Ass.} \\
(2) & \neg B(\phi) \vee \phi & \text{Clause (c)} \\
(3) & \phi & DS\ (1)\text{-}(2) \\
(4) & \neg K(\phi) & \text{Clause (a)} \\
(5) & B(\phi) \wedge \phi \wedge \neg K(\phi) & \wedge\ \text{Intr. (1)-(3)-(4)} \\
(6) & \neg(\neg B(\phi) \vee \neg\phi \vee K(\phi)) & DM\ (5) \\
(7) & \neg B(\phi) \vee \neg\phi \vee K(\phi) & \text{Clause (f)} \\
(8) & \text{Contradiction} & (6)\text{-}(7) \\
\text{Case 2:} & & \\
(1) & B(\neg\phi) & \text{Ass.} \\
(2) & \neg B(\neg\phi) \vee \neg\phi & \text{Clause (d)} \\
(3) & \neg\phi & DS\ (1)\text{-}(2)
\end{array}
$$

$$
\begin{array}{lll}
(4) & \neg K(\neg\phi) & \text{Clause (b)} \\
(5) & B(\neg\phi) \wedge \neg\phi \wedge \neg K(\neg\phi) & \wedge \text{ Intr. (1)-(3)-(4)} \\
(6) & \neg(\neg B(\neg\phi) \vee \phi \vee K(\neg\phi)) \; DM \; (5) \\
(7) & \neg B(\neg\phi) \vee \phi \vee K(\neg\phi) & \text{Clause (g)} \\
(8) & \text{Contradiction} & (6)\text{-}(7)
\end{array}
$$

All cases lead to a contradiction. Therefore, at least one doxastic effect must be true whenever first-order ignorance is present. □

4   Hierarchies of Ignorance

When dealing with hierarchies of ignorance, there are at least two important aspects which require analysis. The first aspect is the one that describes the relation between first-order ignorance and second-order ignorance; the second aspect is the one that describes the relation between second-order ignorance and higher-order levels of ignorance. The importance of those aspects is based on one fundamental fact: first-order ignorance is a common phenomenon of every day life; people are ignorant about many facts and information about the world they live in. Not only common people, but also scientists and curious persons fall victim to the phenomenon of ignoring. It is an indissoluble trait of all human beings. Nonetheless, first-order ignorance is not problematic on its-own; quite the opposite, first-order ignorance is what often stimulates the genuine curiosity that pushes human beings towards making new discoveries and increasing their overall knowledge. What can be considered problematic is the ignorance of ignorance (second-order ignorance), since this phenomenon precludes the possibility of dissipating first-order ignorance, given that people do not have the stimulus to understand something they are not even aware of being ignorant about. This should highlight the importance of understanding and exploring what is the relation between first-order ignorance and second-order ignorance. Once the interplay between the two phenomena is clear, it is possible to design strategies that lock the passage from the former to the latter.

The second aspect (the relation between second-order ignorance and higher-orders of ignorance) is important for similar reasons. Once it is admitted that some forms of second-order ignorance are unavoidable, it might be good to know that such second-order ignorance exists, i.e., to know that one is second-order ignorant about something. At least, such knowledge would stimulate persons to work on their ignorance, in order to avoid it.

While the first aspect is still obscure in the literature on the formal representation of ignorance and will be explored in the next section of this paper, the second aspect has been well explored by Kit Fine in his paper "Ignorance of ignorance" [7].[11] In his paper, Fine shows that second-order ignorance and

---

[11] It should be pointed out that Fine does not use the terms "first-order ignorance", "second-order ignorance" and so on. However, to maintain coherence with the rest of the paper, those terms will be employed when the concepts expressed by Fine are aligned with the meanings attributed to those terms in this paper.

higher-orders of ignorance are tightly tied together. Once second-order ignorance is present, an agent is doomed to the black hole of higher-order levels of ignorance.

Those aspects about the hierarchies of ignorance are especially important when strategies for modelling artificial agents are taken into consideration. This is due to the fact that if modellers do not pay enough attention, those artificial agents might end up falling victims of second-order ignorance and, subsequently, to higher-orders of ignorance; they would thus be unable to recognize that they are missing some information and would not look for it.

Theorem 2 (Fine's Ignorance Theorem). *Second-order ignorance implies higher-orders of ignorance. Specifically, second-order ignorance implies third-order ignorance. Third-order ignorance implies fourth-order ignorance and so forth.*

The proof of this statement is straightforward and only requires a few formal definitions and a few lemmas. First, the notion of *Rumsfeld ignorance* of ϕ is introduced. Intuitively, someone is Rumsfeld ignorant when he is first-order ignorant about ϕ and does not know it.

Definition 6 (Rumsfeld Ignorance). *Rumsfeld ignorance of a formula ϕ is represented by the formula*

$$I_R(\phi) =_{def} I(\phi) \wedge \neg K(I(\phi))$$

where $I(\phi)$ is a first-order ignorance formula.

Lemma 1 (From second-order ignorance to first-order ignorance).
$\vdash_{\mathcal{L}} I(I(\phi)) \rightarrow I(\phi).$[12]

Lemma 2 (From second-order ignorance to Rumsfeld ignorance).
$\vdash_{\mathcal{L}} I(I(\phi)) \rightarrow I_R(\phi).$

Lemma 3 (From Rumsfeld ignorance to second-order ignorance).
$\vdash_{\mathcal{L}} I_R(\phi) \rightarrow I(I(\phi)).$

Lemma 4 (Impossibility of knowing to be Rumsfeld ignorant).
$\vdash_{\mathcal{L}} \neg K(I_R(\phi)).$

A further lemma which will be useful later is the following.[13]

---

[12] Proofs of lemmas will not be provided. If the reader is interested, in [7] it is possible to find all the details concerning the lemmas which are introduced here. The only important detail is that Fine provides proofs in the axiomatic system $S4$, which is a system that defined languages weaker than the one employed in this paper. Therefore, every proof provided by Fine could be easily reproduced inside $\mathcal{L}$.

[13] Fine proves such lemma while proving his main theorem. However, to make the proof easier to read, this lemma will be given separately. The proof of such lemma can be found in appendix A.

Lemma 5 (Second-order ignorance and impossibility of knowing it).
$\vdash_{\mathcal{L}} I(I(\phi)) \rightarrow \neg K(I(I(\phi)))$.

It is now possible to prove Fine's main result about the relationship between second-order ignorance and higher-orders of ignorance.

Proof (Theorem 2).

(1)	$I(I(\phi))$	Ass.
(2)	$I(I(\phi)) \rightarrow \neg K(I(I(\phi)))$	Lemma 5.
(3)	$\neg K(I(I(\phi)))$	*MP* (1)-(2).
(4)	$K(\neg I(I(\phi))) \rightarrow \neg (I(I(\phi)))$	Axiom *T*.
(5)	$I(I(\phi)) \rightarrow \neg K(\neg (I(I(\phi))))$	Contrap. (4).
(6)	$\neg K(\neg (I(I(\phi))))$	*MP* (1)-(5).
(7)	$\neg K(I(I(\phi))) \wedge \neg K(\neg (I(I(\phi))))$	$\wedge$ Intr. (3)-(6).
(8)	$I(I(I(\phi)))$	Definition of (7).

What Theorem 2 shows is that there is a deep connection between second-order ignorance and higher-order levels of ignorance. In fact, as soon as someone is second-order ignorant, there is no possibility that he/she escapes the dark hole of ignorance on his/her own. Once this is well understood, it becomes evident why deep investigations on the relation between first-order ignorance and second-order ignorance are required. Once it is established what causes second-order ignorance in the presence of first-order ignorance, it might be possible to stop agents from crossing the *event-horizon* of the black hole which is second-order ignorance. The rest of the paper will be dedicated to the exploration of such relation.

5   The Birth of Second-Order Ignorance

As it has been shown in the previous section, once an agent steps into second-order ignorance, he/she also enters the black hole of higher-order levels of ignorance, without having much hope to escape, since, formally, this black hole is inescapable employing the resources internal to the language. Assuming that first-order ignorance phenomena are common, it is important, when modelling artificial agents, to avoid possible passages from first-order ignorance to second-order ignorance, so that the black hole of higher-order levels of ignorance is avoided. Interestingly, negative introspection is an incredibly powerful cognitive phenomenon that can block the passage between first-order ignorance and second-order ignorance.[14]

Theorem 3 (Negative introspection and lack of second-order ignorance). *Assume* $\vdash_{\mathcal{L}} \neg K(\phi) \rightarrow K(\neg K(\phi))$, *then* $\vdash_{\mathcal{L}} I(\phi) \rightarrow \neg I(I(\phi))$.

Proof (Theorem 3).
    The proof will be given directly.

---

[14] This result is not novel to this paper, but is well known in the logical literature on formalizing ignorance.

(1)	$I(\phi)$	Ass.
(2)	$\neg K(\phi) \wedge \neg K(\neg\phi)$	Definition of (1).
(3)	$\neg K(\phi)$	$\wedge$ Elim. (2).
(4)	$\neg K(\phi) \rightarrow K(\neg K(\phi))$	Axiom 5.
(5)	$K(\neg K(\phi))$	MP (3)-(4).
(6)	$\neg K(\neg\phi)$	$\wedge$ Elim. (2).
(7)	$\neg K(\neg\phi) \rightarrow K(\neg K(\neg\phi))$	Axiom 5.
(8)	$K(\neg K(\neg\phi))$	MP (6)-(7).
(9)	$K(\neg K(\phi) \wedge \neg K(\neg\phi))$	$\wedge$ Distr. (5)-(8).
(10)	$K(I(\phi))$	Definition of (9).
(11)	$K(I(\phi)) \vee K(\neg I(\phi))$	$\vee$ Intr. (10).
(12)	$\neg(\neg K(I(\phi)) \wedge \neg K(\neg I(\phi)))$	DM (11)
(13)	$\neg I(I(\phi))$	Definition of (12).

It can therefore be safely claimed that negative introspection is an exceptionally effective measure to avoid the black hole of higher levels of ignorance. However, as discussed in Sect. 2, assuming that artificial agents possess the deep introspection that axiom 5 requires might be too much. Unfortunately, a direct negation of negative introspection can become the main culprit in the spread of second-order ignorance. That means that even though it is reasonable to assume that agents are not negatively introspective, it is important to avoid that agents are *completely* non-negatively introspective, as this would tie together first-order ignorance and second-order ignorance, as stated by the following theorem:

Theorem 4 (From first-order ignorance to second-order ignorance).
$\vdash_{\mathcal{L}} I(\phi) \wedge (\neg K(\phi) \wedge \neg K(\neg K(\phi))) \rightarrow I(I(\phi))$.

Proof (Theorem 4).
   The proof will be given by contradiction.

(1)	$I(\phi)$	Ass.
(2)	$\neg K(\phi) \wedge \neg K(\neg K(\phi))$	Ass.
(3)	$\neg I(I(\phi))$	Ass.
(4)	$\neg(\neg K(I(\phi)) \wedge \neg K(\neg I(\phi)))$	Def. of (3).
(5)	$K(I(\phi)) \vee K(\neg I(\phi))$	DM (4).
(6)	$K(I(\phi))$	Ass.
(7)	$I(\phi) \rightarrow \neg K(\phi)$	P. Taut.
(8)	$K(I(\phi) \rightarrow \neg K(\phi))$	Nec. (7).
(9)	$K(I(\phi) \rightarrow \neg K(\phi)) \rightarrow K(I(\phi) \rightarrow K(\neg K(\phi)))$	Axiom K.
(10)	$K(I(\phi) \rightarrow K(\neg K(\phi)))$	MP (8)-(9).
(11)	$K(\neg K(\phi))$	MP (6)-(10).
(12)	$\neg K(\neg K(\phi))$	$\wedge$ Elim. (2).
(13)	Contradiction	(11)-(12).
(14)	$K(\neg I(\phi))$	Ass.
(15)	$K(\neg I(\phi)) \rightarrow \neg I(\phi)$	Axiom T.
(16)	$\neg I(\phi)$	MP (14)-(15)
(17)	Contradiction	(1)-(16).

Since both clauses of $K(I(\phi)) \vee K(\neg I(\phi))$ lead to a contradiction, it must follow that $\neg(K(I(\phi)) \vee K(\neg I(\phi)))$, which is equivalent to $I(I(\phi))$. $\qquad\square$

6 Conclusion and Future Works

In the paper, three possible conditions that make ignorance emerge have been proposed, showing that those conditions are both sufficient and necessary for ignorance to emerge. Those conditions were given in terms of beliefs and thus are employable to enrich previously proposed BDI-frameworks that model intelligent systems. This is especially important, if the modellers want to allow the intelligent system to avoid the black-hole of higher-orders of ignorance. While it might not seem a great improvement, it should be noted that a system which is unaware of being ignorant, will never be in a position to question such ignorance and, thus, will always be unable to produce plans to achieve extra information and make better decisions. In the paper, it has also been shown what can cause the passage between basic ignorance and higher-order levels of ignorance, providing insights on what should be explicitly avoided by intelligent systems. What shall be done in the future is to explore if the negative introspection condition that bridges basic and higher-order levels of ignorance can be expressed through more specific belief conditions that can be tailored to the specific cases of misbelieving, agnosticism and doubting. It would be also interesting to relate ignorance conditions to other knowledge-based aspects, such as trust [21,22] and the spread of fake news.

A Formal Proofs

Proof (Proposition 1).

The proof will be split into two parts.

Case 1: The proof will be given directly.

(1)	$B(\phi) \wedge \neg\phi$	Ass.
(2)	$B(\phi)$	$\wedge$ Elim. (1).
(3)	$\neg\phi$	$\wedge$ Elim. (1).
(4)	$K(\phi) \rightarrow \phi$	Axiom T.
(5)	$\neg\phi \rightarrow \neg K(\phi)$	Contrap. (4).
(6)	$\neg K(\phi)$	MP (3)-(5).
(7)	$\neg(B(\phi) \wedge B(\neg\phi))$	Axiom D.
(8)	$\neg B(\phi) \vee \neg B(\neg\phi)$	DM (7).
(9)	$\neg B(\neg\phi)$	DS (2)-(8).
(10)	$K(\neg\phi) \rightarrow B(\neg\phi)$	Axiom Int_1.
(11)	$\neg B(\neg\phi) \rightarrow \neg K(\neg\phi)$	Contrap. (10).
(12)	$\neg K(\neg\phi)$	MP (9)-(11).
(13)	$\neg K(\phi) \wedge \neg K(\neg\phi)$.	$\wedge$ Intr. (6)-(12).

Case 2: The proof will be given directly.

$$
\begin{array}{lll}
(1) & B(\neg\phi) \wedge \phi & \text{Ass.} \\
(2) & B(\neg\phi) & \wedge \text{ Elim. (1).} \\
(3) & \phi & \wedge \text{ Elim. (1).} \\
(4) & \neg(B(\phi) \wedge B(\neg\phi)) & \text{Axiom } D. \\
(5) & \neg B(\phi) \vee \neg B(\neg\phi) & DM \text{ (4).} \\
(6) & \neg B(\phi) & DS \text{ (2)-(5).} \\
(7) & K(\phi) \rightarrow B(\phi) & \text{Axiom } Int_1. \\
(8) & \neg B(\phi) \rightarrow \neg K(\phi) & \text{Contrap. (10).} \\
(9) & \neg K(\phi) & MP \text{ (6)-(8).} \\
(10) & K(\neg\phi) \rightarrow \neg\phi & \text{Axiom } T. \\
(11) & \phi \rightarrow \neg K(\neg\phi) & \text{Contrap. (10).} \\
(12) & \neg K(\neg\phi) & MP \text{ (3)-(11).} \\
(13) & \neg K(\phi) \wedge \neg K(\neg\phi). & \wedge \text{ Intr. (9)-(12).}
\end{array}
$$

Proof (Proposition 2).

The proof will be given directly.

$$
\begin{array}{lll}
(1) & \neg B(\phi) \wedge \neg B(\neg\phi) & \text{Ass.} \\
(2) & \neg B(\phi) & \wedge \text{ Elim. (1).} \\
(3) & \neg B(\neg\phi) & \wedge \text{ Elim. (1).} \\
(4) & K(\phi) \rightarrow B(\phi) & \text{Axiom } Int_1. \\
(5) & \neg B(\phi) \rightarrow \neg K(\phi) & \text{Contrap. (4).} \\
(6) & \neg K(\phi) & MP \text{ (2)-(5).} \\
(7) & K(\neg\phi) \rightarrow B(\neg\phi) & \text{Axiom } Int_1. \\
(8) & \neg B(\neg\phi) \rightarrow \neg K(\neg\phi) & \text{Contrap. (7).} \\
(9) & \neg K(\neg\phi) & MP \text{ (3)-(8).} \\
(10) & \neg K(\phi) \wedge \neg K(\neg\phi). & \wedge \text{ Intr. (6)-(9).}
\end{array}
$$

Proof (Proposition 3).

The proof will be split into two parts, showing that each disjunct of the antecedent of the conditional implies the consequent of the conditional.

Case 1: The proof will be given directly.

$$
\begin{array}{lll}
(1) & B(\phi) \wedge \phi \wedge \neg K(\phi) & \text{Ass.} \\
(2) & B(\phi) & \wedge \text{ Elim. (1).} \\
(3) & \phi & \wedge \text{ Elim. (1).} \\
(4) & \neg K(\phi) & \wedge \text{ Elim. (1).} \\
(5) & K(\neg\phi) \rightarrow \neg\phi & \text{Axiom } T. \\
(6) & \phi \rightarrow \neg K(\neg\phi) & \text{Contrap. (5).} \\
(7) & \neg K(\neg\phi) & MP \text{ (3)-(6).} \\
(8) & \neg K(\phi) \wedge \neg K(\neg\phi). & \wedge \text{ Intr. (4)-(7).}
\end{array}
$$

Case 2: The proof will be given directly.

(1) $B(\neg\phi) \wedge \neg\phi \wedge \neg K(\neg\phi)$ Ass.
(2) $B(\neg\phi)$ $\wedge$ Elim. (1).
(3) $\neg\phi$ $\wedge$ Elim. (1).
(4) $\neg K(\neg\phi)$ $\wedge$ Elim. (1).
(5) $K(\phi) \rightarrow \phi$ Axiom T.
(6) $\neg\phi \rightarrow \neg K(\phi)$ Contrap. (5).
(7) $\neg K(\phi)$ MP (3)-(6).
(8) $\neg K(\phi) \wedge \neg K(\neg\phi)$. $\wedge$ Intr. (4)-(7).

Proof (Lemma 5). The proof is given by contradiction.

(1) $I(I(\phi))$ Ass.
(2) $I(I(\phi)) \rightarrow I_R(\phi)$ Lemma 2.
(3) $K(I(I(\phi)))$ Ass.
(4) $K(I(I(\phi)) \rightarrow I_R(\phi)$ *Nec.* (2).
(5) $K(I(I(\phi)) \rightarrow I_R(\phi)) \rightarrow (K(I(I(\phi))) \rightarrow K(I_R(\phi))$ Axiom K.
(6) $K(I(I(\phi))) \rightarrow K(I_R(\phi)$ MP (4)-(5).
(7) $K(I_R(\phi)$ MP (3)-(6).
(8) $\neg K(I_R(\phi))$ Lemma 4.
(9) Contradiction   (7)-(8).

Since a contradiction has been reached, one of the assumptions must be rejected. Thus, $I(I(\phi)) \rightarrow \neg K(I(I(\phi)))$ holds in $\mathcal{L}$.   □

References

1. Adam, C., Gaudou, B.: BDI agents in social simulations: a survey. Knowl. Eng. Rev. **31**(3), 207–238 (2016)
2. Artemov, S., Fitting, M.: "Justification Logic", The Stanford Encyclopedia of Philosophy (Summer 2020 Edition), Edward N. Zalta (ed.)
3. Baltag, A., Renne, B.: "Dynamic Epistemic Logic", The Stanford Encyclopedia of Philosophy (Winter 2016 Edition), Edward N. Zalta (ed.)
4. Caillou, P., Gaudou, B., Grignard, A., Truong, C. Q., Taillandier, P.: A simple-to-use BDI architecture for Agent-based Modeling and Simulation. In: Proceedings of the 11th Conference of the European Social Simulation Association (ESSA 2015) (2015)
5. Chellas, B.: Modal Logic: An Introduction. Cambridge University Press, Cambridge (1980)
6. Fano, V., Graziani, P.: A working hypothesis for the logic of radical ignorance, Synthese (2020)
7. Fine, K.: Ignorance of ignorance. Synthese **195**(9), 4031–4045 (2018)
8. Georgeff, M., Pell, B., Pollack, M., Tambe, M., Wooldridge, M.: The belief-desire-intention model of agency. In: Müller, J.P., Rao, A.S., Singh, M.P. (eds) Intelligent Agents V: Agents Theories, Architectures, and Languages. ATAL 1998 (1999)

9. Halpern, J.Y.: A theory of knowledge and ignorance for many agents. J. Logic Comput. **7**(1), 79–108 (1997)
10. Halpern, J., Moses., Y., Fagin, R., Vardi, M.: Reasoning about knowledge. The MIT Press, Cambridge (1995)
11. Hintikka, J.: Knowledge and beliefs: an Introduction to the logic of the two notions. Cornell University Press, Ithaca (1962)
12. van der Hoek, W., Lomuscio, A.: A logic for ignorance. In: Leite, J., Omicini, A., Sterling, L., Torroni, P. (eds) Declarative Agent Languages and Technologies. DALT 2003 (2004)
13. Meyer, C., van der Hoek, W.: Epistemic logic for AI and computer science. Cambridge University Press, Cambridge (1995)
14. Mints, G.: Natural deduction for propositional logic. In: A Short Introduction to Intuitionistic Logic. The University Series in Mathematics, pp. 9–22. Springer, Boston (2002). https://doi.org/10.1007/0-306-46975-8_3
15. Plato, "Theaetetus", traslation by McDowell, J., Oxford University Press, Oxford (2014)
16. Rao, A., Georgeff, M.: Formal models and decision procedures for multi-agent systems. Technical Note, AAII (1995)
17. Rendsvig, R., Symons, J.: "Epistemic Logic". The Stanford Encyclopedia of Philosophy (Summer 2019 Edition), Edward N. Zalta (ed.)
18. Smitha Rao, M.S., Jyothsna, A.N.: BDI: applications and architectures. Int. J. Eng. Res. Technol. **2**(2), 1–5 (2013)
19. Steinsvold, C.: A Note on Logics of Ignorance and Borders. Notre Dame Journal of Formal Logic **49**(4), 385–392 (2008)
20. Souza, M.: Choices that make you change your mind : a dynamic epistemic logic approach to the semantics of BDI agent programming language, Ph.D. Thesis, University of Rio Grande (2016)
21. Aldini, A., Tagliaferri, M.: Logics to reason formally about trust computation and manipulation. In: Saracino, A., Mori, P. (eds.) ETAA 2019. LNCS, vol. 11967, pp. 1–15. Springer, Cham (2020). https://doi.org/10.1007/978-3-030-39749-4_1
22. Tagliaferri, M., Aldini, A.: From knowledge to trust: a logical framework for pre-trust computations. IFIP Adv. Inf. Commun. Technol. **528**, 107–123 (2018)
23. Tripathi, K.P.: A review on knowledge-based expert system: concept and architecture. IJCA Special Issue on "Artificial Intelligence Techniques - Novel Approaches and Practical Applications (2011)
24. Wooldridge, M.: Reasoning about Rational Agents. MIT Press, Cambridge (2000)
25. Wooldridge, M.: Practical reasoning with procedural knowledge. In: Gabbay, D.M., Ohlbach, H.J. (eds.) FAPR 1996. LNCS, vol. 1085, pp. 663–678. Springer, Heidelberg (1996). https://doi.org/10.1007/3-540-61313-7_108

CoSIM-CPS 2020

Organization

CoSim-CPS 2020 – Workshop Chairs

Cinzia Bernardeschi	University of Pisa, Italy
Cláudio Gomes	Aarhus University, Denmark
Paolo Masci	National Institute of Aerospace (NIA), USA
Peter Gorm Larsen	Aarhus University, Denmark

CoSim-CPS 2020 – Programme Committee

Swee Balachandran	National Institute of Aerospace (NIA), USA
Stylianos Basagiannis	Raytheon Technologies, Ireland
Mongi Ben Gaid	IFPEN, France
Joerg Brauer	Verified Systems International GmbH, Germany
Fabio Cremona	Argo.ai, Germany
Julien Deantoni	Inria, France
Paul De Meulenaere	University of Antwerp, Belgium
Joachim Denil	University of Antwerp, Belgium
Marco Di Natale	Scuola Superiore Sant'Anna, Italy
Andrea Domenici	University of Pisa, Italy
Adriano Fagiolini	University of Palermo, Italy
Claudio David López	Delft University of Technology, The Netherlands
Cesar Munoz	NASA Langley Research Center, USA
Maurizio Palmieri	University of Pisa, Italy
Akshay Rajhans	Advanced Research and Technology Office MathWorks, USA
Rudolf Schlatte	University of Oslo, Norway
Neeraj Singh	INPT-ENSEEIHT/IRIT, University of Toulouse, France
Frank Zeyda	Verified Systems International GmbH, Germany

A Case Study on Formally Validating Motion Rules for Autonomous Cars

Mario Henrique Cruz Torres[1], Jean-Pierre Giacalone[2], and
Joelle Abou Faysal[3(✉)]

[1] IVEX.ai Intelligent Vehicle Technology, Leuven, Belgium
[2] Renault SW Labs (RSL), Expert ADAS/AD Software Architecture,
Autonomous Vehicle Algorithms, Sophia Antipolis, France
[3] Renault Software Labs (RSL), Université Cote d'Azur, CNRS, Inria, I3S,
Valbonne, France
joelle.abou-faysal@etu.univ-cotedazur.fr

Abstract. Car motion control is a key functional stage for providing advanced assisted or autonomous driving capabilities to vehicles. Car motion is subject to strict safety rules which are normally expressed in natural language. As such, these natural language rules are subject to potential misinterpretation during the implementation phase of the motion control stage. In this paper, we show a novel approach by which safety rules are expressed in natural language, then in a formal language specification which is then validated and used to generate a car motion checker. We present a case study of using the approach with true road capture data and its associated imperfections. We also show how the approach lowers the validation efforts needed to guarantee that the car motion always respects a desired set of safety rules while other traditional validation methods would be much heavier to deploy and error prone.

Keywords: Formal language · Autonomous drive · Motion safety

1 Introduction

In the past few years, across most regions in the world, there has been a push to improve vehicles driving safety through specific regulations. In Europe, for instance, this has been the case with the issuance of General Safety Regulation (GSR) phase 2 [6]. These regulations tend to mandate the deployment of Advanced Driving Assistance systems in cars, like Automatic Emergency Braking and other car motion control systems such as Autonomous Emergency Steering or Adaptive Cruise Control [5]. These driving features take control of car motion on behalf of the driver, even if the driver still has the ability to take back control, in order to operate within time ranges and with perception reaction time that allows fast action to protect the car occupants. As such, these systems must make sure that passengers safety is guaranteed during their operation.

For such systems, a key element to guarantee safety is the car trajectory control. The trajectory control component is the part that computes the future

© Springer Nature Switzerland AG 2021
L. Cleophas and M. Massink (Eds.): SEFM 2020 Workshops, LNCS 12524, pp. 233–248, 2021.
https://doi.org/10.1007/978-3-030-67220-1_18

trajectory of the vehicle as a function of what is perceived by the car sensors. The trajectory is transformed into 2D positional objectives to follow on the road, from the current position. The problem behind ensuring safety at the current position and along the trajectory is bounded in terms of the expression of the safety rules. Important properties behind such bounding are related to creating priorities (e.g., what is happening at the front of the vehicle versus at the rear) or having parallel conditions expressed (e.g., checks being performed simultaneously on different directions, longitudinally and laterally) [15]. Hence, car motion safety construction is a perfect domain for exploring formal methods to express safety rules.

In the context of proof that is needed to express safety rules described above, one can refer to several specification languages providing design by contract approaches ([1,12] for instance). Usually, though, these languages are not at the right level of abstraction to express key aspects regarding car motion. In this paper, we will explain what that level should be and why. As it will be shown, there are several challenges expressing safety rules about motion control. As a result, we will describe the expected flow to be used between the expression of rules in human language by a car safety engineer and their formalization using the proposed approach. We will also illustrate the application of the formal method depicted in this paper through true road captures of difficult scenarios in traffic jams, with complicated perception situations (at night, for instance). We will devise the impact of those conditions as constraints in the expression of safety rules. Finally, we will indicate some trends in the application of this technique for safe car motion control implementation.

This paper is organized as follows. Section 1.1 briefly discusses related work. Section 2 presents the challenges in implementing safety rules concerning car motion control. Section 3 introduces IVEX tools and details IVEX approach to model safety rules for car motion. Section 4 describes a case study of a Society of Automotive Engineering (SAE) Level 3 car experiments. It also discusses the low speed motion control safety rules set. Finally, Sect. 5 draws conclusions and details some possible future work.

1.1   Related Work

Reachability analysis is used to propose a safety framework to analyse the motion of autonomous vehicles by used at [13]. It is used to guarantee that any feasible future motion of dynamic obstacles around the automated car, also known as Ego car, is taken into account when assessing a planned trajectory. Similar to our technique, [13] define a set of assumptions for the future movement of dynamic obstacles around the Ego car and verifies if the Ego car can reach a future state where it would have a collision. An interesting aspect of the work [13] is to also provide an alternative fail-safe trajectory planning that the autonomous vehicle (AV) could use to avoid a collision.

The work by [16] proposes a mathematical model, called Responsibility-Sensitive Safety (RSS), for safety assurance of autonomous vehicles. The RSS model is explainable and has well defined assumptions. The model tries to formalize common sense human behavior concerning the judgment of "who is responsible for causing an accident". The final goal of the model is to guarantee that an autonomous car never causes an accident, being than considered safe. The main limitation of the model is not taking variability and uncertainty into account. This limitation greatly impacts the ability of RSS to guarantee safety in reality.

2   Challenges in Implementing Safety Rules Around Car Motion Control

When creating safety rules for a car motion control we assume that it is impossible to guarantee there is absolutely no collisions involving the controlled car. The impossibility of zero collisions is due to the fact that the Autonomous Vehicle/Advanced Driver Assistant Systems (AV/ADAS) car cannot control the behaviour of other road users. If another road user is actively trying to cause a collision and has a vehicle capable of very large accelerations/decelerations, it is easy to understand that even if the AV/ADAS executes evasive maneuvers or stops on the side of the road, the other road user can still cause a collision. There is a trade-off between the drive-ability and safety of the controlled car, since there will always be a risk of a collision. We believe that creating the safety rules for the car motion control will always incorporate such trade-off.

The first challenge for creating safety rules for car motion control is identifying the minimum set of assumptions (including hidden assumptions) about the environment, particularly assumptions about other road users, which are measurable and represent reality closely enough. The RSS model [16], for instance, defines a small number of assumptions about the environment, such as the maximum, minimum acceleration/deceleration of other vehicles which leads to a model that can be easily understood by human beings but which also leads to lower drive-ability of a controlled car. For instance, one's assumptions for maximum deceleration may lead to an extremely conservative driving behaviour, thus lowering the drive-ability of the controlled car.

We believe the car motion safety rules should be formally verified for soundness and completeness for a certain environment. This brings the second challenge which is defining the proper abstraction level to formally model the safety rules. Each formalism requires the system (software + hardware) to be specified in a certain way, [9,17]. The modeler has to reduce, simplify, or abstract the system being modeled to be able to use different model checking tools.

Formally modelling the system and its environment has to be done taking into account possible issues, such as:

- the model does not represent the system,
- the model does not properly represent the environment (e.g. wrong assumptions about the environment),
- the model truly represents the system and its environment, but is intractable.

Finding the correct level of abstraction to represent the AV/ADAS system and its environment is challenging because it has to be done in a way that the model is close enough to the reality, but abstract enough to be solvable [7]. Toolboxes like Tulip [18] help to mitigate the problem of having a model which does not represent the system, since it synthesizes controllers, but does not help in defining the proper abstraction level to represent the problem, or mitigating having wrong assumptions about the environment.

Particular attention has to be given to modeling the input information that will be used by the motion control rules. The perception systems which provide information used by the motion controllers may have a great impact into the safety rules. When modeling the car motion rules, it is extremely important to clearly model the assumptions concerning the perception systems used in the car so that limitations of this system can be properly dealt with in the motion of the car.

The focus of this paper is on car motion control under conditions of low speed (lower than 60 km/h) and traffic jams. The control context is either Advanced Driving Assistance where the driver is still under control but is assisted by the electronic system or Autonomous Driving of Level 3, as defined by the SAE [14]. Motion control consists in constructing a trajectory to be followed by the car under control. This trajectory is expressed as 2D positions on the road, provided with a recurrence of 20 to 100 ms into the future from current time. The positions are expressed in the car coordinates (see Fig. 1). Car motion control takes these target positions and transforms them into actions along the longitudinal direction (along the X axis) and the lateral (along the Y axis through the yaw angle) one, for the vehicle. Simply put, these actions relate to defining acceleration or deceleration of the car and steering wheel movement. As a result, there will be rules for checking longitudinal and lateral safety, and these rules are going to be valid simultaneously which is another challenge we have to address with the rules description language.

Fig. 1. Car motion trajectory definition in the car coordinates system.

Car motion control constructs the trajectory based on information reported by sensors mounted in the vehicle. This information is usually named Perception and consists in aggregating different details about moving or static objects around the car, like type, dimensions, position, speed and acceleration. This aggregation is performed by an electronic system called Fusion that materializes and confirms the various detections provided by individual sensors. Depending on the number and type of sensors available, the accuracy and reliability of the information may vary. Challenges regarding obtaining a quality Perception have been highlighted in publications like [8]. Among other issues related to perception conditions (night, rain, fog, as examples), problems of persistence are quite impacting to the definition of safety rules like the ones introduced above. In essence, potential appearance and disappearance of detected obstacles means that safety rules must express a dependency in space (longitudinal, lateral) and time (provided an obstacle is confirmed over a certain time, for instance). And this becomes a constraint to the description language.

As we will see in Sect. 4, the approach for implementing car motion safety rules presented in this paper has been exposed to real road data captured with a car prototype embedding several classes of sensors (cameras, radars, lidars, ultra sonic) and called TRAJAM. We will see clear examples of perception challenges that were faced in order to properly express rules given by a human safety engineer.

3 IVEX Tools Suite and Approach to Model Safety Rules for Car Motion

The IVEX toolchain can be used to model different systems that need safety guarantees and which operate in complex environments. IVEX engineers spent years performing research into the development of safety critical systems for other domains, such as aerial vehicles and Automated Guided Vehicles (AGVs) [2–4]. When doing research, they modelled different systems using varied approaches, exploring diverse ways to specify safety critical autonomous systems. IVEX engineers understood that traditional approaches like creating Finite State Machines (FSM), Behavioral Trees, and traditional formal modeling techniques, like solvers for LTL, had their own limitations to define safe autonomous systems [11]. The systems build by IVEX engineers were used and demonstrated, besides others, in aerial platforms having embedded mission control and autonomous safe behavior, used to fly around electricity towers in Belgium.

At the core of the IVEX toolchain is an engineering process (Depicted in Fig. 2) which supports the creation of the safety rules. The process allows one to automatically transform safety requirements into formally verified software. The toolchain then generates correct-by-construction software, by performing a translation between a solved model specification and a C++ execution policy. The process highlights the limitations of safety requirements (by performing consistency and completeness checks). The process shortens iteration cycles, reuses existing knowledge and is supported by mature toolchain.

The first step into the process is to identify the safety requirements that should be always satisfied by the car motion. Normally, such requirements reflect

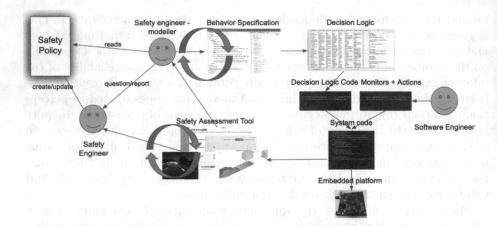

Fig. 2. The IVEX process consists of 3 main steps: 1 - define a behavior specification of the system and its environment, 2 - generation of a decision making logic, 3 - validation of the created system using the Safety Assessment Tool.

a number of safety requirements imposed on the car motion. The safety requirements can even consider what is the expected car behavior in different operational design domains (ODD).

Based on the requirements gathering on system goals, safety and other (as gathered together with the various system stakeholders), IVEX engineers specify the requirements in a behaviour specification. During specification, a concise, formal model of the system is built, including of the perceivable system states, its actuation, rules and constraints that must be fulfilled. The behavior specification is created using a domain specific language (DSL). The DSL has constructs to represent vehicle system properties, such as pre-conditions for action executions, and expected outcomes. The behavior specification is written in the DSL using first-order logic constructs. The specification is declarative (it describes goals, state, actions - as in a traditional planning model - and constraints), it is not imperative (describing for every situation exactly which action to take). This is a fundamental aspect of the approach, which makes it more adaptable and manageable from the ground up. The exact decision making logic is generated later on in the process.

The next step is performing an automated analysis on the specification, by using IVEX verification tool, to check whether the specification is:

– complete (the specification will be able to decide in each possible state)
– consistent (the specification does not contain contradicting requirements)

It could be that (a) the specifications is not covering a combination of states; or (b) that the model is inconsistent. The verification tool will automatically detect and report inconsistencies. If a specification is not verifiable, developers receive clear and punctual feedback on the status of the specification. Through iterative specification and verification steps, developers are guided to unambiguously and completely model the expected system behavior under all circumstances.

Based on the final, verified specification, the toolchain generates the decision making logic as code. At this step the decision logic is translated into a tree-like data structure with a known maximum depth, which is critical for guaranteeing real-time execution deadlines. Besides that, the decision logic is a direct mapping of the behavior specification into C++ code, which lowers the chances of implementation bugs.

The logic implements a mapping for every possible discrete situation - based on the state representation from the requirements & specification - to one or a set of actions to be executed by the system. This logic is guaranteed to cover every possible discrete scenario, and respect all safety requirements and other constraints. A typical specification for a SAE Level 3 car generates around 120.000 safety rules in the decision logic, when considering many aspects of the Operating Design Domain.

The toolchain has a runtime execution environment which is used to perform the integration of the execution policy with the rest of the system, called the IVEX safety co-pilot. The runtime includes specific components for integration in the overall autonomous system. The runtime has well defined inputs and outputs interfaces, to facilitate its integration into the system. For instance:

- The runtime communicates with the rest of the system via a middleware or via direct function invocation (loading new threads to execute continuous controllers).
- Inputs are read by **monitors**. Monitors are runtime components that actively read the perception data from the system and convert this perception data into discrete values. Example of monitor inputs are: static and dynamic obstacle location (with corresponding confidence levels), car velocity, etc. One of the inputs for car motion validation is the planned trajectory created by a path planner.
- Outputs are given by **actions**. Actions represent the actuator components from the autonomous system that should be activated and their parameters. For instance, an action can represent an emergency operation such as a strong longitudinal braking controller.

The process has two distinct parts, being an off-line one, used to model the system, and an on-line one used to verify the behaviour of the system during execution. In order to validate the full system that uses the decision making logic the process has a validation step. The first validation step is done by testing the created system with recorded driving data in the IVEX Safety Assessment Tool (SAT). The SAT tool performs Software-in-the-loop (SIL) tests, using the system created.

The SAT tool allows one to replay recorded driving sensor data into the SIL which will then check its thousands of decision rules to define if a certain car movement is triggering a safety violation. The SAT tool then collects all safety violations occurrences and generates statistics highlighting all critical situations in the driving data. A safety engineer can then proceed to analyse the highlighted safety violations and the safety metrics created by the SAT tool. After analysing thousands of scenarios, the SAT tool indicates how conservative the created system is, allowing safety engineers to proceed to refine the behavior specification or its parameters.

4 Case Study of a SAE Level 3, Low Speed Motion Control Safety Rules Set

In this section, we are entering into real experiments conducted using the tools and method described in the previous section through information available in a re-simulation environment. A re-simulation environment is providing data captured during road trips by a car embedding a set of sensors close to the one used in production and located where they would be installed. Hence, study presented here is based on real data. The environment provides data at various locations in the processing stages through pipes that can be connected to the system to be tested. In our case, these pipes carried kinematic information from sensor fusion outputs for objects, infrastructure and Ego (a.k.a. the automated) car. They also carried the future Ego car trajectory positions as delivered by the planning stage. The infrastructure data consisted of lanes structure information as captured by the perception stage. Kinematic information included positions in 2D as measured in Ego car referential (see Fig. 1) as well as speed and acceleration. The re-simulation environment also provides a situational camera view, towards the front of Ego car, synchronous to the provided data in order to better understand visually a given scenario configuration. Table 1 summarizes the re-simulation data available.

Table 1. Data available through re-simulation.

Type	Content	Sampling
Objects	Position, Speed, Accel., Size	40 ms
Ego car	Position, Speed, Accel	40 ms
Infrastructure	Lines types, Shape	40 ms
Trajectory	Positions	100 ms

With this re-simulation environment, we constructed and verified car motion safety rules corresponding to the SAE Level 3 motion control mode. These were written as a real safety policy, describing longitudinal and lateral situations to be avoided and corresponding expected behavior. Figure 3 shows how these rules are getting exercised with the re-simulation data. The Trajectory Validation function receives the results of the analysis according to safety rules, and apply them on the trajectory data proposed by the planner. This valid trajectory will be passed to the motion control that follows it (as explained earlier). In case checks report

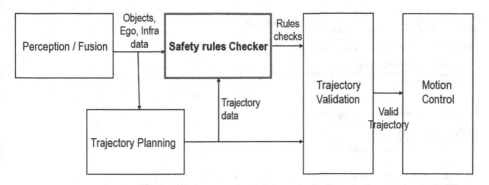

Fig. 3. Safety rules verification high level architecture. Each rectangle represents a functional component in the system. The safety rules are checked at the **Safety rules Checker** component which receives information from the **Perception/Fusion** and **Trajectory Planner** components. The results of the checks performed by the **Safety rules Checker** components are sent to the **Trajectory Validation** component which is then responsible for deciding on following the planned trajectory or not.

a failure in fulfilling the rules then an emergency maneuver could be signaled by this function, as an example.

Safety rules considered here were initially structured and expressed in human text language. Situations covered for L3 control mode were essentially in traffic jam, at low speed and various weather and light conditions as well as infrastructure and slopes. Organized in tables, the rules provide information about the situation being verified, the preconditions, the result being avoided (usually, a collision) and specific aspects to consider. Their definition is owned by a safety engineer from Renault and he was supporting requests for understanding situations to be checked in case there was any ambiguity. An example of longitudinal safety description is depicted in Fig. 4. There were around 20 rules like this that composed the L3 set.

By going over this rule, we can first observe that a traffic jam situation must be the operating case (also known as the Operating Design Domain or ODD). This means Ego car is surrounded by several moving objects. We can also observe that it is following a preceding vehicle with which a safety distance of 2 s is defined. The rule sets a situation by which the preceding vehicle is potentially decelerating with a certain strength (minimal for $1m/s^2$, nominal for $5m/s^2$ or strong for $10m/s^2$). This covers limits of the environment model as discussed in Sect. 2. On his side, the Ego car has the capability to regulate its speed with an Automatic Cruise Control (ACC) deceleration capability up to a strong braking capability of 10 m/s2. In order to model this rule, we need to start discretizing space to separate system states with the preceding vehicle. Figure 5 describes this, based on the rule content. The blue vehicle is Ego car.

Longitudinal states are so that either Ego car is alone or there is a preceding vehicle. And, if there is one, then the situation can be that Ego car is at a safe distance, i.e. 2 s from it, or it is within a range of ACC distance where it should

TYPICAL SCENARIO The EGO vehicle is on the highway in traffic jam. The EGO vehicle is following the preceding vehicle. The preceding vehicle is making: - Deceleration (Strong braking, Nominal deceleration, etc.)	SCHEMA
RISK Front collision with the preceding vehicle	
PASS/FAIL CRITERIA No front accident/If accident the Delta V No rear accident/if accident the Delta V MEASURE The system shall detect the preceding vehicle who is making a deceleration. To avoid the collision with the preceding vehicle, the system shall maintain a safety distance by braking. The system shall brake at the ACC standard for regulation & if necessary the system shall apply to a strong braking	
Obstacle decel = 1m/s2, 5 m/s2, 10 m/s2 ACC Standard Regulation for braking = 0.6 m/s2 to 3 m/s2 Strong braking = 10 m/s2 Safety distance = 2 s	

Fig. 4. Longitudinal rule expressed by safety engineering.

regulate with corresponding deceleration levels or it is within an emergency distance that requires to regulate speed with strong braking capabilities. The resource that is actionable is longitudinal acceleration. The rules brings up 2 system states that matter for verifying it: The traffic jam state (S_traffic_jam, a Boolean, yes/no) and the front car distance state (S_front_car, an enumerated, not_exist, safe_distance, acc_distance, emergency_distance). Each state values are provided by monitors that run at the pace of the re-simulation data. For the front car distance, the state is populated with the equations below, assuming constant velocities within 100 ms trajectory sampling points and the maximum deceleration capability of $3m/s^2$ for ACC:

d is the distance to the front car, as reported by perception,

$Vego$ is Ego car velocity, $Aego$ is $3\ m/s^2$

$Dsafe = 2 \times Vego$ is the safe distance

if $d > Dsafe \rightarrow$ "safe_distance" state

if $d \leq Dsafe$ and $d \geq Dsafe - 6 \rightarrow$ "acc_distance" state

if $d < Dsafe - 6 \rightarrow$ "emergency_distance" state

Finally, the rule expresses three actions to be fulfilled as shown in Table 2.

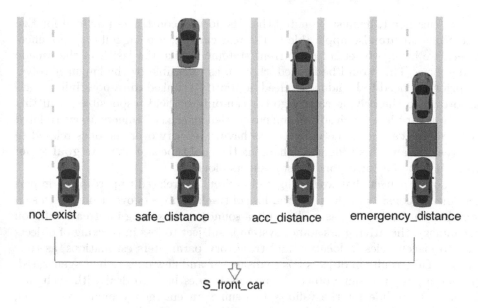

Fig. 5. Example of a state variable used to discretize the different safety distances taken into account between ego car and a front car. Depending on the current sensor readings and the assumptions used in the system, the front car distance can be classified as **not_exist**, meaning there is no visible front car, **safe_distance** meaning that considering current distance, velocities and assumptions for the front car accelerations there is no imminent risk of collision, **acc_distance** means that given the current assumptions and sensor readings, the ego-car would need to reduce its velocity at the ACC rate (3.5 m/s2) to satisfy the 2-s distance rule, while the **emergency_distance** value indicates that given the assumptions and current sensor readings, the ego-car would need to reduce its velocity at a higher rate than (3.5 m/s2) to avoid a collision.

Table 2. Actions to be fulfilled according to the longitudinal rule.

Type	Meaning
A_not_brake	No action
A_brake_acc	ACC braking ($0.6m/s^2$ to $3m/s^2$
A_brake_strong	$10m/s^2$ braking

With these rules elements properly broken down, the following formal description can be constructed to represent the safety rule *goals*. The formal code looks like this:

when

 $S\_traffic\_jam$ **is** *yes*

then

 goal type: constraint

 when S_front_car **is** *acc_ distance* **then goal: executing** $A\_ brake\_ acc$

 when S_front_car **is** *emergency_ distance* **then goal: executing** $A\_ brake\_ strong$

At this point, it must be noted that the formal constructs provided for Ego and front car are also applicable to the rear car (i.e. if a car follows). So, effectively, we have two sets of concurrent statements like the last 2 in the pseudo code above. This would be flagged, though, as infeasible by the language solver as priorities need to be added. Indeed, a priority is linked to responsibility levels according to the driving code: Ego car can only be held responsible for hitting the front car via a longitudinal maneuver. Hence, the statements above (related to the front car) must be indicated as having priority over the ones related to the rear car and this is done by changing the goal type statement to **goal type: priority** for the corresponding statements block.

Another aspect that we highlighted in Sect. 2 is related to perception imperfections. It consists in the potential loss of track of objects over time. With sensors used in cars (cameras, radars, ultra-sound) and existing information fusion technology, the driving assistance system is subject to loss in tracking of objects due to inaccuracies in location and trajectory parameters estimations, as time passes. This results in objects IDs to disappear and new ones to be re-generated, potentially for the same objects. The safety rules have to deal with such case in order to decide for the validity of issuing an emergency maneuver action, for instance. Here, we talk about discretizing in time, collecting, via monitors, the disappearing times statistics from the re-simulation data. The corresponding "tracking" state values are then expressed with a statement like:

> **goal type: constraint**
>
> **when** $S\_front\_car\_tracking$ is *disappeared_ more_than_ t1*
>
> **then goal: executing** $A\_$ *emergency_ operation1*

In the statement above, "t1" is a statistical time value that needs to be evaluated by a monitor from re-simulation data described in this section. The *disappeared_more_than_t1* state is a Boolean created by a comparison with the threshold "t1" in the corresponding monitor.

The whole set of L3 rules, coded with the approach described above, generated a total number of 13500 checking states. These checks were formally compiled, without human intervention, into an executable checker code that was embedded into the Safety Assessment Tool introduced in Sect. 3. This tool provided a global view over several hours of driving under traffic jam or dense traffic conditions, under day light or at night conditions. All safety violations as defined by the formal representation of the rules, i.e. triggering an action as presented above, were reported into a single view, along the timeline. The picture in Fig. 6 shows a graphical representation of an example of such report for a case of ACC braking that was reported as insufficient. The indicators on the left show the Ego car kinematic parameters (speed, acceleration, longitudinal, lateral). The situation is showing a merge to the left, into Ego car lane, of object labeled 14882 (zoom on the upper left), but the lanes structure is not reported yet in this representation. Object 14882 motion intent is depicted, at current time, by its kinematic projection trajectory model shown by a yellow color, in Fig. 6. This trajectory model takes a statistical representation of the longitudinal and lateral speeds evolution over time from the current object position, based on a combination of its current acceleration parameter as well as worst case (strong) deceleration. The Ego car future trajectory is depicted in front of it and consists of 50 points separated by 100 ms. It is colored blue for the points that do not report safety issues and red for points that do. In the case shown below, the safety action is an A_brake_acc action that is required when the trajectory becomes red. For this situation, the insufficient braking level is reported due to the 2-s safe distance definition in the rule that is violated by car 14882 sudden arrival in Ego car lane. This was not anticipated by the motion planner during the road drive. In the functional system shown in Fig. 3, this safety error would be reported to the trajectory validation as a warning of a potential future issue. As we move over time in the Safety Assessment tool we can, hence, deduce whether that situation becomes real when the red color reaches the Ego car position (instantaneous violation), potentially highlighting a critical situation for which an emergency action is required. In the data set captured for this case, the instantaneous violation appeared roughly a second later.

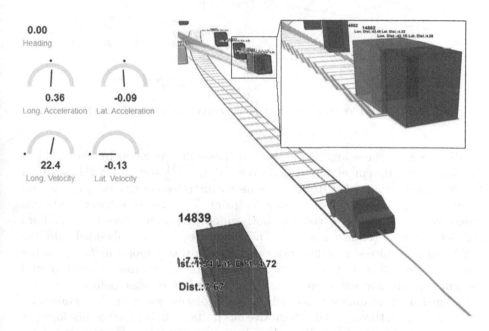

Fig. 6. Graphical representation of an ACC braking violation along Ego trajectory. (Color figure online)

4.1   Notes Regarding Real-Time Performance

For the study presented in this paper, we did not only want to address the formal construction of car motion safety rules and their offline validation with re-simulation data captured on the road. We also wanted to make sure that the safety checker that is automatically built by the method studied was capable to be embedded within the car motion control software. And we wanted this to occur by using a real computing platform as used in the automotive industry. Such platforms are called ECUs in the automotive industry.

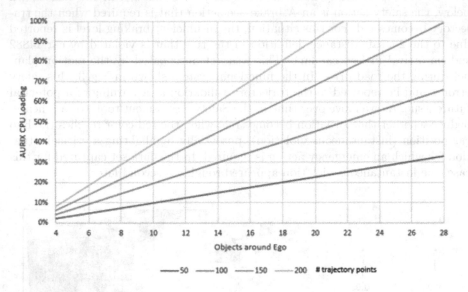

Fig. 7. Motion rules checker performance curves on a single AURIX core.

We chose a processing engine used in those and provided by the Infineon manufacturer [10], embedding a multi-core AURIX^{TM} system (in this study, a TCS397 development board), where we have constrained ourselves to operate on a single core to bound the processing footprint. The checker software code was generated as a single C++ code for both supporting safety assessment (offline) and for our performance analysis. The checker code was embedded into the IVEX Safety Co-Pilot runtime framework, which was responsible for updating the sensor data flowing into the monitors and for maintaining a memory and computationally efficient representation of the safety checker policy. The code was compiled with Infineon tools and run on the platform with re-simulation data patterns injected through the automotive bus ports available on that platform, at speed and synchronized (see Table 1). The various situations found in the data, in terms of objects density around Ego car, and the capability to re-sample in time the trajectory allowed to gather curves like in Fig. 7. Our threshold for considering that the checker is valid to run in a single care was that its execution did not exceed 80% loading at maximum frequency (300 MHz). These curves clearly indicate the fact that the checker can cope with a large number of surrounding objects for a given choice of trajectory sampling in the system.

5 Conclusion

This paper highlights real experiments, conducted over road captures made in the context of advanced driving and autonomous control car prototyping, on a practical approach to formalize the driving rules with an objective of maximal safety, using novel language and tools available from IVEX. Expressing safety rules applied to car motion control in a formal way carry challenges linked to both the imprecise nature of the rules defined by a human as well as the uncertainties related to the motion control process itself. We have presented, in previous sections, the needs for improvements from existing formal methods to address those challenges. And we have shown how a well thought set of language and tools, associated with a practical usage method, can handle all the above concerns together.

The method proposed also drives for a way of considering the safety rules verification in the chain that starts from motion planning and ends in car motion execution via physical actuators. Indeed, if it makes sense to apply safety principle within the various stages of the chain above, this paper has shown that the motion control safety rules generate a formal verification complexity of several thousand states. This shows that a safety checker executing this verification is required, associated with a trajectory validation function in order to cover the full safety complexity (see Fig. 3, above, for an example of this). So, this poses the question for the proposed approach of this paper to be compatible with real-time execution constraints of running within the electronic processing system in the car.

5.1 Next Steps

The motion safety rules used for this study where a preliminary set. Our study allowed to show that, as a whole, they where performing as expected, capturing driving situations that where below the quality requirements. The analysis showed that the proposed method could be used to create indicators of bugs in safety rules coverage and system behavior. This part is worth further studying. Also, and finally, we have noticed that some rules where too "static" in their definition and that some parameters would benefit from being specified according to the driving situation (e.g., the safety distance of 2 s). This is another axis of future study.

References

1. Bezault, E., Howard, M., Kogtenkov, A., Meyer, B., Stapf, E.: Eiffel analysis, design and programming language. ECMA International, Technical report ECMA-367 (2005)
2. De Waen, J., Dinh, H.T., Cruz Torres, M.H., Holvoet, T.: Scalable multirotor UAV trajectory planning using mixed integer linear programming. In: 2017 European Conference on Mobile Robots (ECMR), pp. 1–6 (2017)

3. Dinh, H.T., Cruz Torres, M.H., Holvoet, T.: Dancing uavs: Using linear programming to model movement behavior with safety requirements. In: International Conference on Unmanned Aircraft Systems, pp. 326–335. IEEE (2017). https://doi.org/10.1109/ICUAS.2017.7991352. https://lirias.kuleuven.be/1571693

4. Dinh, H.T., Cruz Torres, M.H., Holvoet, T.: Combining planning and model checking to get guarantees on the behavior of safety-critical UAV systems. In: ICAPS Workshop on Planning and Robotics. ICAPS Workshop on Planning and Robotics (2018)

5. Euro, N.: Euro NCAP 2025 roadmap: in pursuit of vision zero. Belgium, Leuven (2017)

6. European Commission: Revision of the EU General Safety Regulation and Pedestrian Safety Regulation (2018). https://www.unece.org/fileadmin/DAM/trans/doc/2018/wp29grsp/GRSP-63-31e.pdf

7. Fisher, M., Dennis, L., Webster, M.: Verifying autonomous systems. Commun. ACM **56**(9), 84–93 (2013)

8. Giacalone, J., Bourgeois, L., Ancora, A.: Challenges in aggregation of heterogeneous sensors for autonomous driving systems. In: 2019 IEEE Sensors Applications Symposium (SAS), pp. 1–5 (2019)

9. Gu, R., Marinescu, R., Seceleanu, C., Lundqvist, K.: Towards a two-layer framework for verifying autonomous vehicles. In: Badger, J.M., Rozier, K.Y. (eds.) NASA Formal Methods, pp. 186–203. Springer, Cham (2019). https://doi.org/10.1007/978-3-030-20652-9_12

10. Infineon: AURIX^{TM} 32-bit microcontrollers for automotive and industrial applications (2020). https://www.unece.org/fileadmin/DAM/trans/doc/2018/wp29grsp/GRSP-63-31e.pdf

11. Maoz, S., Ringert, J.O.: Gr (1) synthesis for LTL specification patterns. In: Proceedings of the 2015 10th Joint Meeting on Foundations of Software Engineering, pp. 96–106 (2015)

12. Microsoft: AURIX^{TM} 32-bit microcontrollers for automotive and industrial applications (2004). http://research.microsoft.com/en-us/projects/specsharp

13. Pek, C., Koschi, M., Althoff, M.: An online verification framework for motion planning of self-driving vehicles with safety guarantees. In: AAET-Automatisiertes und vernetztes Fahren (2019)

14. SAE International: Automated Driving Levels of Driving Automation are Defined in New SAE International Standard J3016 (2014). http://www.sae.org/autodrive

15. Schwarting, W., Alonso-Mora, J., Rus, D.: Planning and decision-making for autonomous vehicles. Annual Rev. Control, Robot. Autonom. Syst. **1**(1), 187–210 (2018). https://doi.org/10.1146/annurev-control-060117-105157

16. Shalev-Shwartz, S., Shammah, S., Shashua, A.: On a formal model of safe and scalable self-driving cars. CoRR abs/1708.06374 (2017). http://arxiv.org/abs/1708.06374

17. Wolff, E.M., Murray, R.M.: Optimal control of nonlinear systems with temporal logic specifications. In: Inaba, M., Corke, P. (eds.) Robotics Research. STAR, vol. 114, pp. 21–37. Springer, Cham (2016). https://doi.org/10.1007/978-3-319-28872-7_2

18. Wongpiromsarn, T., Topcu, U., Ozay, N., Xu, H., Murray, R.M.: Tulip: a software toolbox for receding horizon temporal logic planning. In: Proceedings of the 14th International Conference on Hybrid Systems: Computation and Control, pp. 313–314 (2011)

Modelling Train Driver Behaviour in Railway Co-simulations

Tomas Hotzel Escardo[1], Ken Pierce[1(✉)], David Golightly[2], and Roberto Palacin[2]

[1] School of Computing, Newcastle University, Newcastle upon Tyne, UK
{T.Escardo,kenneth.pierce}@newcastle.ac.uk
[2] School of Engineering, Newcastle University, Newcastle upon Tyne, UK
{david.golightly,roberto.palacin}@newcastle.ac.uk

Abstract. The performance of a cyber-physical system (CPS) is affected by many factors, however the impact of human performance on a CPS is often overlooked. Modelling and simulation play an important role in understanding CPSs, and co-simulation offers a way to easily incorporate human performance models into co-models of CPSs. This paper demonstrates an initial human performance model in the form of a train driver model in the railway domain. The model is linked to models of the rolling stock and movement authority using the Functional Mock-up Interface (FMI). Initial results are presented and a discussion of future directions for the work.

Keywords: Human performance · FMI · Railways · Co-simulation

1 Introduction

Cyber-Physical Systems (CPSs) are systems constructed of interacting hardware and software elements, with components networked together and distributed geographically [23]. Importantly, humans are a key component of CPS design. For example, Rajkumar et al. call for "systematic analysis of the interactions between engineering structures, information processing, humans and the physical world" [27, p. 734]. Humans may act as operators, acting with or in addition to software controller; or as users, interpreting data from or actions of the CPS.

The railways are a good example of such a CPS, involving physical infrastructure and assets, coupled to increasing levels of digital control systems. Ultimately, this system is operated/supervised by people (drivers, signallers/dispatchers, station staff), for users (the passengers). It is a domain where systematic analysis is important, for technical design, and for delivering the operational framework (i.e. the timetable). This is increasingly relevant to goals such as optimising the network for low-carbon performance [5].

Model-based design techniques offer opportunities to achieve this systematic analysis. When considering the diverse nature of disciplines however, and therefore diverse modelling techniques and even vocabulary, creating models that sufficiently capture all aspects of a CPS present a challenge. Rail modelling is

© Springer Nature Switzerland AG 2021
L. Cleophas and M. Massink (Eds.): SEFM 2020 Workshops, LNCS 12524, pp. 249–262, 2021.
https://doi.org/10.1007/978-3-030-67220-1_19

no exception [7]. Multi-modelling techniques present one solution, where models from appropriate disciplines are combined into a multi-model and are analysed, for example, through co-simulation [16,21].

An open challenge in CPS design is how to accurately reflect human capabilities and behaviours. Without considering such an important part of a CPS, observed performance in an operational context may differ from that predicted by models [13,19]. For example, train systems fail to achieve optimal performance due to drivers' not following eco-driving advice [26], or optimal performance requires unrealistic demands on operators (e.g. challenging peaks or reduced wellbeing) [24].

There is a wealth of modelling of human behaviours within the field of ergonomics—the study of the human role in work and systems design to ensure safety, health and performance—which could be applicable to multi-modelling [15]. In return, multi-modelling would seem to offer an ideal way for these existing models to be incorporated into system-level models of CPSs. In the rail domain, a key contribution would be more realistic simulation of driver behaviour. Rail modelling rarely takes into account human performance characteristics. Either human performance is not considered, or the human operator is assumed to perform perfectly (e.g. the driver always drives to the timetable), or a degree of noise is introduced to reflect variability in operator performance. In practice, drivers are often optimising their performance with knowledge and experience of their route, the wider system state or anticipated implications of their actions (e.g. in conditions of low adhesion). In this sense, operator variability is rational, and describable, rather than stochastic and simply 'noisy' [11,26]. As well as this rational adaptation of behaviour, operational roles are also prone to more general limitations on human performance. For the driver this is often an issue of fatigue and decrements to vigilance [12] or performance change due to underload [10].

A very early model of driver behaviour had been built to control an urban train network [14]. However, this model was limited in a number of ways:

- The model only addressed a very small part of an urban rail network;
- This part of the urban rail network only used a rudimentary two-aspect signalling, which is unrepresentative of the majority of mainline rail in Great Britain;
- Neither the infrastructure model, nor driver behaviour model, had a concept of maximum line speed (i.e. the speed limit): this was not necessary given the short distance in the urban rail model, but is a significant limitation for wider application of the model; and
- The driver model only captured basic performance characteristics; other factors, such as fatigue or workload, were not expressed in the model.

The rest of this paper presents work to develop both a more sophisticated driver model, and relevant enhancement of the initial infrastructure and train models to provide the degrees of freedom necessary for more complex behaviour. By doing so, it gives a demonstration of the possibility to include human operator modelling within CPS multi-model simulation.

The remainder of this paper is structured as follows. Section 2 provides background on behavioural modelling and co-simulation. Section 3 presents the elements of the driver behaviour model and its implementation as a Functional Mockup Unit (FMU). Section 4 presents a railway case study and shows co-simulation results with the driver model included, with results co-simulation results presented in Sect. 5. Finally, Sect. 6 presents conclusions from the work and suggests direction for future work, both in terms of developing the driver model, and wider developments for modelling human performance in CPS with multi-modelling.

2   Background

This section provides background and definitions relating to human behaviour modelling and co-simulation, which serve as a basis for the results presented later in the paper in Sects. 3 and 4.

2.1   Co-simulation

The FMI (Functional Mockup Interface) standard is an emerging standard for co-simulation of multi-models, where individual models are packaged as Functional Mockup Units (FMUs). FMI defines an open standard that any tool can implement, and currently more than 30 tools can produce FMUs, with the number expected to surpass 100 soon, taking into account partial or upcoming support[1]. INTO-CPS is a tool chain based on FMI for the modelling and analysis of CPSs [21]. At the core of the tool chain is Maestro [28], an open-source and fully FMI-compliant co-simulation engine supporting variable- and fixed-step size Master algorithms across multiple platforms. Maestro includes advanced features for simulation stabilisation and hardware-in-the-loop simulation. INTO-CPS also provides a graphical front end for defining and executing co-simulations.

The Vienna Development Method (VDM) [22] is a family of formal languages based on the original VDM-SL language for systematic analysis of system specifications. The VDM-RT language allows for the specification of real-time and distributed controllers [29], including an internal computational time model. VDM-RT is an ex-tension of the VDM++ object-oriented dialect of the family, which itself extends the base VDM-SL language. VDM is a state-based discrete-event (DE) language, suited to modelling system components where the key abstractions are state, and modifications of that state through events or decisions. Overture is an open-source tool for the definition and analysis of VDM models, which supports FMU export.

The 20-sim tool[2] supports modelling and simulation of physical formula based on differential equations. 20-sim can represent phenomena from the mechanical, electrical and even hydraulic domains, using graphs of connected blocks. Blocks may contain further graphs, code or differential equations. The connections represent channels by which phenomena interact; these may represent signals (one-way) or bonds (two-way). Bonds offer a powerful, compositional and domain-independent way to model physical phenomena, as they carry both effort and

---

[1] http://fmi-standard.org/tools/.
[2] http://www.20sim.com/.

flow, which map to pairs of familiar concepts, e.g. voltage and current. 20-sim is a continuous-time tool which solves differential equations numerically to produce high-fidelity simulations of physical components.

2.2   Human Behaviours

There are many aspects to human behaviour that require understanding or study when humans are included in a CPS. Several of these are relevant to a potential train driver model.

Reaction Time. At high speed, slow reaction times can easily lead to overrunning and potential accidents, particularly on high-speed rail. Reaction time in humans can be broken down into simple reaction time (the minimal time needed to respond to a stimulus) and choice response time (time taken for a decision). Simple reaction time has been shown to increase with age [31]. The range of the mean is between 0.55 ms to 1.7 ms increase per year after the age of 18. Error rates also showed a difference between ages; the 18–24 bracket showed the highest error rate, at around 8% vs 3% for other age groups [31].

Fatigue. Railway drivers are usually shift-workers, so often have night shifts, as well as continuously varying periods of the day when they work [12]. Shift workers have been shown to have up to 50% higher (264 ms up from 182 ms) visual reaction time, as a simple reaction time response [18]. Despite being only performed with only young male participants, this research is relevant for rail study due to rail driver jobs being heavily male dominated, at 93.5% male in the UK, though this picture is changing [8].

There are studies on how sleep deprivation affects error rates in various domains. Railway specific studies show similar patterns of fatigue in shift-workers in terms of work length, despite showing that severe sleepiness (higher fatigue) decreased with age [17]. Accident rates on both passenger and freight rail increased with the number of consecutive hours worked, and higher fatigue levels can show higher rates of lapses in attention leading to delayed responses, which in the tested case was of changing speed restrictions [6,9]. Other factors may also affect train drivers, but which are more difficult to account for, include:

- Noise—"High continuous noise levels increase arousal, reducing error in repetitive, monotonous tasks. However, the performance of complex tasks decreases with reduced accuracy, poor response to the unexpected, increased annoyance and induction of fatigue" [3].
- Vibration—"Vibration may cause significant changes in arterial pressure, reduce tactile feedback, induce fatigue and motion sickness" [3].

Modelling driver behaviour correctly is vital to the accuracy of the model; differences in acceleration and brake usage will produce different results such as energy usage. Some alternative strategies lead to similar energy usage, such as low but consistent acceleration vs high acceleration with coasting [11]. However, the variation in driver approach can lead to inconsistent and contradictory energy consumption, especially around the use of train control notches, since throttle notches do not necessarily show linear power increases, due to differences between trains. This error might not have shown up in real rail usage simply due to assuming that drivers are following the appropriate control advice, showing the need for driver models covering all variables and variations of driver style [26], but can affect simulation modelling. Even minor differences between simulated and observed performance can have a significant impact, for example at very high capacity sections of track, junctions and stations.

Powell et al. [25] produce a model of different sections of urban rail (also of sections of the Tyne & Wear Metro), superseding previous the use of spreadsheets for such calculations, with the specific goal of optimising the driving profiles to mirror real-life drivers, producing a reasonably well-fitting model compared to the collected data. Errors were mainly due to the perceptions of when to brake being different (data was gathered during autumn when drivers are told to brake more gently to avoid sliding due to low wheel adhesion caused by leaves). Other issues which could not be accounted for included other random conditions such as wind (influencing drag) and lack of adhesion caused by other weather conditions.

Given the various aspect of human behaviours that can affect the operation of cyber and physical elements, it is important to model such behaviours and

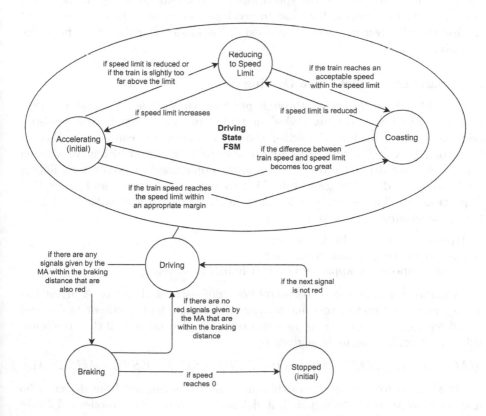

Fig. 1. Finite state machine diagram showing the primary states of the driver model.

ensure that they can be incorporated with models of the other elements, to predict better the real-world performance of CPSs.

3   Driver Behaviour Model

A driver model was developed using VDM-RT encoding the basic operation of a driver, assuming perfect response time and no delay. The main behaviours of a driver can be represented in four distinct states [26]:

1. Accelerating—throttling to reach the speed limit
2. Cruising—short combined phases of:
   (a) Throttling
   (b) Free-running—period of no throttle
   (c) Applying brake
3. Coasting—Long period of free-running
4. Braking—apply the brake to follow the new lower speed limit, stop at a station or for a signal.

The state machine in Fig. 1 shows how these are encoded in the model, where accelerating, cruising and coasting are part of the driving state. Although the driver must respond to the speed limits and signals, it us up to the driver how to drive, for example how fast to accelerate or brake. These is the core of the individual driver behaviours encoded in the model which we describe in the following sections.

3.1   Baseline vs. Defensive Driving

Drivers differ in style, either through practice, through specific training (e.g. in more fuel-efficient driving styles) or under guidance from Driver Advisory systems. In terms of how the driver chooses to drive, this can be described in terms of a 'baseline' driving style, and a more fuel efficient or 'defensive' driving style. Defensive driving involves using lower application of power with lighter acceleration, and earlier, gentler braking to reduce energy use and generation of particulates. The model encodes these in the following way, with the default being the baseline driving style:

- Higher throttle and brake values;
- Shorter braking distances set; and
- Slower response to applying brake or reducing power.

A parameter is included to control how 'defensive' a driver is. A driver has a maxi-mum and minimum value to which it can set the throttle or brake, and this defensiveness parameter is used as a ratio for what the actual throttle/brake value is set to. The formula is given by:

$$((MAX\_VAL - MIN\_VAL) * (1.0 - DEFENSIVENESS)) + MIN\_VAL)$$

This allows for flexibility in exploring behaviours ranging from drivers who may be slow to react to signals but defensive in terms of throttle and brake values, or any other combination.

3.2   Response Time

Response time this is the most basic factor in terms of the variability of humans in a CPS. As described above, this is affected by a variety of factors. Response time is encoded in the driver model by separating input events, in this case signals and speed limit signs, and adding them to a delay queue through which they must proceed before being processed. The time events spend in the delay queue is dependent on four parameters:

- A 'base human reaction time' constant;
- A notional age of the driver;
- The number of hours the driver has been 'on shift'; and
- Whether it is night or day.

The default values of these parameters were calculated based on values and patterns described in Härmä et al. [17], Hemamalini et al. [18], and Woods et al. [31].

3.3   Fatigue

Fatigue is a major factor in the performance of rail drivers in terms of awareness and failure. Fatigue leading to driver mistakes is a probabilistic process: tiredness makes mistakes more likely. Within this driver model, mistakes relate to the processing of events and there are two types. The first are 'errors', where an event (signal or speed limit) is completely ignored. The second are 'lapses', where reduced concentration induces additional delay. Both can be realised using the delay queue, by not adding an event to the queue or adding an additional delay, respectively.

Errors and lapses are controlled by a by a pseudo random number generator. To make the model deterministic, a random seed parameter is included if a specific run needs to be repeated. The 'error rate' is controlled variable, in the range (0, 1), and is calculated based on how long the driver has been on shift and the time of day, using assumptions from Härmä et al. [17] and Volná and Šonka [30]. The 'lapse rate' works in a similar way but has random numbers controlling both the length and severity of lapses. Different lapse rates are encoded for different degrees of fatigue after Woods et al. [31], as below:

Length of lapse (s)	Probability of lapse (lapse rate)		
	Low fatigue	Medium fatigue	High fatigue
0	0.67	0.58	0.3
0–0.5	0.33	0.34	0.4
0.5–2.0	0.01	0.08	0.2
2.0–4.0	0	0	0.1
4.0–6.0	0	0	0.05
6.0–8.0	0	0	0.05

4   Railway Co-simulation

The driver model described above was exported to an FMU and incorporated into a railway multi-model for co-simulation. The multi-model is based on the urban rail model described by Golightly et al. [14], with the train and signalling model updated to mainline speeds, and the basic driver model replaced with the more sophisticated implementation. An overview of the multi-model structure is given in Fig. 2, showing the relationship between the FMUs, the main data exchanged between them, and the parity between them, i.e. multiple driver-train pairs are controlled by a single movement authority and power system. The FMUs in the multi-model are:

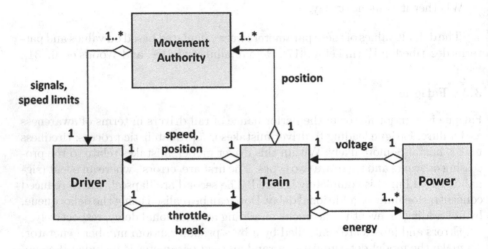

Fig. 2. Abstract UML diagram showing the relationship between the FMUs in the railway case study, their data exchange and parity.

- Movement Authority (VDM-RT): controls the signalling and provides the driver FMU with the signals and speed limits which they can 'see' up ahead.
- Train (20-sim): a model of an Intercity 125 train composed of three carriages, including variations for a lighter model with regenerative braking.
- Power (20-sim): a simple power model providing line voltage to the train; does not model voltage drop or other phenomena.
- Driver (VDM-RT): the model described in the previous section.

Though there are exceptions, the majority of Great Britain's mainline rail network uses a signalling method called track circuit block signalling, where only one train is allowed in a block of track, with each block being protected by colour light signals. In conjunction with this, GB rail also functions with 'route' signalling, meaning that signals inform the driver of the state of the signals that follow that signal. While signals can have up to seven states (or 'aspects'), the four most common configurations are:

- Green: clear; proceed subject to track or train speed limits.
- Double-yellow: preliminary caution; the next signal is yellow.
- Yellow: caution, the next signal is danger/stop.
- Red: danger/stop. Stop at this signal unless given explicit permission to proceed by a signaller.

Green and yellow signals are known as 'proceed aspects' since the driver may pass the signal. These signals are controlled by a 'movement authority' which knows the state of the system. Drivers must respect these signals, along with any speed limits on the track which affect how fast the train can go despite a green signal. The movement authority model used signal type and position, collected from open source rail models, to create a realistic infrastructure model that capture both signal position and line speeds for a 25 km stretch of the East Coast Main Line, between Newcastle and Durham.

5   Results

A co-simulation was run with a single train performing a journey of a few minutes from Newcastle to Durham in the North East of England. Three driver profiles were used, a baseline driver, a fully defensive driver and an intermediate driver with 50% defensiveness (a defensive parameter of 0.5). Plots for different levels of fatigue are shown in Fig. 3. This allows a comparison of the number of and length of different lapses. These show the intended smaller gap between 'low' and 'medium' fatigued driver models, and a large difference between them and the 'high' fatigued driver, as intended to reflect lapses described by Dorrian et al. [9]. The 'medium' fatigued driver still has more longer lapses than the 'low' fatigued driver. One issue that is immediately obvious in the high fatigued driver is that the lapses are both long and repeated consecutively. One possible way to fix this would be for the randomness around laps-es to be pseudo-random (reduced likelihood after a long lapse has occurred), as research has shown train drivers will be more alert in the period after making a mistake [20].

Plots for the velocity, acceleration and energy usage for these three drivers are shown in Fig. 4. When comparing velocity, the baseline driver is much closer to the speed limit at all times than both more defensive drivers, despite with the defensive drivers braking earlier in response to speed limits compared to the baseline driver. The drawbacks of using the higher throttle and braking values is shown by the much larger changes in acceleration as the defensiveness of a driver decreases. This highlights some of the issues with the parameters used in this model. The values acceleration and jerk (rate of change of acceleration) are unlikely to be comfortable for either the driver or passengers, showing an important avenue in which driver models could be improved in the future.

The cost of baseline driving is further demonstrated when considering energy consumption, as shown in Fig. 4. This reflects what has been shown in previous research, including and an earlier study with this model [26], with less defensive drivers showing much larger energy consumption. The following table shows the energy consumption and distance travelled by each driver:

Low Fatigue Lapses

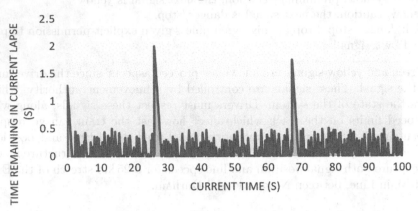

Medium Fatigue Lapses

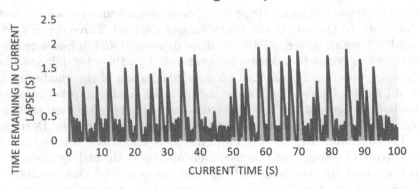

High Fatigue Lapses

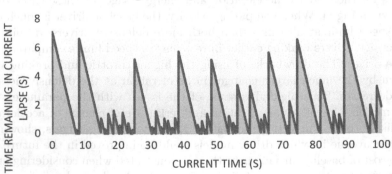

Fig. 3. Plots showing the rate and duration of lapses during a co-simulation for low, medium and high fatigue drivers.

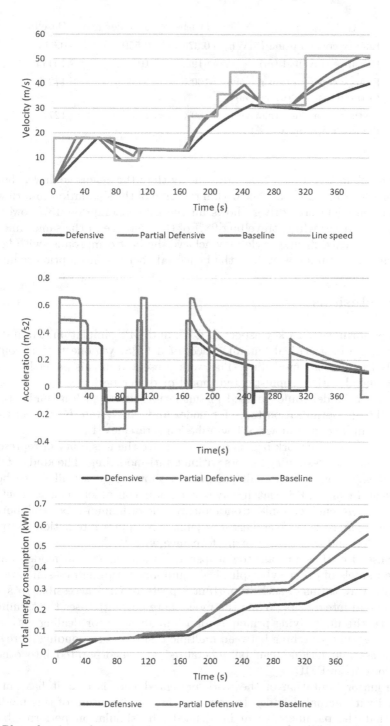

Fig. 4. Plots showing velocity, acceleration and energy usage for different driving styles.

	Defensive	50% defensive	Baseline
Energy consumption (MWh)	0.375	0.559	0.643
Distance travelled (m)	8192	10197	10444
Energy consumption relative to the defensive driver (%)	100%	149%	172%
Distance travelled relative to the defensive driver (%)	100%	124%	127%

The baseline driver uses 72% more energy than the defensive driver, however the train also travels 27% farther within the time of the simulation run. However, the somewhat defensive driver shows an energy consumption 13% lower than the base-line driver whilst travelling 98% of the distance in the same time. This shows how a mixed driving style may achieve the best compromise with keeping to current travel timing as well as the beneficial energy consumption reductions.

6 Conclusions

Human performance is a key aspect of CPS, and CPS simulation. Train driving is an example of a critical human aspect of a CPS, yet one that is routinely overlooked in systems modelling. The work presented in this paper has started to move ahead with a sophisticated model of driver performance. In practical terms, such a model could be used in simulation efforts such as understanding the viability of optimisation tools for nodes and rail system bottlenecks [2], or for understanding the impact of low-adhesion strategies [1].

More broadly, the work begins to demonstrate the feasibility of representing human performance-shaping factors within multi-modelling. The kind of fatigue, workload, and competence/skill parameters shown here are likely to be just as relevant in any CPS that involves a human controller such as control of an aircraft/aero-engine, semi-autonomous vehicle or human-robot collaboration. Moreover, there is scope to add error or human-computer interaction parameters to the model, and thus represent performance with ICT.

The use of FMI as a means to encapsulate human behaviour models and link them to models of other cyber-physical components through co-simulation can serve as a way to make such important aspects of CPS accessible to systems engineers and integrators. While the model here currently uses the terminology of driving, the under-lying principles of a human operator dealing with events and actions while switching between multiple tasks, and responding more slowly under stress and fatigue, can be generalized and incorporated into other co-simulations as an FMU.

One major limitation of the work presented here is that it has not been validated with actual driving. To some extent, the intention of this model has shown that the parameters can be defined, the simulation performs broadly within expectations, and the human performance can be modelled to influence system performance. A next critical step will be to consult both with actual drivers, and with sources such as On-Train Data Recorders [4].

Acknowledgements. This work was partially supported by the Rail Safety and Standards Board (RSSB) as the Digital Environment for Collaborative Intelligent Decarbonisation (DECIDe) project (COF-IPS-06).

References

1. Alturbeh, H., Stow, J., Tucker, G., Lawton, A.: Modelling and simulation of the train brake system in low adhesion conditions. Proc. Inst. Mech. Eng. Part F J. Rail Rapid Transit **234**(3), 301–320 (2020)
2. Armstrong, J., Preston, J.: Capacity utilisation and performance at railway stations. J. Rail Transp. Plan. Manage. **7**(3), 187–205 (2017)
3. Arnstein, F.: Catalogue of human error. Br. J. Anaesth. **79**(5), 645–656 (1997)
4. Balfe, N.: Human factors applications of on-train-data-recorder (OTDR) data: an exploratory study. Cogn. Technol. Work, 1–15 (2020). https://doi.org/10.1007/s10111-019-00622-y
5. Basile, D., Di Giandomenico, F., Gnesi, S.: Statistical model checking of an energy-saving cyber-physical system in the railway domain. In: Proceedings of the Symposium on Applied Computing, pp. 1356–1363 (2017)
6. Chang, H.L., Ju, L.S.: Effect of consecutive driving on accident risk: a comparison between passenger and freight train driving. Accid. Anal. Prev. **40**(6), 1844–1849 (2008)
7. Chen, L., James, P., Kirkwood, D., Nguyen, H.N., Nicholson, G.L., Roggenbach, M.: Towards integrated simulation and formal verification of rail yard designs-an experience report based on the UK east coast main line. In: 2016 IEEE International Conference on Intelligent Rail Transportation (ICIRT), pp. 347–355. IEEE (2016)
8. Davies, C.: Rail union in push for more female and BAME train drivers. The Guardian (2019). https://www.theguardian.com/uk-news/2019/jun/17/female-bame-train-drivers-aslef
9. Dorrian, J., Roach, G.D., Fletcher, A., Dawson, D.: The effects of fatigue on train handling during speed restrictions. Transp. Res. Part F Traffic Psychol. Behav. **9**(4), 243–257 (2006)
10. Dunn, N., Williamson, A.: Driving monotonous routes in a train simulator: the effect of task demand on driving performance and subjective experience. Ergonomics **55**(9), 997–1008 (2012)
11. Ellis, R., et al.: Observations of train control performance on a camshaft-operated DC electrical multiple unit. Proc. Inst. Mech. Eng. Part F J. Rail Rapid Transit. **230**(4), 1184–1201 (2016)
12. Filtness, A.J., Naweed, A.: Causes, consequences and countermeasures to driver fatigue in the rail industry: the train driver perspective. Appl. Ergon. **60**, 12–21 (2017)
13. Flach, J.M.: Complexity: learning to muddle through. Cogn. Technol. Work **14**(3), 187–197 (2012)
14. Golightly, D., Gamble, C., Palacin, R., Pierce, K.: Multi-modelling for decarbonisation in urban rail systems. Urban Rail Transit **5**(4), 254–266 (2019)

15. Golightly, D., Gamble, C., Palacin, R., Pierce, K.: Applying ergonomics within the multi-modelling paradigm with an example from multiple UAV control. Ergonomics **63**, 1–17 (2020)
16. Gomes, C., Thule, C., Broman, D., Larsen, P.G., Vangheluwe, H.: Co-simulation: a survey. ACM Comput. Surv. **51**(3), 49:1–49:33 (2018)
17. Härmä, M., Sallinen, M., Ranta, R., Mutanen, P., Müller, K.: The effect of an irregular shift system on sleepiness at work in train drivers and railway traffic controllers. J. Sleep Res. **11**(2), 141–151 (2002)
18. Hemamalini, R., Krishnamurthy, N., Saravanan, A.: Influence of rotating shift work on visual reaction time and visual evoked potential. J. Clin. Diagn. Res. JCDR **8**(10), BC04 (2014)
19. Hollnagel, E., Woods, D.D.: Joint Cognitive Systems: Foundations of Cognitive Systems Engineering. CRC Press, Boca Raton (2005)
20. Itoh, K., Tanaka, H., Seki, M.: Eye-movement analysis of track monitoring patterns of night train operators: effects of geographic knowledge and fatigue. In: Proceedings of the Human Factors and Ergonomics Society Annual Meeting, vol. 44, pp. 360–363. SAGE Publications, Sage CA: Los Angeles, CA (2000)
21. Larsen, P.G., Fitzgerald, J., Woodcock, J., Gamble, C., Payne, R., Pierce, K.: Features of integrated model-based co-modelling and co-simulation technology. In: Cerone, A., Roveri, M. (eds.) SEFM 2017. LNCS, vol. 10729, pp. 377–390. Springer, Cham (2018). https://doi.org/10.1007/978-3-319-74781-1_26
22. Larsen, P.G., et al.: The VDM-10 language manual. Technical report TR-2010-06, The Overture Open Source Initiative, April 2010
23. Lee, E.A.: Cyber physical systems: design challenges. Technical report. UCB/EECS-2008-8, EECS Department, University of California, Berkeley, January 2008. http://www.eecs.berkeley.edu/Pubs/TechRpts/2008/EECS-2008-8.html
24. de Mattos, D.L., Neto, R.A., Merino, E.A.D., Forcellini, F.A.: Simulating the influence of physical overload on assembly line performance: a case study in an automotive electrical component plant. Appl. Ergon. **79**, 107–121 (2019)
25. Powell, J.P., Fraszczyk, A., Cheong, C.N., Yeung, H.K.: Potential benefits and obstacles of implementing driverless train operation on the Tyne and Wear Metro: a simulation exercise. Urban Rail Transit **2**(3–4), 114–127 (2016)
26. Powell, J., Palacín, R.: A comparison of modelled and real-life driving profiles for the simulation of railway vehicle operation. Transp. Plan. Technol. **38**(1), 78–93 (2015)
27. Rajkumar, R., Lee, I., Sha, L., Stankovic, J.: Cyber-physical systems: the next computing revolution. In: 2010 47th ACM/IEEE Design Automation Conference (DAC), pp. 731–736 (2010)
28. Thule, C., Lausdahl, K., Larsen, P.G., Meisl, G.: Maestro: The INTO-CPS Co-Simulation Orchestration Engine (2018, submitted to Simulation Modelling Practice and Theory)
29. Verhoef, M., Larsen, P.G., Hooman, J.: Modeling and validating distributed embedded real-time systems with VDM++. In: Misra, J., Nipkow, T., Sekerinski, E. (eds.) FM 2006. LNCS, vol. 4085, pp. 147–162. Springer, Heidelberg (2006). https://doi.org/10.1007/11813040_11
30. Volná, J., Šonka, K.: Reaction time measurement in different periods of shift work at nurses. České Budějovice 2010, p. 147 (2010)
31. Woods, D.L., Wyma, J.M., Yund, E.W., Herron, T.J., Reed, B.: Factors influencing the latency of simple reaction time. Front. Hum. Neurosci. **9**, 131 (2015)

Cross-level Co-simulation and Verification of an Automatic Transmission Control on Embedded Processor

Cinzia Bernardeschi[1]([✉]), Andrea Domenici[1], Maurizio Palmieri[1],
Sergio Saponara[1], Tanguy Sassolas[2], Arief Wicaksana[2], and Lilia Zaourar[2]

[1] Department of Information Engineering, University of Pisa, Pisa, Italy
cinzia.bernardeschi@unipi.it
[2] CEA, LIST, Gif-sur-Yvette CEDEX, France

Abstract. This work proposes a method for the development of cyber-physical systems starting from a high-level representation of the control algorithm, performing a formal analysis of the algorithm, and co-simulating the algorithm with the controlled system both at high level, abstracting from the target processor, and at low level, i.e., including the emulation of the target processor. The expected advantages are a smoother and more controllable development process and greater design dependability and accuracy with respect to basic model-driven development. As a case study, an automatic transmission control has been used to show the applicability of the proposed approach.

1 Introduction

Simulation is an essential activity in model-driven development (MDD), as it enables developers to implement virtual prototypes of their designs at all required levels of abstraction, until the design has been refined and validated to the point that it can be prototyped in hardware and code.

The existence of design models at different levels of abstraction makes it convenient to use different tools and formalisms for each model. Let us consider, for example, the control part of a cyber-physical system (CPS). This component must implement a high-level control algorithm that can be defined mathematically and modelled and simulated with the well-known tools together with a model of the controlled plant, usually built with the same tools, e.g., with Simulink. From now on, it is tacitly assumed that simulations include a plant model built with Simulink. Further refinements lead to a lower-level design including programming code and a hardware platform of the target processor(s), including system software/microcode. In this work, the terms *hardware* or *platform* refer to the physical and software infrastructure that executes the control

Work partially supported by the EPI (European Processor Initiative) project, EU-H2020 and by the Italian Ministry of Education and Research (MIUR) in the framework of the CrossLab project (Department of Excellence).

L. Cleophas and M. Massink (Eds.): SEFM 2020 Workshops, LNCS 12524, pp. 263–279, 2021.
https://doi.org/10.1007/978-3-030-67220-1_20

algorithm. At this level of abstraction, it is possible to run the developed programming code on a simulated or real processor architecture. At this point, the hardware platform is the critical issue, as it affects significantly system performance and dependability. Accurate simulation of the platform makes it possible to evaluate hardware from different vendors, compare different architectural solutions, and choose optimal parameter configurations.

Processor simulation, however, requires formalisms and tools that are quite different from those used for high-level design. This mismatch is both conceptual and organisational, since the two levels require different fields of expertise, and is a potential source of issues ranging from project delays to design errors.

This paper introduces the concept of *cross-level simulation*, an approach to MDD aimed at bridging the gap between high- and low-level models, preserving coherence between them, and furthermore enabling formal verification of the control algorithm. A key point in this concept is that the implementation of the control algorithm is the same for both levels of simulation, and that the implementation is produced automatically from a formally verifiable model. Depending on application characteristics or project constraints, verification may be performed upfront on the formal model, or concurrently with simulation, the two activities cross-checking each other. This approach, summarised in Fig. 1, is an extension to the common development flows based on Simulink-like tools, and relies on various tools for model construction, transformation, and simulation. More precisely, (i) a prototyping environment is used to create a high-level, automaton-based model and generate both a logic-based specification and C code; (ii) the specification is used to verify the control algorithm with a theorem-proving environment; (iii) high-level simulation executes the controller code together with a plant model, e.g., in Simulink; and (iv) low-level simulation executes the same code on simulated hardware, built in the SESAM/VPSim environment to account for timing behaviour.

In summary, this work extends the common MDD process by (i) starting with an abstract formal model; (ii) automatically generating an executable and a verifiable model; (iii) using formal verification side by side with simulation; (iv) relying on co-simulation to achieve modularity and flexibility of system models; and (v) using the same control code in high- and low-level simulation. The expected advantages are (i) a smoother and more controllable process and (ii) greater design dependability and accuracy with respect to basic MDD, relying on tools that enforce coherency among models at different levels of abstraction. In particular, the same code is used in both high-level and low-level co-simulations.

The rest of the paper is structured as follows: a selection of related works is presented in Sect. 2, the methods and tools for virtual prototyping/verification are introduced in Sect. 3, Sect. 4 illustrates the proposed approach, and Sect. 5 shows its application to a case study. Section 6 concludes the paper.

2   Related Work

Model-driven development relies mainly on simulation to analyse the system behaviour [25]. In cyber-physical systems, simulation often takes the form of

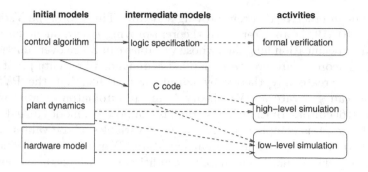

Fig. 1. Overview of cross-modelling.

co-simulation [11], which integrates simulation of heterogeneous sub-systems, modelled and simulated with the appropriate tools.

Due to the complexity of such systems, formal verification can help to assess compliance to safety requirements. Hybrid model checking, which relies on the formalism of Hybrid Automata [12], is used for the analysis of cyber-physical systems with a model-checking approach. One example of a model checking tool is HYCOMP [7] which also relies on Satisfiability Modulo Theories [8]. A complementary approach to model checking is theorem proving. Dynamic Logic [5] is used with the KeyMaeraX [22] theorem prover, which has been integrated with the SPIRAL environment [23] as reported by Franchetti et al. [10]. A framework that integrates simulation and theorem proving is PVSio-web [18], which uses higher-order logic as modelling language, as reported in [3,9,20].

None of these works integrates the processor emulation in the co-simulation.

In the field of electronic systems design, virtual prototypes are extensively used to simulate the behaviour of a system to be built. This allows hardware/software co-design to be better assessed and provides fast software development, reducing time to market. Many tools from academic work [6,14] or electronic design automation vendors [26] address this need. The cited tools are all based on the IEEE SystemC standard [13] meant for model sharing. The standard defines a C++ library providing both a full discrete event simulation environment and design specific architectural constructs to enable hardware design at this level of abstraction. SystemC is further extended by the TLM 2.0 standard [1] which abstracts complex communication channels and protocols into simple function calls for faster simulation.

3   Background

This section provides details on the methods and tools used in this work.

3.1   PVS, Emucharts, and PVSio-web

Theorem proving consists in describing a system as a theory in some logic language, expressing its requirements as theorems, and verifying them with automatic or interactive theorem-proving software. The Prototype Verification System (PVS) [19] is an interactive theorem-proving environment based on a higher-order specification language whose variables can range over functions and predicates. Theorems are proved by issuing commands to execute proof steps.

Users can create any theory by editing a text file, but the PVSio-web toolkit [18] can generate a PVS theory from an automaton created with the Emucharts [16] editor. An Emucharts graph is composed of modes linked by transitions. The graph is complemented by a set of variables whose values, together with the current mode, define the current state. Transitions are defined by a trigger (an event), a guard (an enabling condition on variables), and an action (updates on variables). The variables may range over discrete or continuous domains, they may represent time, state variables, values of time derivatives, and updated values in difference equations, so the Emucharts are a form of hybrid automata as defined in [12]. An example is shown in Sect. 5.1, Fig. 5.

The PVSio-web toolkit can translate Emucharts into various specification and programming languages, including Misra C, a dependability-oriented version of C [17]. It is then possible to create a high-level automaton-based system model and from it generate a PVS theory to assert its properties, and use executable code automatically produced from the same model.

3.2   INTO-CPS

Co-simulation [11] is a technique to couple different simulation units together. A complex system can then be divided in many simpler submodels, and each submodel can be simulated using its specific language and tools. The Functional Mock-up Interface (FMI) [4] is an emerging standard for co-simulation, in which different simulation units, called Functional Mock-up Units (FMU), are orchestrated by a master algorithm in charge of synchronisation and data exchange among the FMUs. The master algorithm adopted in this work is the Co-simulation Orchestration Engine (COE), developed by the INTO-CPS Association [15]. The COE requires as input the logical connections between FMUs, the parameter values and the constraints on the co-simulation time step size. The INTO-CPS application also collects and graphically displays data produced by the co-simulation experiments.

3.3   SESAM/VPSim Environment

Within the SESAM [27] CPS design framework, the VPSim [6] tool targets the fast assembly and simulation of SoC architecture for both design space exploration and hardware/software co-design and validation. VPSim uses Python scripts to define architectures composed of SystemC modules from an extensive library of simulated commercial components including CPUs, interconnects, peripherals, and external controllers from various vendors (Xilinx, Renesas, Cadence, etc.). VPSim relies on the QEMU [2] processor emulator to provide a rich and fast CPU library model. As it targets fast simulation, VPSim is based on a loosely-timed model in compliance with TLM 2.0.

As all SESAM tools, VPSim supports FMI co-simulation for tool interoperability throughout the design stages. It is a fully automated solution for exporting a hardware/software virtual prototype as an FMU. This enables the co-simulation of a whole CPS as detailed in [24]. Therefore, it can easily interface with other FMI-compliant simulators. An FMU encapsulating the virtual platform can be automatically generated based on a high-level description of the hardware/software platform.

4   Cross-level Modelling, Co-simulation and Verification

Cross-level simulation, introduced in Sect. 1, is discussed below.

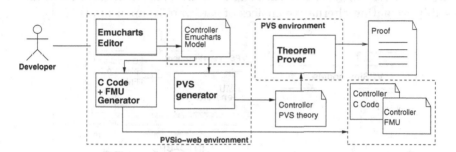

Fig. 2. Model generation and formal verification.

4.1   Development Process

The initial steps of the development process are depicted in Fig. 2: first, the developer uses the PVSio-web environment to generate the Emucharts model of the algorithm under analysis and then the developer uses the PVS and C code generators to generate the PVS theory and the Misra C code.

In the verification activity, the theory is used for two forms of verification: First, the well-formedness of the system model is assessed with the PVS type checker [21], then its compliance to requirements is checked with the theorem prover. The type checker may generate *type checking conditions* (TCC), i.e., statements that must be proved to ensure type correctness. Many TCCs are discharged automatically, others can be proved by the user with one or few commands, but unprovable TCC reveal incompleteness or inconsistency in a theory. The specification of requirements involves translating the desired property from natural or mathematical language to a PVS theorem.

In the high-level simulation phase, the controller is co-simulated to validate it in connection with the plant model (Fig. 3). The latter, built with a tool such as Simulink, is packaged in an FMU.

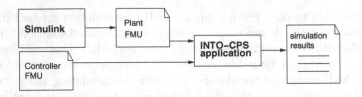

Fig. 3. High-level co-simulation.

In the low-level simulation phase, the controller implementation is compiled and executed on a simulated platform including accurate models of real hardware, such as processors, memories, and controllers. In this phase, performance-related properties are assessed, such as execution time, latencies, or cache misses, possibly evaluating alternative choices of hardware components (Fig. 4).

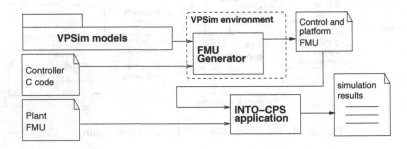

Fig. 4. Low-level co-simulation.

The verification or simulation activity (or both) may fail, e.g., because the results of the co-simulation are not the expected ones or because the discharge of some TCC failed; In this case, the Emucharts model should be refined, using the results of the failed activities, then a new Misra C code and a new PVS theory should be generated, and the two activities iterated until both succeed.

When type checking and co-simulation succeed, it is then possible to specify and prove safety properties of the submodel under analysis.

4.2   Emulation of Processors with VPSim

Any architecture can be simulated by VPSim using the CPU models provided by the QEMU, an open-source hardware emulation solution, although it also allows the integration of model providers that have a SystemC/TLM interface, such as ARM Fast Models. By using QEMU for CPU modelling, we can obtain a very high simulation speed. Such a high performance is achieved mainly by abstracting the architectural aspects of CPUs while maintaining the functional accuracy in the execution. To provide the essential performance statistics to users, the QEMU models is enriched by VPSim in the SystemC simulation domain to model architectural aspects. To that end, all models that are backed by QEMU, including the VirtIOs, are encapsulated in SystemC modules and executed in the context of SystemC threads. Accordingly, QEMU models are controlled by the SystemC kernel like any native module and are transparently exposed to the user like any other VPSim component. For CPS validation purposes, the VPSim virtual platform can be packaged as an FMU by adding the definition of necessary FMI interfaces used in the Python front-end interface. VPSim supports models such as CAN controllers in its hardware library.

5   Automatic Transmission Control Case Study

The case study of this work is based on the Automatic Transmission Controller example from the Matlab documentation[1]. This example is a Simulink/Stateflow model composed of five high-level blocks: the Engine, the ShiftLogic, the Transmission, the Vehicle, and the ManeuvresGUI, which drives the simulation by producing the throttle and brake signals for a passing manoeuvre. The ShiftLogic block is a hybrid automaton, defined in Stateflow, that produces the upshift and downshift commands to the Transmission, according to a *shift schedule* that takes into account the current gear, the vehicle speed, and the throttle position.

The ShiftLogic controller is in a steady state if the vehicle is driving at an intermediate speed between the upshift and downshift thresholds for the current gear and throttle. If the throttle or speed cross a threshold, the controller moves to either of two waiting states (for upshift or downshift). If the speed remains beyond the threshold for a given time, the corresponding shift command is issued and the controller moves to the steady state of the new gear.

5.1   High-Level Virtual Prototyping

The ShiftLogic block has been re-designed as outlined in Sect. 4.1 and packed in an FMU. Another FMU, generated by Simulink, contained the other four blocks. The two FMUs were then co-simulated in INTO-CPS.

Emucharts Model for ATC. The ATC behaviour is specified by the shift schedule. Using the data from the cited example, the shift schedule is defined by the functions represented in Tables 1 and 2. Each row labelled as $n - m$ shows the threshold speed value (in miles per hour) for a shift from gear n to gear m in consecutive intervals of throttle position $t_\%$ (in percent).

The shift schedule has been modelled as an Emucharts diagram. Figure 5 shows the diagram fragment relative to the lower two gears, while the complete diagram (Fig. 13), drawn in a more compact form, is in the Appendix with the transition definitions. Each transition is identified with a label followed by a guard in square brackets and possibly an action in braces.

---

[1] https://www.mathworks.com/help/simulink/slref/modeling-an-automatic-transmission-controller.html.

Table 1. Shift schedule, speed thresholds for upshifts.

Shift	$t_\% \leq 25$	$25 < t_\% \leq 35$	$35 < t_\% \leq 50$	$50 < t_\% \leq 90$	$90 < t_\% \leq 100$
1-2	10	$0.5t_\% - 2.5$	$0.53333t_\% - 3.6667$	$0.425t_\% + 1.75$	40
2-3	30	30	$0.73333t_\% + 4.3333$	$0.725t_\% + 4.75$	70
3-4	50	50	$0.66667t_\% + 26.6667$	$t_\% + 10$	100

Table 2. Shift schedule, speed thresholds for downshifts.

Shift	$t_\% \leq 5$	$5 < t_\% \leq 40$	$40 < t_\% \leq 50$	$50 < t_\% \leq 90$	$90 < t_\% \leq 100$
4-3	35	$0.1429t_\% + 34.28571429$	$t_\%$	$0.75t_\% + 12.5$	80
3-2	20	$0.1429t_\% + 19.2857$	$0.5t_\% + 5$	$0.5t_\% + 5$	50
2-1	5	5	5	$0.625t_\% - 26.25$	30

In the diagram, *stdy* modes represent steady conditions of the ATC, while *up* and *down* modes represent the waiting phases before the ATC is going to issue an upshift or downshift command, respectively, if the speed remains beyond the corresponding threshold for a long enough time.

The transitions depend on variables: the discrete variables *clock* and *gear* and the continuous variables *tht* (throttle), *up_th* (upshift threshold), *dw_th* (downshift threshold), and *speed*. Variable *clock* is a timer that can be incremented by one step or reset, and *gear* is the controller output. The flow conditions [12] for the input variables *tht* and *speed* are defined externally by the Vehicle and ManeuvresGUI models, while the flow conditions for *up_th* and *dw_th* are given by the shift schedule in Tables 1 and 2, respectively.

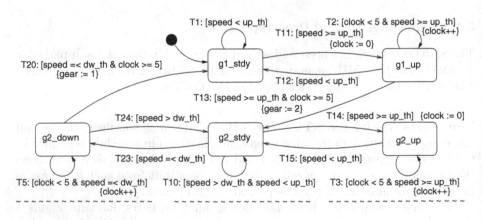

Fig. 5. Fragment of the Emucharts diagram for the shift logic automaton.

Discharging the TCCs for the ATC. The Emucharts model of the ATC was derived from the Stateflow machine in the cited Matlab example. Typechecking the PVS theory generated from the first Emucharts version produced unprovable

TCCs. One is a *coverage* TCC, stating that the disjunction of the guards of all transitions is identically true, i.e., at least one transition is enabled. The complementary *mutual exclusion* TCC requires that at most one transition is enabled. The problem was that some transitions were implicit in the Stateflow model, and was fixed by explicitly adding the needed transitions to the Emucharts model. These transitions are labelled in boldface as T1, T8, T9, and T10 in Fig. 13. This is an example of how a sophisticated type system, together with automatic checking, helps spotting hidden assumptions that are often sources of errors.

Co-Simulation for ATC. Figure 6 shows the co-simulation architecture for the high-level simulation. The co-simulations are executed using a fixed time step of 0.1 s and last 80 s of simulated time. Figure 7 shows the results of a co-simulation run. These results are consistent with those obtained with the original Simulink model, i.e., the shapes of the throttle and speed curves between the first and third upshifts match those of the plot in the MATLAB documentation, except for the initial speed value and the initial transient as discussed in Sect. 5.2 below.

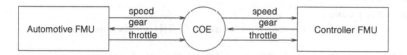

Fig. 6. High-level co-simulation architecture.

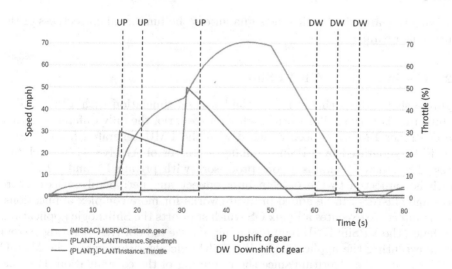

Fig. 7. High-level control co-simulations.

Verification Process for ATC. As an example, one of the safety properties of the ATC algorithm is *"it is never possible to move in one step from a state where gear equals g to a state where gear equals $g \pm 2$"*. This natural language statement can be translated in PVS with the *main_th* theorem, where abs is the absolute value function:

```
gear_T: TYPE = {x: posnat | x<=4}

main_th: THEOREM
FORALL (N:nat, g:gear_T):
  kth_step(N)'gear = g => abs(kth_step(N+1)'gear - g) < 2
```

where kth_step(N) is a data structure containing the values of all variables at step N, and 'gear selects the value of *gear*. The proof has been done by induction on the number of steps and by analysing separately the different values of gear. The proof relies on a few lemmas. For example, it must be proved that in any gear, the computed threshold speed for an upshift (compute_UP_TH(s)'up_th) is greater than the one for a downshift (compute_DW_TH(s)'dw_th):

```
UPgtDW: LEMMA
FORALL (s:State):
  compute_UP_TH(s)'up_th > compute_DW_TH(s)'dw_th
```

Another lemma excludes direct transitions between two *steady* states:

```
gear1: LEMMA
FORALL(N:nat):
  kth_step(N)'mode = g1_steady => kth_step(N+1)'mode /= g2_steady
```

Verifying this and similar theorems guarantees the functional correctness of the control algorithm.

5.2   Co-simulation with VPSim

Figure 8 shows the architecture for the low-level simulation with VPSim. This architecture is very similar to the high-level scenario, the only difference is that the controller FMU has been replaced with the FMU generated from VPSim. The FMU generated by VPSim emulates a cluster of ARMv8 64-bit architectures. The cluster contains 1-core processors with private L1 and L2 caches, which is connected to the on-chip interconnect and peripheral devices. More cores and clusters can be added in future works for more complex applications. The architecture executes a Linux OS which supports the ShiftLogic application.

Hence, the VPSim FMU requires an initial time to boot the operating system before executing the application of the ShiftLogic algorithm, while the MisraC FMU executes the algorithm since the beginning of the co-simulation. For sake of comparison between the two architectures, the value of throttle is always kept close to zero for the first seconds of the co-simulations.

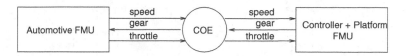

Fig. 8. Low-level co-simulation architecture.

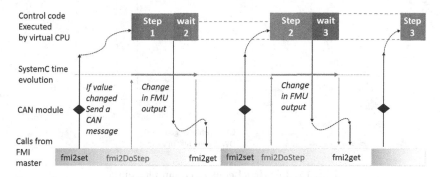

Fig. 9. Co-simulation of VPSim-generated FMU with fast ATC execution.

VPSim enables the timing behaviour of a system to be captured. Hence the duration of the applicative code has a direct impact on the evolution of the FMU outputs that it may change. Simulations have been performed with different execution times of the ATC, obtained by adding delay loops to the original code. When the ATC executes faster than the FMI simulation step demanded by the FMI master (Fig. 9), the behaviour is similar to what is achieved with high-level simulation. (Functions fmi2set and fmi2get are write and read operations, while function fmi2DoStep triggers the execution of one simulation step.) In that case, Fig. 11 shows the behaviour of the co-simulation with the VPSim FMU: The resulting behaviour is consistent with the one previously obtained with the high-level co-simulation. On the contrary (Fig. 10), when the ATC executes slower than the interval between invocations of fmi2DoStep, output value changes are differed to after future simulation steps.

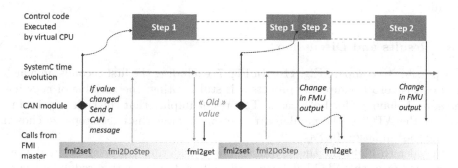

Fig. 10. Co-simulation of VPSim-generated FMU with slow ATC execution.

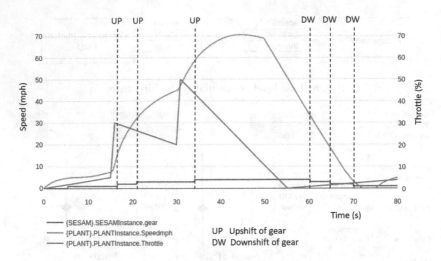

Fig. 11. Low-level control co-simulations.

Having both simulation levels together allows inadequate execution speed of the code under scrutiny to be better underlined. Indeed, when the application execution speed is appropriate, the behaviour is consistent throughout the validation levels as expected. However, it is worth noting that, even if one may set the co-simulation step to an arbitrarily small value, this will create new discrepancies due to processing delay in the low-level simulation compared to the high-level one. If achievable, users should be advised to keep a simulation step duration in line with the expected control decision deadline.

To further improve the alignment of models without this constraint, it could also be beneficial in future work to model the target execution time during high-level simulation by delaying control decisions accordingly. This would render the high-level model behaviour invariant to simulation step choices. Similarly the low-level simulation results shall not be made available too early, even if the control code execute too fast. This would likely be taken into account in real-time code where control decision would never be output before the target deadline.

5.3   Results and Discussion

It is possible to compare Fig. 11 with Fig. 7: excluding the first seconds, in which gear equals zero because the processor is still booting, the variable of gear has the same behaviour for both cases. This result implies that the time required to execute the ATC on the emulated processor is lower than the step-size chosen for the co-simulation (0.1 s).

In order to highlight the advantages of considering low-level co-simulation, the algorithm of the ATC has been artificially extended with a redundant code that increases its computation time so that the execution time of the ATC

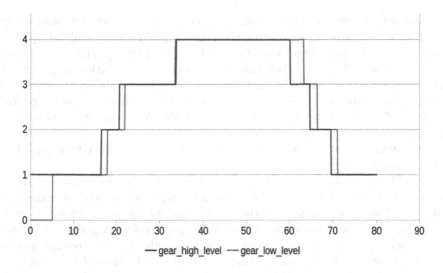

Fig. 12. Gear shifts in high- and low-level simulations.

becomes greater than the co-simulation step-size. Figure 12, shows the comparison between the low- and high-level behaviour of the gear variable with the extended code. It is possible to notice a small time delay in the behaviour of the variable which is due to the different time management: The high-level co-simulation always executes the whole ATC algorithm within a co-simulation step while the low-level now requires more co-simulation steps. Please notice that the delay in the first two gear transitions of the low-level co-simulation has affected the value of speed, increasing it, in such a way that the next gear transition occurs earlier with respect to the high-level simulation and, apparently, with no delay.

The co-simulation results show that the underlying hardware performance (e.g., computation speed), must be taken into account to ensure that the plant can be controlled within the step-size. High-level simulation hides system performance issues that the virtual prototype can highlight.

In all the co-simulation runs, both high-level and low-level, the results obtained with PVS hold, as it is never the case that two gear transitions are executed in two adjacent steps, thus validating the results obtained in Sect. 5.1. Of course, time-related properties will be affected by the different time management and so require an additional step in the verification process, i.e., the specification of the processor in PVS, but this will be subject of future work.

6   Conclusions

This work proposes an approach to the analysis of control algorithms deployed on automotive systems. The approach uses models with different levels of abstraction: a more abstract, high-level, model composed of the executable code of the control algorithm, and a more accurate, low-level, model that also includes the emulation of the hardware executing the code. The high-level analysis provides information on the functional correctness of the model by exploiting both formal reasoning tools such as the PVS theorem prover and simulation tools such as INTO-CPS, while the low-level analysis provides information on the execution performances related to the chosen hardware for the low-level model by exploiting VPSim. The proposed approach also uses FMI co-simulation to include the physical components of the car in the analysis.

Both levels of analysis are needed in the development of CPSs, especially safety- or mission-critical ones.

It would not make sense to jump to software/hardware integration before validating and possibly verifying the controller design, as it would not make sense to choose a hardware platform without assessing its adequacy with respect to timing constraints and evaluating its performance. This work strives to provide a framework to maintain as much coherence as possible to the three key aspects of development, i.e., formal verification, high-level, and low-level modelling.

A case study of an automatic transmission controller algorithm is used as a proof of concept for the methods and tools involved in a safety-critical area like automotive applications. The results highlight that it is possible to assess the performance of the chosen hardware: If the emulated processor is fast enough to accommodate the execution of the algorithm within a co-simulation time step, then the behaviour of the low-level co-simulation is the same as the one of high-level co-simulation, otherwise the co-simulation shows a different behaviour. The proposed methodology was applied to an ARMv8 64-bit single-core processor with private L1 and L2 caches, used in many application automotive processors (e.g. NXP S32V). More cores will be added in future work to consider high-end processors (like the Rhea1, expected as output of the European Processor Initiative project) and for more complex safety-critical applications like autonomous driving and model predictive control of vehicle dynamics.

Acknowledgements. The authors would like to thank the reviewers for their useful comments and suggestions.

Appendix

Table 3 below shows the transition definitions of the simplified Emucharts diagram shown in Fig. 13. Functions up_th and dw_th implement the shift schedule specifications from Tables 1 and 2, respectively.

Table 3. Shift logic transitions.

Transition	Source	Target	Guard	Action
T1	g1_stdy		speed < up_th(1, $t_\%$)	
T2	g1_up		clock < 5 AND speed ≥ up_th(1, $t_\%$)	clock++
T3	g2_up		clock < 5 AND speed ≥ up_th(2, $t_\%$)	clock++
T4	g3_up		clock < 5 AND speed ≥ up_th(3, $t_\%$)	clock++
T5	g2_down		clock < 5 AND speed ≤ dw_th(2, $t_\%$)	clock++
T6	g3_down		clock < 5 AND speed ≤ dw_th(3, $t_\%$)	clock++
T7	g4_down		clock < 5 AND speed ≤ dw_th(4, $t_\%$)	clock++
T8	g4_stdy		speed > dw_th(4, $t_\%$)	
T9	g3_stdy		speed > dw_th AND speed < up_th(3, $t_\%$)	
T10	g2_stdy		speed > dw_th AND speed < up_th(2, $t_\%$)	
T11	g1_stdy	g1_up	speed ≥ up_th(1, $t_\%$)	clock := 0
T12	g1_up	g1_stdy	speed < up_th(1, $t_\%$)	
T13	g1_up	g2_stdy	speed ≥ up_th(1, $t_\%$) AND clock ≥ 5	gear := 2
T14	g2_stdy	g2_up	speed ≥ up_th(2, $t_\%$)	clock := 0
T15	g2_up	g2_stdy	speed < up_th(2, $t_\%$)	
T16	g2_up	g3_stdy	speed ≥ up_th(2, $t_\%$) AND clock ≥ 5	gear := 3
T17	g3_stdy	g3_up	speed ≥ up_th(3, $t_\%$)	clock := 0
T18	g3_up	g3_stdy	speed < up_th(3, $t_\%$)	
T19	g3_up	g4_stdy	speed ≥ up_th(3, $t_\%$) AND clock ≥ 5	gear := 4
T20	g2_down	g1_stdy	speed ≤ dw_th(2, $t_\%$) AND clock ≥ 5	gear := 1
T21	g3_down	g2_stdy	speed ≤ dw_th(3, $t_\%$) AND clock ≥ 5	gear := 2
T22	g4_down	g3_stdy	speed ≤ dw_th(4, $t_\%$) AND clock ≥ 5	gear := 3
T23	g2_stdy	g2_down	speed ≤ dw_th(2, $t_\%$)	clock := 0
T24	g2_down	g2_stdy	speed > dw_th(2, $t_\%$)	
T25	g3_stdy	g3_down	speed ≤ dw_th(3, $t_\%$)	clock := 0
T26	g3_down	g3_stdy	speed > dw_th(3, $t_\%$)	
T27	g4_stdy	g4_down	speed ≤ dw_th(4, $t_\%$)	clock := 0
T28	g4_down	g4_stdy	speed > dw_th(4, $t_\%$)	

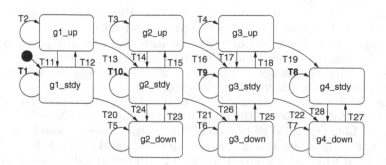

Fig. 13. Simplified Emucharts diagram for the shift logic automaton.

References

1. Accelera: TLM-2.0 Language Reference Manual (2009). https://www.accellera. org/images/downloads/standards/systemc/TLM_2_0_LRM.pdf
2. Bellard, F.: QEMU, a fast and portable dynamic translator. In: Proceedings of the Annual Conference on USENIX Annual Technical Conference, ATEC 2005, p. 41. USENIX Association, USA (2005)
3. Bernardeschi, C., Domenici, A., Masci, P.: A PVS-simulink integrated environment for model-based analysis of cyber-physical systems. IEEE Trans. Softw. Eng. **44**(6), 512–533 (2018)
4. Blochwitz, T., et al.: Functional mockup interface 2.0: the standard for tool independent exchange of simulation models. In: Proceedings of the 9th International MODELICA Conference, pp. 173–184. No. 76 in Linköping Electronic Conference Proceedings (2012)
5. Bohrer, B., Rahli, V., Vukotic, I., Völp, M., Platzer, A.: Formally verified differential dynamic logic. In: Proceedings of the 6th ACM SIGPLAN Conference on Certified Programs and Proofs, CPP 2017, pp. 208–221. ACM (2017). https://doi. org/10.1145/3018610.3018616
6. Charif, A., Busnot, G., Mameesh, R.H., Sassolas, T., Ventroux, N.: Fast virtual prototyping for embedded computing systems design and exploration. In: Chillet, D. (ed.) Proceedings of the Rapid Simulation and Performance Evaluation: Methods and Tools, RAPIDO 2019, Valencia, Spain, 21–23 January 2019, pp. 3:1–3:8. ACM (2019). https://doi.org/10.1145/3300189.3300192
7. Cimatti, A., Griggio, A., Mover, S., Tonetta, S.: HyComp: an SMT-based model checker for hybrid systems. In: Baier, C., Tinelli, C. (eds.) TACAS 2015. LNCS, vol. 9035, pp. 52–67. Springer, Heidelberg (2015). https://doi.org/10.1007/978-3-662-46681-0_4
8. De Moura, L., Bjørner, N.: Satisfiability modulo theories: introduction and applications. Commun. ACM **54**(9), 69–77 (2011)
9. Domenici, A., Fagiolini, A., Palmieri, M.: Integrated simulation and formal verification of a simple autonomous vehicle. In: Cerone, A., Roveri, M. (eds.) SEFM 2017. LNCS, vol. 10729, pp. 300–314. Springer, Cham (2018). https://doi.org/10. 1007/978-3-319-74781-1_21
10. Franchetti, F., et al.: High-assurance spiral: end-to-end guarantees for robot and car control. IEEE Control Syst. **37**(2), 82–103 (2017). https://doi.org/10.1109/ MCS.2016.2643244
11. Gomes, C., Thule, C., Broman, D., Larsen, P.G., Vangheluwe, H.: Co-simulation: a survey. ACM Comput. Surv. (CSUR) **51**(3), 1–33 (2018)
12. Henzinger, T.A.: The theory of hybrid automata. In: Inan, M.K., Kurshan, R.P. (eds.) Verification of Digital and Hybrid Systems. NATO ASI Series (Series F: Computer and Systems Sciences), vol. 170, pp. 265–292. Springer, Heidelberg (2000). https://doi.org/10.1007/978-3-642-59615-5_13
13. IEEE: IEEE Standard for Standard SystemC Language Reference Manual. IEEE Std 1666–2011 (Revision of IEEE Std 1666–2005), pp. 1–638 (2012)
14. Imperas Ltd.: Open Virtual Platforms (2020). http://www.ovpworld.org/
15. Larsen, P.G., et al.: Integrated tool chain for model-based design of Cyber-Physical Systems: the INTO-CPS project. In: 2016 2nd International Workshop on Modelling, Analysis, and Control of Complex CPS (CPS Data), pp. 1–6, April 2016. https://doi.org/10.1109/CPSData.2016.7496424

16. Masci, P., et al.: Combining PVSio with Stateflow. In: Badger, J.M., Rozier, K.Y. (eds.) NFM 2014. LNCS, vol. 8430, pp. 209–214. Springer, Cham (2014). https://doi.org/10.1007/978-3-319-06200-6_16

17. Mauro, G., Thimbleby, H., Domenici, A., Bernardeschi, C.: Extending a user interface prototyping tool with automatic MISRA C code generation. In: Dubois, C., Masci, P., Méry, D. (eds.) Third Workshop on Formal Integrated Development Environments. Electronic Proceedings in Theoretical Computer Science, vol. 240, pp. 53–66. Open Publishing Association (2017). https://doi.org/10.4204/EPTCS.240.4

18. Oladimeji, P., Masci, P., Curzon, P., Thimbleby, H.: PVSio-web: a tool for rapid prototyping device user interfaces in PVS. In: FMIS 2013, 5th International Workshop on Formal Methods for Interactive Systems, London, UK, 24 June 2013 (2013)

19. Owre, S., Rushby, J.M., Shankar, N.: PVS: a prototype verification system. In: Kapur, D. (ed.) CADE 1992. LNCS, vol. 607, pp. 748–752. Springer, Heidelberg (1992). https://doi.org/10.1007/3-540-55602-8_217

20. Palmieri, M., Bernardeschi, C., Masci, P.: A framework for FMI-based co-simulation of human-machine interfaces. Softw. Syst. Model. 19(3), 601–623 (2020)

21. Palmieri, M., Macedo, H.D.: Automatic generation of functional mock-up units from formal specifications. In: Camara, J., Steffen, M. (eds.) SEFM 2019. LNCS, vol. 12226, pp. 27–33. Springer, Cham (2020). https://doi.org/10.1007/978-3-030-57506-9_3

22. Platzer, A., Quesel, J.-D.: KeYmaera: a hybrid theorem prover for hybrid systems (system description). In: Armando, A., Baumgartner, P., Dowek, G. (eds.) IJCAR 2008. LNCS (LNAI), vol. 5195, pp. 171–178. Springer, Heidelberg (2008). https://doi.org/10.1007/978-3-540-71070-7_15

23. Püschel, M., et al.: SPIRAL: code generation for DSP transforms. Proc. IEEE 93(2), 232–275 (2005). https://doi.org/10.1109/JPROC.2004.840306

24. Saidi, S.E., Charif, A., Sassolas, T., Le Guay, P.G., Souza, H.V., Ventroux, N.: Fast virtual prototyping of cyber-physical systems using SystemC and FMI: ADAS use case. In: Proceedings of the 30th International Workshop on Rapid System Prototyping (RSP 2019), pp. 43–49 (2019)

25. Selic, B.: The pragmatics of model-driven development. IEEE Softw. 20(5), 19–25 (2003). https://doi.org/10.1109/MS.2003.1231146

26. Synopsys: Virtualizer (2020). https://www.synopsys.com/verification/virtual-prototyping/virtualizer.html

27. Ventroux, N., et al.: SESAM: an MPSoC simulation environment for dynamic application processing. In: 2010 10th IEEE International Conference on Computer and Information Technology, pp. 1880–1886 (2010). https://doi.org/10.1109/CIT.2010.322

A Semantic-Aware, Accurate
and Efficient API for (Co-)Simulation
of CPS

Giovanni Liboni[1,2](✉) [ID] and Julien Deantoni[1] [ID]

[1] Université Cote d'Azur, CNRS/INRIA Kairos, Sophia Antipolis, France
julien.deantoni@univ-cotedazur.fr
[2] Safran Tech, Modeling and Simulation,
Rue des Jeunes Bois, 78114 Magny-Les-Hameaux, France
giovanni.liboni@safrangroup.com

Abstract. To understand the behavior emerging from the coordination
of heterogeneous simulation units, co-simulation usually relies on either
a time-triggered or an event-triggered Application Programming Interface (API). It creates bias in the resulting behavior since time or event
triggered API may not be appropriate to the behavioral semantics of the
model inside the simulation unit. This paper presents a new semantic-
aware API to execute models. This API is a simple and straightforward
extension of the Functional Mock-up Interface (FMI) API. It can be
used to execute models in isolation, to debug them, and to co-simulate
them. The new API is semantic aware in the sense that it goes beyond
time/event triggered API to allow communication based on the behavioral semantics of internal models. This API is illustrated on a simple
co-simulation use case with both Cyber and Physical models.

Keywords: Co-simulation · API · Behavioral semantics

1 Introduction

Cyber-Physical Systems are a class of systems where computation parts (cyber)
and the plant parts (physical) can not be developed in isolation since the behavior of one impacts the others. In this context, it is of prime importance that
orchestration of software and physical processes are based on semantic models that reflect properties of interest in both [14]. Echoing this, such systems are
usually developed by multiple stakeholders, which use domain-specific languages,
tailored both syntactically and semantically to the domain of expertise [10].

One solution to address the orchestration of different processes is the use of
co-simulation, where processes are computed/solved by dedicated tools and kept
in synchronization by a coordination algorithm. Usually, co-simulation makes use
of a common Application Programming Interface (API) to communicate more
easily with the various tools. For instance, the Functional Mockup Interface
(FMI [18]) proposes a homogeneous time-driven interface to realize a computation step on a solver; this is well suited to the simulation of continuous physical

© Springer Nature Switzerland AG 2021
L. Cleophas and M. Massink (Eds.): SEFM 2020 Workshops, LNCS 12524, pp. 280–294, 2021.
https://doi.org/10.1007/978-3-030-67220-1_21

processes. Another well-known co-simulation approach is HLA [2], which proposes a homogeneous publish-subscribe event-based API well suited to discrete event processes.

However, in order to keep during the orchestration the behavioral semantics specificity of each modeling language, it is important to avoid using an API that hides semantic specificity in order to homogenize. Not taking care of internal semantics during the orchestration of such processes may lead to wrong results, lack of accuracy, and bad simulation performance [9,11,16,19,25].

In this paper, we presented a versatile API, which can be tailored to the internal model under simulation. By using this API, it is possible to communicate with the simulation unit according to different semantics (e.g., time-triggered, event-triggered, mix); according to the internal semantics of the simulation unit. Additionally, it is also possible to ask the simulation unit to stop under specific conditions like for instance when crossing a threshold or when it reaches breakpoint (for debugging purpose).

The next section explains the problems and overviews of the solutions proposed by other approaches. In Sect. 3, we present the semantic-aware API and in Sect. 4, we illustrate its use and benefits on a case study. Finally, before to conclude in Sect. 6, we propose a small discussion about the approach in Sect. 5.

2   Problem Statement and Related Work

Collaborative simulation of Cyber Physical Systems relies on the data and time synchronizations between various simulation units; where a simulation unit is, generally speaking, an encapsulation of a system part execution. Depending on the co-simulation, a simulation unit can encapsulate different entities amongst which (but not limited to): a model and its solver, a program together with its virtual machine, a compiled executable code, a proxy to an existing system part (hardware/software in the loop). A co-simulation actually implements the coordination that should ensure a *correct*[1] synchronization between the data produced and required by different simulation units. The goal is to be able to understand the behavior emerging from the coordination of the different parts of the system; either for simulation or analysis. Usually, such parts were developed by different domain experts, who are using domain specific languages and tools tailored syntactically and semantically to their needs. Also, most of the time, a simulation unit is seen as a black box by the coordination to ensure Intellectual Property preservation.

In this context, various algorithms were proposed to realize the coordination; well known classics like the Jacobi or Gauss-Siedel algorithms but also many others variants [3,4,7,8,20,22,23,28,29]. It is worth noticing that all the proposed algorithms are time-triggered, i.e., simulation units are all executed for a specific predefined time-step. This is a surprising fact since from the 90's work about coordination languages and architecture description languages (ADLs)

---

[1] This notion is defined later in this section.

proposed more sophisticated techniques for the correct and efficient coordination among software components [13,17,21]. Actually, using the same API, being time-triggered or event-triggered on all the simulation units creates bias in the understanding of what the simulation units do. For instance, [9,16,25,26] identified the problem introduced by the use of a time-triggered API on *Cyber* simulation units where sampling the behavior at specific point in times creates artificial delays and loss of information from the coordinator point of view. Such delay can lead to error in results, to bad performances of the co-simulation, or both [16].

In this paper, *we consider a coordination algorithm as correct if it does not introduce any delays or lose information during the communication with the simulation unit.* Consequently, delays and information loss that appear when using a time-triggered API on a piece-wise constant data are considered incorrect (see Fig. 1). Three important things must be noticed at this point. First, sampling a piece-wise constant value can make sense and does not necessarily introduce major problem; however, this should be done on purpose and not be the result of an inappropriate API. Second, there exists in many API (e.g., the FMI standard [18]) the possibility to avoid such delay, typically by roll-backing the simulation to a previous state and trying to locate the actual value change. This can be done only if the simulation can actually be rolled-backed; also this is costly in terms of simulation-time. Finally, third, it is worth noticing that the problem is broader than the simple illustrative case. As illustrated in [26], the coordination algorithm can have an impact on the correctness of the system.

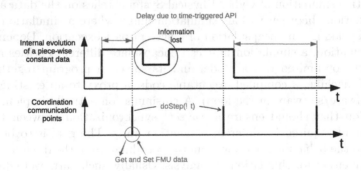

Fig. 1. Time-triggered simulation considered as incorrect.

The core of the problem was identified in several papers: it is not appropriate for any simulation unit to communicate only through a time-triggered or event-triggered API. In the literature, some approaches proposed to extend some existing API to fix a particular problem. This was for instance the case in [25] where they proposed to add a new parameter to the FMI time-triggered *doStep(Δt)* function. The new parameter is *nextEventTime*, a placeholder to store the time at which unpredictable events occurred. [16] went further by proposing to extend the FMI API with new functions that simulate until input and output ports are respectively ready to be read or just written. Finally, the new features of FMI3.0 for hybrid co-simulation tries to aggregate such propositions (see Chapter 5 of FMI3.0 development version https://fmi-standard.org/docs/3.0-dev/#fmi-for-hybrid-co-simulation).

However, in all these related works, the problem is not handled in its generality and they make specific cases of something that should be straightforward. In order to speak *correctly* with a simulation unit, you should be aware of its behavioral semantics and adapt the way to realize the *doStep* accordingly. As an abstraction of a simulation unit behavioral semantics, previous works proposed to focus on the nature of the inputs and outputs of the simulation units [8,24] like for instance *continuous*, *piecewise-continuous*, *piecewise-constant* or *spurious*. We believe this abstraction is very interesting and can be used as a basis for a semantics-aware API.

3 Proposition

We propose to consider the FMI time-triggered interface *doStep(Δt)* as a specific case where we ask a simulation unit to simulate until a specific predicate characterized by an amount of time spent in the simulation unit[2]. Following the same rationale, the proposed semantics-aware API can ask to a simulation unit to execute until a specific coordination predicate holds. The predicate must be expressed according to the information from the simulation unit behavioral interface (typically containing input/output nature, time representation, and solver capability [8,15,16,24]). Consequently, the general form of the proposed doStep API is:

StopCondition doStep(CoordinationPredicate p);

where p expresses a condition under which the execution should pause, *i.e.*, the condition under which the doStep function returns. For instance, considering the input and output nature as defined in [24] (i.e., continuous, piece-wise constant, piece-wise continuous or spurious), the concept of predicate for a correct coordination can be defined as shown in the class diagram Fig. 2.

If the simulation unit supports only temporal predicates, then it corresponds to the FMI API. However, other coordination predicates have been defined. Here is a brief description of their meaning and their typical use case.

1. `TemporalPredicate` is a predicate that becomes true when the internal time of the simulation unit reaches the value of the predicate. This is the classical FMI predicate.
2. `UpdatedPredicate` is a predicate that becomes true as soon as the referenced variable, which must be a piece-wise constant output, has been assigned. It typically corresponds to example from Fig. 1, which can then be managed without data loss, delays, or very small communication step size; i.e., in a correct way.

---

[2] Note that, in reference to study on Model of Computations [27] that this may be done only for timed simulation units.

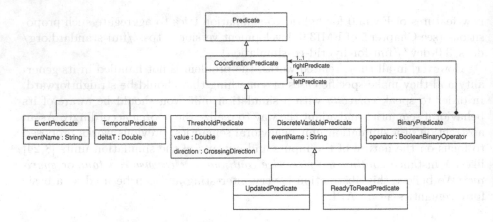

Fig. 2. Minimal but extendable set of predicates.

3. `ReadyToReadPredicate` is a predicate that becomes true just before the simulation units actually read the referenced variable. It is typically used if there is a need to provide an input to a simulation unit that actually reads (non necessarily in a deterministic way) this input at specific points in time. Instead of periodically providing the input data (consequently with unavoidable delays), the data is provided only when needed by the simulation unit.
4. `ThresholdPredicate` is a predicate that becomes true when the referenced variable crosses the defined threshold (according to the crossing direction[3]). It is typically used when a simulation unit is waiting for a specific threshold on a value from another simulation unit. Instead of periodically providing the input data (consequently with unavoidable delays) to be tested and possibly using rollback for more precision, the data is provided only when the condition is reached.
5. `EventPredicate` is a predicate that becomes true when the referenced event occurs. While this is in our implementation only used for cyber events, it may also be extended to encompass discontinuities or other kinds of events on (piecewise) continuous signals.
6. `BinaryPredicate` defines the disjunction of other predicates.

Finally, the proposed API also provides the classical function like for instance `loadModel`, `get/set Variable`, `get/set State` and `terminate`.

What is important is the (preliminary) definition of the coordination predicate, which is, according to our experiments, the minimal set of predicates to have an accurate coordination *i.e.*, without loosing any data, events or signals. Note that for now, we are only using the disjunction of predicates since it is not clear about the meaning of their conjunction. For instance, existing works about Event constraints suggest using Union or Inf/Sup constraints instead of AND since they intrinsically embed a notion of order which is not existing into the classical Boolean operators [1,12].

---

[3] It can be either from above to below, from below to above or both.

To these predicates, many others could be added like for instance a discontinuity predicate that stops when a discontinuity is detected on a piece-wise continuous variable (see description of the Event predicate). Another more complex predicate could be a Bchi predicate, which is verified when a specific state-based observation occurs. There is no real reason to limit the kind of predicate that can be defined, as long as it makes sense according to the simulation unit execution semantics.

In other words, based on the simulation unit behavioral interface, one can speak about the simulation in terms of predicates which are relevant in the particular simulation units used in the co-simulation. For instance, considering a simulation unit interface of an untimed simulation unit, no temporal predicate can be used. In the same idea, if the simulation unit exposes only (piece-wise) continuous variable, then it should not be possible to refer to these variable updates (since it creates an undesired connection with the internal simulation unit discretization step). In short, the acceptable predicates for a specific simulation unit can be inferred from the simulation unit behavioral interface of such simulation unit. However, it is also important that each tool specifies the predicates it supports.

The value returned by the `doStep` function must allow the coordinator to understand why the simulation was actually paused, so that it can do the appropriate action. For instance, if the simulation unit was paused due to an `UpdatedPredicate`, then the variable that has been updated should be communicated to the appropriate simulation unit input (after being sure that the receiving simulation unit is at the same time than the emitting simulation unit, aligning the time if needed). For now, we used a simple form a `StopCondition` but it might be aligned with the `Predicate` class diagram. The Fig. 3 shows a minimal proposition for a simple `StopCondition`. The *StopReason* is a predicate type defining why the simulation was paused; the *elementName* defines the referenced element link with the stop reason and the *stopTime* stores the internal time of the simulation unit when paused.

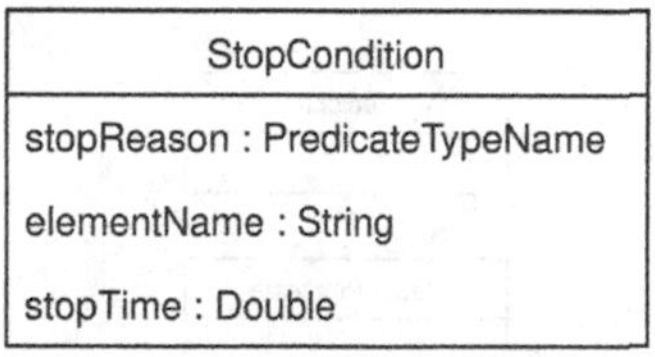

StopCondition
stopReason : PredicateTypeName
elementName : String
stopTime : Double

Fig. 3. Simple `StopCondition`, returned by the `doStep` function.

Remarque 1: This is not clear yet how the link should be made between the name of an exposed variable in the simulation unit behavioral interface and the actual variable inside the model under simulation. For now, we are using qualified names instead of simple names like in the simulation unit behavioral interface.

Similarly, for experimental facilities, we are using a Double to encode time in the co-simulation. It does not mean that the time is internally a double (since it may be encoded by super dense time for instance) but it provides a helpful homogenization of the time from the coordination point of view.

Remarque 2: According to our definition, FMI is a specific mold of our interface since it defines only (piece-wise) continuous variables and (and it does not allow for Threshold predicate injection). Consequently, the only acceptable predicate is a Temporal predicate.

We show in the next section how this API, implemented for language developed in the GEMOC studio [6], provides a simple way to gain in term of accuracy and performance during the coordination of multiple simulation units. However, in the next subsection, we overview how it can be used for other usages, typically debugging.

Example of Extension of the API for Debugging

In this subsection, we show an implementation experimented in the GEMOC studio to use the very same API for debugging. Our goal was to implement the functionality of an API as defined in the usual debugger. We consider this useful for the developer of one simulation unit when she/he wants to debug the simulation unit in the context of the other simulation units. For this reason, we considered that breakpoints are defined with another interface and considered only the way to execute the simulation unit. To define the new use of the interface, we simply defined the necessary Predicate for debugging (see Fig. 4) and implemented the corresponding management of the Predicate in a wrapper. Details can be found here: https://github.com/jdeantoni/cosimulationOfCpuHeatManagement. Furthermore, it is interesting to realize that debugging equational simulation units could use a totally different notion of breakpoint. For instance, one could want to pause the simulation when the derivative of a specific output reaches a symptomatic threshold, in order to check different values in the system and try to understand what actually happens. In this case, Predicates should be defined accordingly.

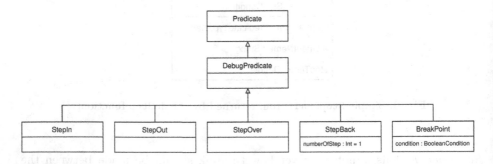

Fig. 4. Simple StopCondition, return by the doStep function.

Once again, we tried to provide an extendable simulation API, focused on co-simulation but suitable for different activities.

4   Case Study

We used the management of a CPU temperature as a simple but representative case study[4]. This system is made up of 3 simulation units (see Fig. 5). *CPUinBoxWithFan* and *fanControler* have been developed in the OpenModelica tool[5] to respectively define the CPU in a box which is cooled by a fan and the controller of the fan speed (a simple Proportional controller). The heat between the box and the CPU is transferred according to the fan speed. The *overHeatController* has been developed as a state machine in the GEMOC studio[6].

In the *CPUinBoxWithFan* simulation unit, the CPU is activated as long as the *stopWorking* input is equal to false. When activated, the CPU produces heat, which is exchanged with the air of its box more or less rapidly depending on the *fanSpeedCommand* input ($\in [0..10]$ where at 0 the fan is stopped and at 10 the fan is at full speed).

In the *overHeatController* simulation unit, a state machine is defined. It monitors periodically (every 3 s) the *cpuTemperature* and if it exceeds a specific threshold, the *switch* event occurs and the state machine enters in a new state where it monitors the CPU temperature every 5 s. If it goes above a specific threshold, the *switch* event occurs and the state machine enters the first state (see Fig. 6).

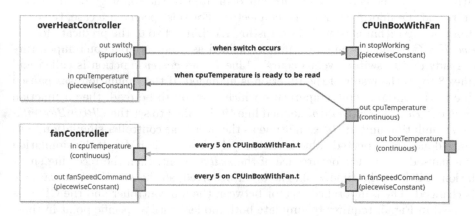

Fig. 5. Simple but representative case study for co-simulation.

---

[4] The associated code can be retrieved from http://i3s.unice.fr/~deantoni/cosim-cps2020.

[5] https://openmodelica.org.

[6] http://eclipse.org/gemoc.

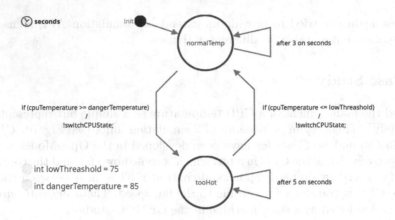

Fig. 6. Over Heat Controller state machine.

To connect the different simulation units we relied on strategies defined in [16]. Consequently the temperature from *CPUinBoxWithFan* to *overHeatController* is only exchanged when the later simulation unit is ready to read the data. Similarly, the change of the *stopWorking* input is only done only when the *switch* event occurs. Between the two simulation units obtained from Modelica, the connectors define classical time trigger communication.

Of course, we handled these different cases by using different Predicates in the *doStep* function call. However, one can notice that the coordination algorithm will not be generic anymore but dedicated to the topology of simulation units and the information on the connectors. For this specific use case, the coordination algorithm is provided on Listing 1.1. Lines 4 to 6, the predicate for the *overHeatController* simulation unit is defined as "the variable cpuTemperature is ready or the switch event occurs". Line 7, the *dostep* function is called and lines 8 to 16 the result of the function is managed. If the simulation was paused due to the variable cpuTemperature which is ready to be read, then a function (*simulateBoxAndFanControl* defined line 19) is called to set the *CPUinBoxWithFan* simulation unit at the same time as the over heat controller simulation unit. Once done, the expected value is exchanged between the FMU. If the simulation was paused due to the occurrence of the *switch* event, then the receiving simulation unit is at the time when the event occurred, so the *stopWorking* variable is changed. The temporal connector between the fan controller and the CPU, as defined in Fig. 5, requires to simulate both models until a specific point in time. In lines 21 to 36, the simulation units must reach an *expectedTime*. If there is one (or several) intermediate temporal steps in between now and the expected time (i.e., $now\%5 = 0$ in our case), then the simulation units are simulated until this point in time and data are exchanged as expected.

Listing 1.1. Coordination Algorithm dedicated to the example on Figure 5 using the proposed interface

```
1   public void coSimulate(double endtime) {
        //now = 0; localIsStopped = false;
3   while (now < endTime){
        ReadyToReadPredicate r2rp("cpuTemperature");
5       EventPredicate ep ("switch");
        BinaryPredicate bp (r2rp, ep);
7       StopCondition sc = controlerSU.doStep(bp);
        if (sc.stopReason == READYTOREAD) {
9           simulateBoxAndFanControl(sc.stopTime);
            double cpuTemperature = c.boxSU.read("cpuTemperature");
11          controlerSU.setVariable("cpuTemperature", cpuTemperature);
        } else { //event occured
13          simulateBoxAndFanControl(sc.stopTime);
            localIsStopped = !localIsStopped;
15          boxSU.write("stopWorking").with(localIsStopped);
        }
17  }

19  public void simulateBoxAndFanControl(double expectedTime) {
        double delta = expectedTime - now;
21      while (delta + (now % 5) >= 5) { //\Delta t == 5 for each
            ↪ connector from boxSU and fanControllerSU
            double stepToDo = (5-(now % 5));
23          boxSU.doStep(stepToDo);
            fanControllerSU.doStep(5);
25          double cpuTemperature = boxSU.read("cpuTemperature");
            fanControllerSU.write("cpuTemperature").with(cpuTemperature);
27          int fanCommand = fanControllerSU.read("fanSpeedCommand");
            boxSU.write("fanSpeedCommand").with(fanCommand);
29          double boxTemperature = boxSU.read("BoxTemperature");
            now += stepToDo;
31          delta = expectedTime - now;
        }
33      if (delta > 0) {
            boxSU.doStep(delta);
35          now += delta;
        }
37  }
```

The results from the beginning of the co-simulation obtained with this setup are provided in Fig. 7. The reader should notice that the points are only retrieved as specified in Fig. 5, i.e., at the exact time it is needed to have a correct co-simulation. For instance on Fig. 7, we can see that a first paused was realized by the overheat controller at time 2, i.e., which is the non deterministic time spent for the state machine to enter in the *normalTemp* state, where the guard of output transition is evaluated and consequently the CPU temperature is read. Then, pauses are realized every 5 s and every multiple of 3 (the reading period in the first state of *overHeatController*). This way, we reduce the number of communication points to their strict minimum to have a correct co-simulation and we avoid the delays introduced by the classical sampling strategy.

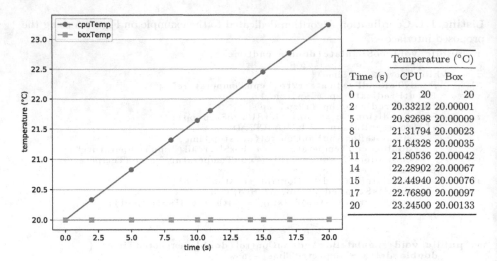

Time (s)	Temperature (°C)	
	CPU	Box
0	20	20
2	20.33212	20.00001
5	20.82698	20.00009
8	21.31794	20.00023
10	21.64328	20.00035
11	21.80536	20.00042
14	22.28902	20.00067
15	22.44940	20.00076
17	22.76890	20.00097
20	23.24500	20.00133

Fig. 7. Results obtained at the begin of the co-simulation.

In the Fig. 8, the first point in time is the one when the state machine switch from the *normalTemp* state to the *tooHot* state. It occurred at time 14679. Consequently, as long as the state machine remains in this state, data are retrieved every 5 s as specified in the temporal connectors and in the reading period from the state machine. However, since the state machine entered in the *tooHot* state at time 14679, then the simulation unit was paused after 5 s, i.e., at 14684, while the temporal connectors induce a pause every 5 s. We can see here that the internal semantics of the simulation is consistently exposed and took into consideration.

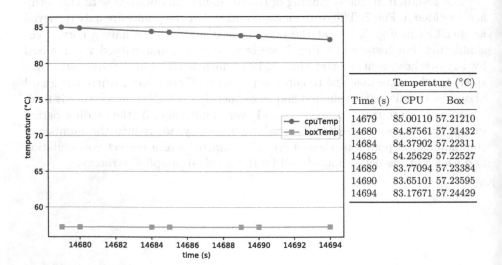

Time (s)	Temperature (°C)	
	CPU	Box
14679	85.00110	57.21210
14680	84.87561	57.21432
14684	84.37902	57.22311
14685	84.25629	57.22527
14689	83.77094	57.23384
14690	83.65101	57.23595
14694	83.17671	57.24429

Fig. 8. Results obtained when the controller enters in the *tooHot* state.

Finally, in Fig. 9, the simulation is run for 8 h and 20 min (30000 s). For this simulation, we obtained 15023 communication steps without sacrificing accuracy over performance. If we were using a time-triggered interface and allowed an error up to 100 ms, then we would have 300'000 communication step and a loss of accuracy. Additionally, we believe that the proposed interface is intuitive to use and may be extended for different purposes. In the next section, before to conclude, a small discussion about implementation is made.

5   Discussion

We argued that the proposed interface is extendable, efficient, and intuitive to use. In this section, we discuss some of these points according to our experiment in implementing the API in the GEMOC studio.

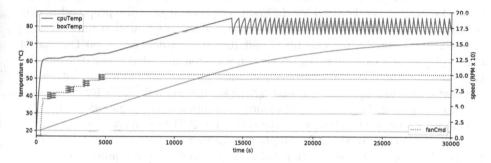

Fig. 9. Results obtained when running the coordination algorithm.

Concerning the implementation of the predicates, two main points can be addressed. First, its efficiency strongly relies on how the API is internally implemented. In our case we modified the code generation to generate a pause when needed. For instance, for the *Updated* predicate, all assignments are instrumented to create a pause. This has only a minor impact on performance. However, if the implementation is done in a wrapper where all micro steps are checked to see if a variable has been updated, then the execution may suffer from a slowdown. The same phenomenon happens for the Threshold predicate. If one sample the variable to check the crossing, the execution will be slow down and the exact point in time when the crossing occurs may be missed. It is better to inject the actual zero crossing in the model (typically in the equation set) to ensure better performance and accuracy. This is what is expected to be done in collaboration with Safran. Also, the implementation of the predicates must actually follow the semantics of the simulation unit. For instance, if a simulation unit is executing a model developed in a synchronous language [5], then all the assignments should NOT be caught since according to the synchronous semantics, data are latched at specific points in time. In our implementation, we relied on annotations to provide flexibility on the exposed semantics. Consequently, the tool developer is in charge of providing the expected semantics.

Concerning the extension of the predicate, there are two minors points to take care of. First, it is important to rely on a mechanism to clearly specify which predicate is supported for a specific simulation unit. This may for instance be done in an artefact equivalent of the FMI model description. Second, there is a risk of an uncontrolled evolution of predicates, leading to a predicate tower of Babel. This is a long term issue and we believe there are few risks it happens. If the road to this situation is taken, it may be interesting to provide an official set of predicate extension repository, where people can look for existing predicate before to create their own and where all predicates are put together.

6   Conclusion

We presented in this paper a new API, initially thought as a co-simulation interface. However, it can be used for different purposes. It uses the proposed notion of predicate to represent the condition under which a simulation unit must be paused. These predicates can be of different nature depending on the use of the API. In each case, it relies on the information provided about the simulation unit. In the co-simulation case, it relies on the data nature (continuous, piece-wise continuous, etc.) exposed by the simulation unit. This information is an abstraction of the internal behavioral semantics of the simulation unit. We developed a case study where we showed how the API can be used in a semantic-aware way. The use of the API adapts the number of co-simulation steps to the internal behavior of the simulation units, keeping only the communication points required for a correct co-simulation. We believe such communication between the coordination algorithm and the simulation unit provides the basis for an analysis of a co-simulation.

In future works, we first want to focus on the automated generation of the coordination algorithm. As shown in Listing 1.1, the coordination is dedicated to a specific simulation and it may be tricky to write it by hand for a more complex system. Additionally, it becomes important to allow for the distribution of co-simulation. Our approach, by limiting the number of co-simulation steps to the minimum, is well appropriate to distribution. This is why we are actually finishing the development of the generator of distributed coordination algorithm based on our interface. Another future work concern the integration of such approach into a system engineering approach but this is a longer term work.

References

1. André, C.: Syntax and semantics of the clock constraint specification language. Technical report 6925, INRIA (2009)
2. IEEE Standards Association and others: IEEE Standard for Modeling and Simulation (M&S) High Level Architecture (HLA)—Framework and Rules. Institute of Electrical and Electronics Engineers, New York. IEEE Standard (1516-2010), pp. 10–1109 (2010)

3. Awais, M.U., Palensky, P., Elsheikh, A., Widl, E., Matthias, S.: The high level architecture RTI as a master to the functional mock-up interface components. In: 2013 International Conference on Computing, Networking and Communications (ICNC), pp. 315–320. IEEE (2013)
4. Bastian, J., Clauß, C., Wolf, S., Schneider, P.: Master for co-simulation using FMI. In: Proceedings of the 8th International Modelica Conference, March 20th–22nd, Technical Univeisity, Dresden, Germany, pp. 115–120, No. 63. Linköping University Electronic Press (2011)
5. Benveniste, A., Caspi, P., Edwards, S.A., Halbwachs, N., Le Guernic, P., De Simone, R.: The synchronous languages 12 years later. Proc. IEEE **91**(1), 64–83 (2003)
6. Bousse, E., Degueule, T., Vojtisek, D., Mayerhofer, T., Deantoni, J., Combemale, B.: Execution framework of the GEMOC studio (tool demo). In: Proceedings of the 2016 ACM SIGPLAN International Conference on Software Language Engineering, pp. 84–89. ACM (2016)
7. Broman, D, et al.: Determinate composition of FMUS for co-simulation. In: Proceedings of the Eleventh ACM International Conference on Embedded Software, p. 2. IEEE Press (2013)
8. Broman, D., Greenberg, L., Lee, E.A., Masin, M., Tripakis, S., Wetter, M.: Requirements for hybrid cosimulation standards. In: Proceedings of the 18th International Conference on Hybrid Systems: Computation and Control, HSCC 2015, pp. 179–188. Association for Computing Machinery, New York (2015)
9. Centomo, S., Deantoni, J., De Simone, R.: Using SystemC cyber models in an FMI co-simulation environment. In: 19th Euromicro Conference on Digital System Design 31 August - 2 September 2016. 19th Euromicro Conference on Digital System Design, Limassol, Cyprus, vol. 19, August 2016. https://doi.org/10.1109/DSD.2016.86. https://hal.inria.fr/hal-01358702
10. Combemale, B., Deantoni, J., Baudry, B., France, R.B., Jézéquel, J., Gray, J.: Globalizing modeling languages. Computer **47**(6), 68–71 (2014). https://doi.org/10.1109/MC.2014.147
11. Cremona, F., Lohstroh, M., Broman, D., Di Natale, M., Lee, E.A., Tripakis, S.: Step revision in hybrid co-simulation with FMI. In: 14th ACM-IEEE International Conference on Formal Methods and Models for System Design. IEEE, Kanpur, November 2016
12. Deantoni, J., André, C., Gascon, R.: CCSL denotational semantics. Research report RR-8628, Inria, November 2014. https://hal.inria.fr/hal-01082274
13. Garlan, D., Shaw, M.: An introduction to software architecture. In: Advances in Software Engineering and Knowledge Engineering 1(3.4) (1993)
14. Lee, E.A.: Cyber physical systems: design challenges. In: 2008 11th IEEE International Symposium on Object and Component-Oriented Real-Time Distributed Computing (ISORC), pp. 363–369 (2008)
15. Liboni, G., Deantoni, J.: WIP on a coordination language to automate the generation of co-simulations. In: 2019 Forum for Specification and Design Languages (FDL), pp. 1–4. IEEE (2019)
16. Liboni, G., Deantoni, J., Portaluri, A., Quaglia, D., De Simone, R.: Beyond Time-triggered co-simulation of cyber-physical systems for performance and accuracy improvements. In: 10th Workshop on Rapid Simulation and Performance Evaluation: Methods and Tools, Manchester, United Kingdom, January 2018. https://hal.inria.fr/hal-01675396

17. Medvidovic, N., Taylor, R.N.: A framework for classifying and comparing architecture description languages. ACM SIGSOFT Softw. Eng. Notes **22**(6), 60–76 (1997)
18. Modelisar: FMI for Model Exchange and Co-Simulation, July 2014. https://fmi-standard.org/downloads#version2
19. Mustafiz, S., Gomes, C., Vangheluwe, H., Barroca, B.: Modular design of hybrid languages by explicit modeling of semantic adaptation. In: 2016 Symposium on Theory of Modeling and Simulation (TMS-DEVS), pp. 1–8, April 2016. https://doi.org/10.23919/TMS.2016.7918835
20. Neema, H., et al.: Model-based integration platform for FMI co-simulation and heterogeneous simulations of cyber-physical systems. In: Proceedings of the 10th International Modelica Conference, Lund, Sweden, pp. 235–245, No. 096. Linköping University Electronic Press (2014)
21. Papadopoulos, G.A., Arbab, F.: Coordination models and languages. Adv. Comput. **46**, 329–400 (1998)
22. Savicks, V., Butler, M., Colley, J.: Co-simulating event-B and continuous models via FMI. In: Proceedings of the 2014 Summer Simulation Multiconference, p. 37. Society for Computer Simulation International (2014)
23. Schierz, T., Arnold, M., Clauß, C.: Co-simulation with communication step size control in an FMI compatible master algorithm. In: Proceedings of the 9th International MODELICA Conference, Munich, Germany, pp. 205–214, No. 076. Linköping University Electronic Press (2012)
24. Tavella, J.P., et al.: Toward an Hybrid Co-simulation with the FMI-CS Standard, Research report, April 2016. https://hal-centralesupelec.archives-ouvertes.fr/hal-01301183
25. Tavella, J.P., et al.: Toward an accurate and fast hybrid multi-simulation with the FMI-CS standard. In: 21st IEEE International Conference on Emerging Technologies and Factory Automation (ETFA), pp. 1–5. IEEE, Berlin, September 2016. https://doi.org/10.1109/ETFA.2016.7733616
26. Thule, C., Gomes, C., Deantoni, J., Larsen, P.G., Brauer, J., Vangheluwe, H.: Towards the verification of hybrid co-simulation algorithms. In: Workshop on Formal Co-Simulation of Cyber-Physical Systems (SEFM satellite), Toulouse, France, June 2018. https://hal.inria.fr/hal-01871531
27. Tripakis, S.: Bridging the semantic gap between heterogeneous modeling formalisms and FMI. In: 2015 International Conference on Embedded Computer Systems: Architectures, Modeling, and Simulation (SAMOS), pp. 60–69. IEEE (2015)
28. Van Acker, B., Denil, J., Vangheluwe, H., De Meulenaere, P.: Generation of an optimised master algorithm for FMI co-simulation. In: Proceedings of the Symposium on Theory of Modeling & Simulation: DEVS Integrative M&S Symposium, DEVS 2015, pp. 205–212. Society for Computer Simulation International, San Diego (2015)
29. Wang, B., Baras, J.S.: Hybridsim: a modeling and co-simulation toolchain for cyber-physical systems. In: Proceedings of the 2013 IEEE/ACM 17th International Symposium on Distributed Simulation and Real Time Applications, DS-RT 2013, pp. 33–40. IEEE Computer Society, Washington, DC (2013). https://doi.org/10.1109/DS-RT.2013.12. http://dx.doi.org/10.1109/DS-RT.2013.12

An FMI-Based Initialization Plugin
for INTO-CPS Maestro 2

Simon Thrane Hansen$^{(\boxtimes)}$, Casper Thule , and Cláudio Gomes

DIGIT, Department of Engineering, Aarhus University, Aarhus, Denmark
{sth,casper.thule,claudio.gomes}@eng.au.dk

Abstract. The accuracy of the result of a co-simulation is dependent
on the correct initialization of all the simulation units. In this work, we
consider co-simulation where the simulation units are described as Functional Mock-up Units (FMU). The Functional Mock-up Interface (FMI)
specification specifies constraints to the initialization of variables in the
scope of a single FMU. However, it does not consider the initialization
of interconnected variables between instances of FMUs. Such interconnected variables place particular constraints on the initialization order
of the FMUs.

The approach taken to calculate a correct initialization order is based
on predicates from the FMI specification and the topological ordering
of both internal connections and interconnected variables. The approach
supports the initialization of co-simulation scenarios containing algebraic
loops using fixed point iteration. The approach has been realized as a
plugin for the open-source INTO-CPS Maestro 2 Co-simulation framework. It has been tested for various scenarios and compared to an existing *Initializer* that has been validated through academic and industrial
application.

Keywords: Co-simulation · Initialization · Algebraic loop ·
Topological ordering · FMI

1 Introduction

Cyber-physical systems (CPS) are becoming ever more sophisticated, while market pressure shortens the available development time. One of the tools to manage
the increasing complexity of such systems is co-simulation since it tackles their
heterogeneous nature. Co-simulation is a technique to combine multiple black-
box simulation units to compute the combined models' behavior as a discrete
trace (see, e.g., [12,14]). The simulation units, often developed independently
from each other, are coupled using a master algorithm, often developed independently, that communicates with each simulation unit via its interface. This

We are grateful to the Poul Due Jensen Foundation, which has supported the estab-
lishment of a new Centre for Digital Twin Technology at Aarhus University. Finally,
we thank the reviewers for the thorough feedback.

L. Cleophas and M. Massink (Eds.): SEFM 2020 Workshops, LNCS 12524, pp. 295–310, 2021.
https://doi.org/10.1007/978-3-030-67220-1_22

interface comprises functions for setting/getting inputs/outputs and computing the associated model behavior over a given time interval. The Functional Mock-up Interface (FMI) standard [4,7] is such an interface prescribing how to communicate with each simulation unit. The interface is used to connect different simulation units, called Functional Mock-up Units (FMUs), exchange values between them, and make them progress in time.

A typical co-simulation consists of three phases: initialization, simulation, and teardown [22]. This work concentrates on the first. The FMI standard specifies criteria for how a single FMU shall be initialized. However, FMI is not concerned with how a connected system of multiple FMUs is initialized correctly as a whole.

The way a system of multiple FMUs should be initialized and interacted with depends on each FMU's implementation and interconnections to other FMUs [9], since these place precedence constraints between the FMU variables. These precedence constraints can introduce algebraic loops between the FMU variables. An algebraic loop places particular requirements on the strategy for both the order of initialization and the method used to calculate the correct initial values of the variables in the algebraic loop [3]. Algebraic loops occur whenever an interconnected FMU variable indirectly depends on itself. Not solving an algebraic loop can lead to a prohibitively high error in the co-simulation result [2], and invalid results, as shown in Sect. 4. It is crucial for all interconnected variables that the initialization procedure ensures that a variable is never read before it is set. For variables within an algebraic loop, the initialization must ensure that all initial values have converged to a fixed point before entering the next phase of the co-simulation.

Other approaches for the generation of co-simulation algorithms have avoided co-simulation scenarios containing algebraic loops since their presence reduces the chance of obtaining a deterministic co-simulation result[1,5,10]. This choice is driven by the fact that not all co-simulation scenarios containing algebraic loops are valid since those algebraic loops never converge, or might converge to unexpected solutions. However, as shown in Sect. 4, solving algebraic loops can be essential to obtaining valid simulation results, and a well-established co-simulation framework should be able to handle these scenarios.

Contribution: This paper describes an approach for calculating the initialization order of an FMI-based co-simulation in linear time of the number of interconnected variables, even when algebraic loops are present. The approach does not put any constraints on the master algorithm chosen to carry out the simulation. The approach is realized as a plugin to the co-simulation framework called INTO-CPS Maestro 2 (Maestro 2), introduced in [22]. The realized plugin has been tested for various co-simulation scenarios and compared to an existing Initializer that has been validated through academic and industrial applications. Furthermore, the calculated initialization order is systematically verified by the semantics of co-simulation introduced in [9,10].

Structure: The paper is structured as follows: Sect. 2 gives a brief background of the formalization of FMUs and Maestro 2. Section 3 describes the approach taken to calculate the initialization order. It is followed by Sect. 5, where the realization of the approach is presented. Finally, Sect. 7 provides concluding remarks and describes future work.

2 Background

In this section, we provide a formalization of FMI co-simulation and a brief background on INTO-CPS Maestro 2.

2.1 FMU Definitions

To describe the formalization of FMUs, we adopt the vocabulary from [9]. The main definitions of relevance to this paper will be presented, but readers are referred to the original publications for more information. This paper is only concerned with the initialization-phase of a co-simulation, making time of an FMU irrelevant. The formalization from Gomes et al.[9] is extended with new definitions regarding algebraic loops, and convergence of fixed point iteration.

Definition 1 (FMU). *An FMU with identifier c is represented by the tuple*

$$\langle S_c, U_c, Y_c, \mathtt{set}_c, \mathtt{get}_c \rangle,$$

where: S_c represents the state space of FMU c; U_c and Y_c the set of input and output variables, respectively; $\mathtt{set}_c : S_c \times U_c \times \mathcal{V} \to S_c$ and $\mathtt{get}_c : S_c \times Y_c \to \mathcal{V}$ arc functions to set the inputs and get the outputs, respectively (we abstract the set of values that each input/output variable can take as $\mathcal{V}$).

Definition 2 (Scenario). *A scenario is a structure $\langle C, L \rangle$ where each identifier $c \in C$ is associated with an FMU, as defined in Definition 1, and $L(u) = y$ means that the output y is connected to input u. Let $U = \bigcup_{c \in C} U_c$ and $Y = \bigcup_{c \in C} Y_c$, then $L : U \to Y$.*

Note a single output can connect to multiple inputs, but a single input can only rely on a single output. The following definitions correspond to the operations that are permitted in the initialization phase of a co-simulation.

Definition 3 (Output Computation). *The $\mathtt{get}_c(\_, y_c)$ represents the calculation of output y_c of $c \in C$. Given a co-simulation state, it checks whether all inputs that feed-through to y_c are defined.*

Definition 4 (Input Computation). *The $\mathtt{set}_c(\_, u_c, v)$ represents the setting of input u_c of $c \in C$. Given a co-simulation state, it checks whether all outputs connected to u_c are defined.*

Definition 5 (Fixed Point). *The $\mathtt{fixedpoint}_l$ represents an ordered sequence of the setting or getting of all variables of a given SCC l, see Definition 9 for a definition of SCC. The $\mathtt{fixedpoint}_l \subseteq \bigcup_{c \in C} \{\mathtt{get}_c, \mathtt{set}_c\}$*

Definition 6 (Initialization). *Given a scenario $\langle C, L \rangle$, we define the initialization procedure $(I_i)_{i \in \mathbb{N}}$ as is a finite ordered sequence of FMU function calls that needs to be performed in the initialization of a co-simulation scenario. The ordered sequence is defined as: $(f_i)_{i \in \mathbb{N}} = f_0, f_1, \ldots$ with $f_i \in I = \bigcup_{l \in loops}$ fixedpoint$_l$, and i denoting the order of the function call. Loops is defined as the set of all SCC see Definition 9.*

It should be noted that a trivial SCC (see Definition 9) is only a single get or set action and is not regarded as a fixed point outside Definition 6, but just a simple computation.

Definition 7 (Feed-through). *The input $u_c \in U_c$ feeds through to output $y_c \in Y_c$, that is, $(u_c, y_c) \in D_c$, when there exists $v_1, v_2 \in V$ and $s_c \in S_c$, such that $\mathtt{get}_c(\mathtt{set}_c(s_c, u_c, v_1), y_c) \neq \mathtt{get}_c(\mathtt{set}_c(s_c, u_c, v_2), y_c)$.*

A graph of the dependencies of a co-simulation scenario is established from the interconnected variables by Definition 8. The graph is the foundation for the calculation of the initialization procedure and is therefore referred to as the Initialization Graph. The graph construction is similar to the one in [10], except the later focuses on a general co-simulation step, while this work focus on the initialization phase.

Definition 8 (Initialization Graph). *Given a co-simulation scenario $\langle C, L \rangle$, and a set of feed-through dependencies $\bigcup_{c \in C} \{D_c\}$, we define the Initialization Graph where each node represents a port $y_c \in Y_c$ or $u_c \in U_c$ of some fmu $c \in C$. The edges are created according to the following rules:*

1. *For each $c \in C$ and $u_c \in U_c$, if $L(u_c) = y_d$, add an edge $y_d \rightarrow u_c$ (output to input).*
2. *For each $c \in C$ and $(u_c, y_c) \in D_c$, add an edge $u_c \rightarrow y_c$ (input to output).*

The interconnections of FMU variables can lead to circular dependencies between the variables. An example of this behavior is the car suspension system that is presented in Sect. 4. Figure 1 shows the co-simulation scenario of the example and the Initialization Graph of the system. The Initialization Graph in Fig. 1 is annotated with the strongly connected components of the graph.

The following definitions formalize the concept of an algebraic loop in a co-simulation scenario and define the problem these algebraic loops are introducing. The definition of strongly connected components is adapted from the semantics of Causal Block Diagrams (see [8] for an overview).

Definition 9 (Algebraic loops). *An algebraic loop is defined as a non-trivial, strongly connected component of the graph in Definition 8. Formally, a strong connected component satisfies $\{a, b \in SCC : Path(a, b)\}$, where $Path(a, b)$ is true when there's a path (including an empty path from a node to itself) between nodes a and b ($Path(a, a)$ is always true). An SCC is non-trivial when it has more than one node.*

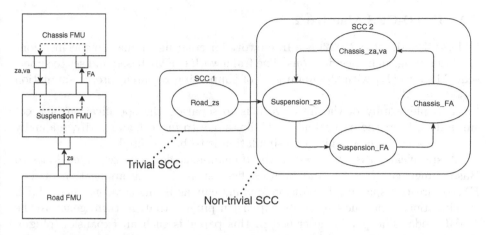

Fig. 1. An FMU co-simulation scenario of the Quarter car and its Initialization Graph denoted with SCCs.

Since the edges of the graph represent dependencies between the variables, the value of every variable in a non-trivial strong component depends on itself. Let X denote a vector of one or more variables whose value depends on itself. The non-trivial strong component forms an equation with the form $F(X, U) = X$, where F denotes the relations between the variables in the loop and U denotes the variables whose values are calculated elsewhere. This means that algebraic loops need to be handled using fixed point iterations[12].

An example of a co-simulation scenario where fixed point iteration is needed can be seen in Fig. 1 where the Initialization Graph of the quarter car system from Sect. 4 is shown.

A fixed point iteration technique is not guaranteed to convergence if the system is unstable. The fixed point is as a numerical fixed point that approximates a limit if such a value exist (the system is stable). It means that an upper bound of the number of repetitions needs to be established to ensure termination. In the case of a non-converging algebraic loop, the simulation should be stopped since the result of the co-simulation scenario would not be trustworthy. The criteria of a valid co-simulation scenario are specified in Definition 10.

Definition 10 (Convergence of Fixed point iteration). *A fixed point iteration converges if a finite number of iterations will make the difference of the output value of the same operation between two following iterations within a certain threshold ϵ.*
Formally, $\exists n \in \mathbb{N} : |F(X^{n+1}, U) - F(X^n, U)| \leq \epsilon$.

2.2   INTO-CPS Maestro 2

INTO-CPS Maestro 2^1 [22] is a framework for creating simulation specifications and executing such specifications. The framework is FMI-based and set to supersede Maestro [21] with the main goal of supporting research into co-simulation based on FMI.

The philosophy of the framework is to separate the specification of a co-simulation from the execution. This allows one to inspect and verify, manually or automatically, how a given co-simulation is to be executed.

A specification is expressed in the domain-specific language called Maestro Base Language (MaBL), and it is explicit, such that the application of, i.e., FMUs are transparent. Expansion plugins can assist in creating such MaBL specifications, and one can apply expansion plugins that, in turn, generate the MaBL code. The plugin described in this paper is such an expansion plugin. The application of a plugin is evident in a MaBL specification. Upon processing of the specification, a new specification is created where the application of a plugin is replaced by the MaBL code generated by the plugin. This process is known as expansion, and a specification without any expansions remaining is a fully expanded MaBL specification. An example of a part of the folded MaBL specification of the case study example of Sect. 4 can be seen below.

```
1  simulation
2  import Initializer;
3  {
4  FMI2 chassis = load("FMI2", "{8c4e810f-3df3-4a00-8276-176
      fa3c9f000}", "src/chassis-c.fmu");
5  ...
6  IFmuComponent components[3]={chassis,suspension, road};
7  expand initialize(components,START_TIME, END_TIME);
8  ...
9  }
```

To conduct a co-simulation, Maestro2 also features an interpreter that can execute a fully expanded MaBL specification, resulting in the execution of the co-simulation.

3   Calculation of an Initialization Order

The FMI specification defines certain information about the initialization order described through different states of a co-simulation. The initialization phase covers the two states (in chronological order) defined in the FMI specification:

– *Instantiated*
– *Initialization Mode*

---

[1] Currently in alpha https://github.com/INTO-CPS-Association/maestro/tree/2.0.0-alpha.

In each of the two states, different groups of FMU variables and parameters are potentially assigned a value. The groups are defined by FMI based on the characteristics of the FMU variables. The rules have been extracted as predicates and used in the implementation. Some groups consist of variables and parameters whose value does not depend on other variables. These independent variables and parameters can be set in the *Instantiated* phase of the Initialization. Since these variables have no connections to other FMU variables - meaning they are not represented in the graph of Definition 8, the order their value is set in is insignificant. The setting and getting operations of each FMU are grouped to perform the fewest possible FMU-operations during the Initialization.

In the *Initialization Mode* state all the interconnected variable is being defined, but as stated by the Definitions 3, 4 and 7 the operations *get* and *set* **require** that the operations are performed in a specific order. Furthermore, algebraic loops place even more requirements on the initialization strategy. Since each non-trivial strongly connected component (algebraic loop) needs to be isolated from the other variables of the system to calculate their initial values using fixed point iteration as described in Definition 5. After the *Initialization Mode* state, all variables of all FMUs in the co-simulation scenario should be defined, and the co-simulation should be ready to enter the *modelInitialized* state.

3.1   Method to Calculate the Initialization Order

This section describes the approach taken to calculate the initialization order of the interconnected FMU variables. The approach is based on the strategy proposed in Gomes et al. [5,10], but the approach in this work is extended with the ability to handle the Initialization of algebraic loops.

The initialization algorithm starts by building a directed graph of the dependencies between the interconnected variables of the FMUs. The graph is constructed based on the interconnected variables and internal connections (feedthrough); please see Definition 8 for a formal definition of the graph.

The topological ordering of the strongly connected components of the graph defined in Definition 8 is the initialization order of the interconnected FMU variables. The non-trivial strongly connected components are algebraic loops of the system. The trivial ones are standard interconnected FMU variables, whose port operation should be performed only once during the initialization procedure. The calculation of an initialization order is performed in linear time based on the number of external and internal connections using Tarjan's algorithm [20].

As described in earlier sections, it is essential to handle algebraic loops by a particular initialization strategy since the loops otherwise would invalidate the co-simulation result. The procedure for initializing algebraic loops is identifying and initializing them using a fixed point iteration strategy until convergence. Since convergence is not guaranteed, this property is monitored using Definition 10 to see if the difference between all the output variables of two successive iterations is below a defined threshold. Suppose convergence is not established within a finite number of iterations[2], the co-simulation scenario is rejected to avoid running an invalid simulation.

3.2 Optimization of a Initialization Procedure

An initialization procedure can sometimes be optimized since the FMI specification allows multiple *set* or *get* operations of the same FMU to be performed in bulk by grouping them together to a single operation over multiple variables with similar characteristic. This criteria of optimization is formalized in Definition 11.

Definition 11 (Optimization of a Initialization procedure). *Given an initialization procedure* $(I_i)_{i \in \mathbb{N}}$ *with a finite ordered sequence of FMU function calls* $f_i \in F = \bigcup_{c \in C} \{\text{set}_c, \text{get}_c\}$, *and i denoting the order of the function call. It can be optimized if* $\exists f_i, f_{i+1} \in F : \exists c \in C : (f_i \in \text{set}_c \wedge f_{i+1} \in \text{set}_c) \vee (f_i \in \text{get}_c \wedge f_{i+1} \in \text{get}_c)$

The correctness of the optimization in Definition 11 is established by the proof of using the Initialization Graph's topological ordering as the initialization order by Gomes et al. [11]. Their proof is trivially shown to cover this approach since the optimization does not change the structure of the Initialization Graph. A limitation of this optimization strategy is that it is not guaranteed to find all potentially valid optimizations of a co-simulation scenario. Considering it works only on a specific co-simulation step (a topological order of a graph), which is not necessarily unique for a given co-simulation scenario. A more advanced optimization strategy needs to be developed to perform all viable optimizations of a co-simulation step. Another solution is to apply this optimization strategy on the set of all valid co-simulation steps - yielding a potential very inefficient initialization algorithm. The initialization of a co-simulation is typically not the most time consuming or computational heaviest part of the co-simulation. However, it is still considered a low hanging fruit to apply this optimization to optimize the initialization.

3.3 The Complete Initialization Strategy

The pseudo-code in Algorithm 1 formulates the entire initialization strategy of the interconnected variables of a co-simulation scenario.

---

[2] 5 iterations is the default in our approach. This number is based on experience.

Algorithm 1. Initialization strategy for Interconnected variables

1: $InitializationGraph \leftarrow createGraph(connections)$
2: $SCCS \leftarrow Tarjan(InitializationGraph)$
3: $OptimizeInitializationOrder(SCCS)$
4: **for each:** $SCC \in SCCS$ **do**
5:    **if** $isAlgebraicLoop(SCC)$ **then**
6:        $applyFixedPointIteration(SCC)$;
7:    **else**
8:        $initializeVariable(SCC)$;
9:    **end if**
10: **end for**

As seen from the algorithm in Algorithm 1, the algebraic loops are handled using a different initialization strategy compared to the other trivial SCC of a single interconnected FMU variable.

4   Case Study

In this section, we give a simple example of a co-simulation whose correct initialization demands the solution to an algebraic loop.

We consider a co-simulation of a quarter car model [19, Section 6.4], illustrated in Fig. 2. We omit the equations that each FMU is solving but note that gravity acts on both wheel and chassis masses and that the origin of each mass is when the springs are not displaced. The equations and simulation model for this example are available online[3].

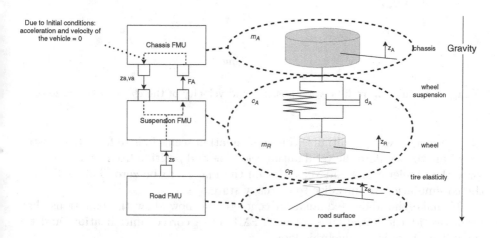

Fig. 2. Quarter car model co-simulation. Adapted from [19, Section 6.4].

---

[3] https://github.com/SimplisticCode/QuarterCarCaseStudy.

The FMUs need initial conditions specified by equations that restrict the possible initial values for the position and velocity of the wheel and chassis masses. Figure 3 illustrates what happens when we set those positions and velocities to zero. Note that, because of gravity, the car chassis bounces on the suspension wheel, with a maximum compression of about 17cm compared to when the system's springs are uncompressed. This is most likely an invalid scenario, as the car's suspension might not be rated to be displaced that much. In any case, the purpose of simulation studies involving quarter car models is to understand how well a suspension system absorbs shock when the car goes over a bump, not when the car *falls on the road*, which is what the simulation results in Fig. 3 resemble.

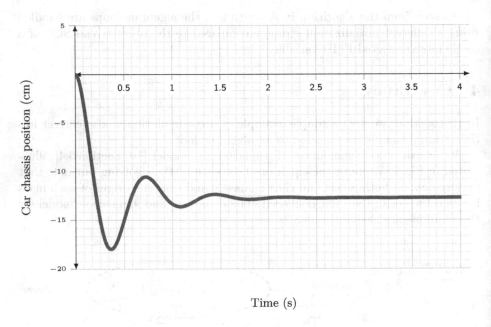

Fig. 3. Simulation results when position and velocity of the chassis mass is zero.

The correct way to initialize this co-simulation scenario is to force the master algorithm to calculate the valid initial velocities and position from equations that force the accelerations and velocities on the masses to be zero. This will force the co-simulation to initialize to a steady state.

To make the above explanation concrete, we now show the equations that are active at the initial time for each FMU for a correct initialization, and we show that there is an algebraic loop.

For the road FMU, the initial equation is simply the initial height of the road surface, which in this case is zero, i.e., $z_s = 0$. For the suspension FMU, the following equations are active:

$$a_R = 0.0 \qquad \text{Acceleration of tire} \qquad (1)$$
$$v_R = 0.0 \qquad \text{Velocity of tire} \qquad (2)$$
$$F_{gR} = 9.81 * m_R \qquad \text{Gravity on the tire} \qquad (3)$$
$$F_R = -c_R * z_R \qquad \text{Rubber force acting on tire} \qquad (4)$$
$$F_A = c_A * (z_A - z_R) + d_A * (v_A - v_r) \qquad \text{Suspension force acting on tire} \qquad (5)$$
$$F_{total} = F_R + F_A - F_{gR} \qquad \text{Total forces acting on tire} \qquad (6)$$
$$a_R = (1/m_R) * F_{total} \qquad \text{Acceleration of tire.} \qquad (7)$$

Finally, for the Chassis FMU, the following equations are active at the initial time:

$$a_A = 0.0 \qquad \text{Acceleration of chassis} \qquad (8)$$
$$v_A = 0.0 \qquad \text{Velocity of chassis} \qquad (9)$$
$$F_{gA} = 9.81 * m_A \qquad \text{Gravity on the chassis} \qquad (10)$$
$$a_A = (1/m_A) * (-F_A - F_{gA}) \qquad \text{Acceleration of chassis.} \qquad (11)$$

To see that there is an algebraic loop, note that the output z_A of the chassis FMU is not restricted directly, but instead has to be computed from the acceleration equations $a_A = 0 = (1/m_A) * (-F_A - F_{gA})$. The later contains the output F_A of the Suspension FMU. This output, in turn, depends on z_A, thus yielding

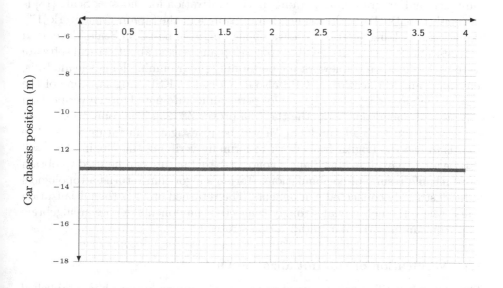

Time (s)

Fig. 4. Simulation results starting from a correct initial state (a steady state).

an algebraic loop. Figure 4 shows the simulation results when the algebraic loop is properly solved during initialization.

5   Realization of a Maestro 2 Plugin

The presented approach has been realized as a Maestro 2 expansion plugin that generates the *Initialization*-phase of a co-simulation specification expressed in MaBL. The plugin calculates the MaBL-specification based on the FMUs of a co-simulation scenario and a specific plugin-configuration to let the user supply the initial values of FMU parameters and fine-tune the initialization of the system. The plugin can calculate a correct initialization specification if the co-simulation scenario adheres to the behavior dictated by the definition given in Definition 10 meaning all algebraic loops in the scenario convergences within a finite number of iterations.

The plugin optimizes the initialization order by grouping operations that can be executed in *parallel* to take advantage of FMI's ability to *set* or *get* multiple variables of a single FMU in bulk. The criteria for this optimization is defined in Definition 11. The developed plugin has been tested on numerous co-simulation scenarios from the INTO-CPS universe[21] and compared with the existing *Initializer* of Maestro. The plugin has been tested as a part of the complete Maestro 2 pipeline.

5.1   Realization of the Topological Sorting

The topological sorting algorithm (Tarjan's Algorithm) is implemented in Scala [18], an object-oriented programming language incorporating many features from the functional programming paradigm. The motivation for choosing Scala [18] is its relation to JVM and the connection to Slang and the Sireum framework [17]. Slang (Sireum Language) is a programming language based on Scala, developed at Kansas State University (KSU), to develop and reason about critical software systems. Sireum is a framework for performing programming language analysis, reasoning, and verification of CPS also developed at KSU. Logika is one of the tools in the Sireum framework used for performing automated formal verification of a piece of Slang code using the theorem prover Z3 [23]. The connection of the implementation to Slang and Logika will be investigated in future work. The plan is to use the Logika framework to formally verifying the plugin. This will also be used to explore how Slang's contract-based nature can be used to obtain more reliable results of co-simulations. Tarjan's algorithm returns a topological order of strongly connected components. The returned order is the initialization order, where the non-trivial strongly connected components denote an algebraic loop requiring a particular initialization strategy.

5.2   Verification of the Initialization Order

The plugin is verified using several methods. The plugin approach is established using traditional proof methods, and the plugin has been practically verified against an established co-simulation step verifier. Gomes et al. have verified the approach in [9]; they proved the correctness of using the topological order of a dependency graph of the interconnected FMU-variables as the order of the operations in a co-simulation step (both the initialization procedure and an arbitrary step). Gomes et al. [9] used a graph of FMU-operations (*Set, Get, doStep*) in their proof instead of interconnected FMU-variables, which is the approach of this paper. The simplification of using the interconnected FMU-variables is valid and preserves the properties proved by Gomes et al. since this approach only considers the initialization phase of a co-simulation. This makes it possible to omit all the *doStep* nodes from Gomes et al.'s graph, eventually ending up with a graph similar to the initialization graph described in Definition 8. This approach is a subgraph of the graph by Gomes et al.[9], which allows their proof to be modified to the approach presented in this paper.

Practical Verification Against an Established Verifier. Gomes et al.'s [9] main contribution is a Prolog implementation of the principles for a valid FMI based co-simulation step [4]. Gomes et al. use the Prolog implementation in their research to verify their approach for generating different co-simulation algorithms. The Prolog realization encapsulates all the rules of a valid co-simulation step (both master-algorithm and an initialization algorithm). The Initializer includes an integration to the Prolog Verifier. The integration is a Java program based on JIProlog [13] - a library that allows calling Prolog predicates directly from Java. The integration is used to check the initialization order against the rules in the Prolog database. The integration performs all the necessary transformations of the dependency graph (see definition 8) used in the Maestro plugin to a graph of FMU operations used in the Prolog database. The transformation is based on the definitions 3 and 4. The integration has been realized to systematically verify the calculated initialization order's correctness against an established and recognized co-simulation Algorithm Verifier. The Prolog implementation does support co-simulation scenarios containing algebraic loop, so these scenarios are not tested against the Prolog database.

6   Related Work

Prior work [5,11] is looking into the generation of co-simulation algorithms (both master and initialization algorithms) for FMI-based scenarios. Their generation technique is like ours, based on a dependency graph of the operations of a co-simulation step. Both Gomes and Broman present an approach for using the topological order of a dependency graph to establish a correct order of operations in a co-simulation step of a given co-simulation scenario. The work by Gomes et al. [11] does also define the criteria for a correct co-simulation step. Their work has many similarities with ours. However, their work is mostly concerned with the theoretical aspect of co-simulation algorithm generation and

---

[4] http://msdl.cs.mcgill.ca/people/claudio/projs/PrologCosimGeneration.zip.

verification, while our work has a more practical nature. Gomes et al. do also not consider the handling of algebraic loops, which is a key feature of our approach. Furthermore, the approach taken in our work is only concerned with the initialization procedure of a co-simulation.

Broman et al. [5] also suggest to use the topological sorting of a dependency graph of the interconnected variables to detect algebraic loops and discover the partial order of port-operations. Nevertheless, they explicitly specify the requirement for cycle freedom in the dependency graph as a precondition for generating a valid co-simulation. It means they refuse all co-simulation scenarios containing algebraic loops. It is a significant difference to our approach that applies a fixed point iteration strategy to handle these scenarios. Also, the approach in this paper is more specialized because it only considers the initialization of a co-simulation, which means it deals with non-interconnected variables.

Amalio et al. [1] investigate how to avoid algebraic loops in FMU based co-simulation scenarios by statically checking the architectural design of a CPS. The publication's purpose is like ours, to avoid invalid co-simulation scenarios. Nevertheless, they achieve this by excluding co-simulation scenarios containing algebraic loops. Their method is realized in a co-simulation tool, INTO-SysML [15]. Formal methods form the basis of their work (Theorem Proving and Model-checking). It will be an inspiration for the future work of formally verifying the plugin and other parts of Maestro 2.

The work by Gomez et al. [6] is similar to ours. They use Tarjan's SCC algorithm to generate a sorted DAG of strongly connected components to solve the initialization problem. Even though their work is very similar to ours, we extend their approach with the verification against the simulation semantics resulting in a formally more sound approach. However, further work will look into further improvements and formal verification of the current approach.

7   Concluding Remarks

This work uses a topological ordering of a dependency graph of the interconnected FMUs variable and internal FMU connections along with predicates from the FMI specification to calculate a correct initialization order for a co-simulation scenario potentially containing algebraic loops. The initialization procedure optimizes the initialization order by grouping variables with similar characteristics to perform the fewest possible operations in the initialization procedure. This approach supports the initialization of a co-simulation scenario containing algebraic loops by using fixed point iteration. The approach is suitable to combine with well-established master algorithms like Gauss-Seidel and Jacobi [16]. The approach is realized as an expansion plugin for the open-source INTO-CPS Maestro 2 tool and verified against the existing *Initializer* and the calculated initialization order was verified against an established co-simulation Algorithm Generator and Verifier implemented in Prolog [9].

Future work includes formal verification of the plugin using the Logika framework[17]. We will also look into the generation of a verification strategy for the whole Maestro 2 framework to examine how different forms of verification jointly can extend the trust of the correctness of the result of a co-simulation.

Acknowledgements. We would like to thank Stefan Hallerstede, Christian Møldrup Legaard, and Peter Gorm Larsen for providing valuable input to this paper and the developed plugin.

References

1. Amálio, N., Payne, R., Cavalcanti, A., Woodcock, J.: Checking SysML models for co-simulation. In: Ogata, K., Lawford, M., Liu, S. (eds.) ICFEM 2016. LNCS, vol. 10009, pp. 450–465. Springer, Cham (2016). https://doi.org/10.1007/978-3-319-47846-3_28
2. Arnold, M., Clauß, C., Schierz, T.: Error analysis and error estimates for co-simulation in FMI for model exchange and co-simulation v2.0. In: Schöps, S., Bartel, A., Günther, M., ter Maten, E.J.W., Müller, P.C. (eds.) Progress in Differential-Algebraic Equations. DEF, pp. 107–125. Springer, Heidelberg (2014). https://doi.org/10.1007/978-3-662-44926-4_6
3. Bastian, J., Clauß, C., Wolf, S., Schneider, P.: Master for co-simulation using FMI. In: 8th International Modelica Conference, pp. 115–120. Linköping University Electronic Press, Linköpings universitet (2011). https://doi.org/10.3384/ecp11063115
4. Blockwitz, T., et al.: Functional mockup interface 2.0: the standard for tool independent exchange of simulation models. In: 9th International Modelica Conference, pp. 173–184. Linköping University Electronic Press (2012). https://doi.org/10.3384/ecp12076173
5. Broman, D., et al.: Composition of FMUs for Co-Simulation. Technical report, University of California, Berkeley (2013). http://www.eecs.berkeley.edu/Pubs/TechRpts/2013/EECS-2013-153.html
6. Évora Gómez, J., Hernández Cabrera, J.J., Tavella, J.P., Vialle, S., Kremers, E., Frayssinet, L.: Daccosim NG: co-simulation made simpler and faster. In: The 13th International Modelica Conference, Regensburg, Germany, March 4–6, 2019, pp. 785–794, February 2019. https://doi.org/10.3384/ecp19157785
7. FMI.: Functional Mock-up Interface for Model Exchange and Co-Simulation (2014). https://fmi-standard.org/downloads/
8. Gomes, C., Denil, J., Vangheluwe, H.: Causal-block diagrams: a family of languages for causal modelling of cyber-physical systems. Foundations of Multi-Paradigm Modelling for Cyber-Physical Systems, pp. 97–125. Springer, Cham (2020). https://doi.org/10.1007/978-3-030-43946-0_4
9. Gomes, C., Lucio, L., Vangheluwe, H.: Semantics of co-simulation algorithms with simulator contracts. In: 2019 ACM/IEEE 22nd International Conference on Model Driven Engineering Languages and Systems Companion (MODELS-C) (2019). https://doi.org/10.1109/models-c.2019.00124
10. Gomes, C., Thule, C., Lúcio, L., Vangheluwe, H., Larsen, P.G.: Generation of co-simulation algorithms subject to simulator contracts. In: Camara, J., Steffen, M. (eds.) SEFM 2019. LNCS, vol. 12226, pp. 34–49. Springer, Cham (2020). https://doi.org/10.1007/978-3-030-57506-9_4
11. Gomes, C., et al.: HintCO - hint-based configuration of co-simulations. In: International Conference on Simulation and Modeling Methodologies, Technologies and Applications, pp. 57–68 (2019). https://doi.org/10.5220/0007830000570068

12. Gomes, C., Thule, C., Broman, D., Larsen, P.G., Vangheluwe, H.: Co-simulation: a survey. ACM Comput. Surv. **51**(3), 1–33, Article 49 (2018). https://doi.org/10.1145/3179993

13. JIProlog: JIProlog, October 2016. http://www.jiprolog.com. Accessed 20 Aug 2020

14. Kübler, R., Schiehlen, W.: Two methods of simulator coupling. Math. Comput. Model. Dyn. Syst. **6**(2), 93–113 (2000). https://doi.org/10.1076/1387-3954(200006)6:2;1-M;FT093

15. Miyazawa, U.Y.A., Woodcock, U.J.: Integrated tool chain for model-based design of CPSs foundations of the SysML profile for CPS modelling (2016). https://www.semanticscholar.org/paper/INtegrated-TOol-chain-for-model-based-design-of-of-Miyazawa-Woodcock/3042572251aba18ab21ced9cc2fb49223dea2a2c. Accessed 13 Nov 2020

16. Palensky, P., Van Der Meer, A.A., Lopez, C.D., Joseph, A., Pan, K.: Cosimulation of intelligent power systems: fundamentals, software architecture, numerics, and coupling. IEEE Ind. Electron. Mag. **11**(1), 34–50 (2017). https://doi.org/10.1109/MIE.2016.2639825

17. Robby Hatcliff, J., Belt, J.: Model-based development for high-assurance embedded systems. In: Margaria, T., Steffen, B. (eds.) Leveraging Applications of Formal Methods, Verification and Validation. Modeling, pp. 539–545. Lecture Notes in Computer Science, Springer International Publishing (2018). https://doi.org/10.1007/978-3-030-03418-4_32

18. Scala: The Scala Programming Language, August 2020. https://www.scala-lang.org. Accessed 19 Aug 2020

19. Schramm, D., Hiller, M., Bardini, R.: Force components. Vehicle Dynamics, pp. 207–224. Springer, Heidelberg (2018). https://doi.org/10.1007/978-3-662-54483-9_9

20. Tarjan, R.: Depth-first search and linear graph algorithms. SIAM J. Comput. **1**(2), 146–160 (1972). https://doi.org/10.1137/0201010

21. Thule, C., Lausdahl, K., Gomes, C., Meisl, G., Larsen, P.: Maestro: the INTO-CPS co-simulation framework. Simul. Model. Pract. Theor. **92**, 45–61 (2019). https://doi.org/10.1016/j.simpat.2018.12.005

22. Thule, C., et al.: Towards reuse of synchronization algorithms in co-simulation frameworks. In: Camara, J., Steffen, M. (eds.) SEFM 2019. LNCS, vol. 12226, pp. 50–66. Springer, Cham (2020). https://doi.org/10.1007/978-3-030-57506-9_5

23. Z3prover: z3, September 2020). https://github.com/Z3Prover/z3/wiki. Accessed 13 Sept 2020

Introducing Regression Tests and Upgrades to the INTO-CPS Application

Prasad Talasila[(⊠)], Armine Sanjari, Kristoffer Villadsen, Casper Thule,
Peter Gorm Larsen, and Hugo Daniel Macedo

DIGIT, Department of Engineering, Aarhus University, Aarhus, Denmark
{prasad.talasila,casper.thule,pgl,hdm}@eng.au.dk,
{201607125,201607406}@post.au.dk

Abstract. In this paper, we report on the progress made to upgrade
and develop a stable upgrading process to the INTO-CPS Application,
an Electron.js based desktop application providing a front-end to an
INtegrated TOolchain, which is used to develop Cyber-Physical Systems
models. We added regression tests to the codebase and for the first time
can detect the loss of functionality of the application and its accompanying training tutorials using an automated process. The tests were developed on top of the Mocha, Chai and Spectron frameworks and cover
all the tutorials steps performed in the desktop application (approximately 33% of the app and other tools total). The testing process is not
yet ready to be deployed in the also recently developed GitHub Actions
automated workflow, but this is a possibility to be considered in future
developments. We expect this work to improve the stability and security
of the code, thus improving user experience.

Keywords: Integrated toolchain · Front end · Automated regression
test

1 Introduction

The INTO-CPS Application is a front-end used by engineers and students in
several projects and universities. It enables users to harness the backend co-
simulation toolchain in the development of Cyber-Physical Systems models, and
its typical use-cases are illustrated in a set of tutorials. The front-end was created during the INtegrated TOolchain for Model-based Design of Cyber-Physical
Systems (INTO-CPS) project [2], and intends to reduce the entry barriers to
newcomers interested in the modelling and co-simulation of CPS, by integrating
other tools like Overture, Modelio, and Maestro.

The front-end was developed on top of *Electron*[1] and *Angular*[2] among other
web-based dependencies. Given their nature, the dependencies and in particular

---

[1] See https://www.electronjs.org/.
[2] See https://angular.io/.

© Springer Nature Switzerland AG 2021
L. Cleophas and M. Massink (Eds.): SEFM 2020 Workshops, LNCS 12524, pp. 311–317, 2021.
https://doi.org/10.1007/978-3-030-67220-1_23

Electron and Angular evolve fast with short release cycles. Furthermore, newer versions of any of the frameworks are not necessarily backwards compatible. Thus applications built on top of such frameworks require frequent upgrades with non-trivial code changes. In addition, the back-end third-party tools evolve and such upgrades need to be accommodated in the front-end. For instance, the new releases of tools such as Maestro require frequent upgrades to the application.

Upgrades are likely to break the app features, and in particular the performance of its tutorial steps, thus frustrating precisely the most vulnerable in the userbase. To prevent that, the tool-induced and framework-induced upgrades to the codebase require significant manual regression testing to ensure that all the expected functional and non-functional requirements are satisfied. Manual regression testing of the application requires hundreds of tedious and error-prone steps. The only guaranteed way to make successful upgrades to the application is to have automated regression tests. Automation of the regression tests paves way for more frequent testing of the application. An added advantage of automated tests is the ability to take advantage of Continuous Integration/Continuous Delivery pipelines during the software development process.

In this paper we describe two improvements – automated regression tests and framework upgrades – carried out recently. With these two improvements in place, it is now possible to make successful and frequent releases of the application. This paper starts off with background information on the application in Sect. 2. Afterwards, Sect. 3 provides an overview of the initiatives taken for automated regression testing of the application. This is followed by Sect. 4 explaining the systematic steps taken to upgrade the dependencies of the application to the latest stable versions. Finally, Sect. 5 provides concluding remarks and an overview of the future work.

2   Background

The INTO-CPS toolchain has been used to develop CPS case studies in various fields and features co-simulation [3], the combination of models simulating system components to obtain a joint whole system simulation, as a paradigm in the development of CPS models. Any model living up to the *Functional Mockup Interface (FMI)* standard [4] is compatible with the toolchain, and is referred to as a *Functional Mockup Unit (FMU)* [7]. The INTO-CPS toolchain consists of the following baseline tools:

Modelio[3] is a modeling environment for generating the co-simulation scenario, i.e. connections between FMUs, the interface of the FMUs and related parameters, from a SysML profile for CPS modeling [1].

---

[3] See http://www.modelio.org/.

Overture[4], **20-sim**[5], **OpenModelica**[6], **and RT-Tester**[7] are tools that export FMUs.

Maestro [9] is a co-simulation orchestration engine to simulate selected FMUs based on a given co-simulation scenario.

Because the INTO-CPS Application brings together different tools, and each party whose tool is to be available within the application can develop a tool-specific UI component that is plugged into the application, *Electron.js* was chosen as a framework. Applications built using Electron framework use web technologies yet work like native desktop applications, and simplify the support of Windows, Linux, and Mac versions. With most of the codebase written using web technologies, the desktop version of the application can be easily converted into a web application [6,8].

3 Introducing Regression Tests

We developed tests for the tutorials[8] that are performed by trainees and engineering students while learning how to explore the INTO-CPS toolchain [5]. Each tutorial consists of a sequence of steps, each requiring a user to interact with the application or an external tool. See Fig. 1 for an illustration of an application state expected to be visible at the beginning of a tutorial step.

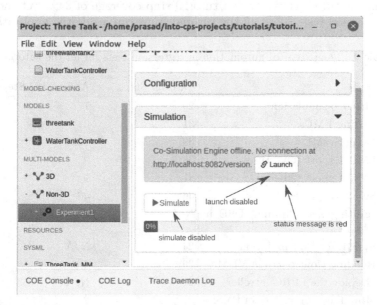

Fig. 1. A tutorial fig. with the expected state before launching the co-simulation engine.

---

[4] See http://overturetool.org/.

[5] See http://www.20sim.com/.

[6] See https://www.openmodelica.org/.

[7] See http://www.verified.de/products/rt-tester/.

[8] See https://github.com/INTO-CPS-Association/training.

We use *Spectron*[9] as a test framework and wrote the tests resourcing to primitives available in the *mocha*[10] and *chai*[11] testing libraries. As an example, the test checking the conformance of the app state to the expected (Fig. 1) is shown in Listing 1.1.

```
1  it('Co-Simulation Engine offline', function () {
2
3      this.app.client.$('coe-simulation').waitForVisible()
4      .then(() => {
5          return this.app.client.$('coe-simulation')
6              .$('.alert.alert-danger')
7              .getText()
8              .then(function (text) {
9                  expect(text)
10                 .contain('Co-Simulation Engine offline')
```

Listing 1.1. Testing if the app state is according to the expectation (Figure 1).

Although the INTO-CPS Application and its tutorials steps involve the usage of external tools like Overture, our tests do not cover such steps, because such interactions happen beyond the scope of the Spectron framework. Table 1 shows the coverage achieved by the automated regression tests. The covered steps correspond to steps that are part of the application. Based on the number of steps covered in the tutorials, we have a tutorial step coverage of 34%, yet the automated regression tests cover all the tutorial steps run inside the application.

Table 1. The extent of automation in regression tests for all the tutorials.

Tutorials	Tutorial Steps	Automated Steps
1 – First Co-simulation	20	7
2 – Adding FMUs	31	21
3 – Using SysML	39	10
4 – FMU Export (Overture)	38	9
5 – FMU Export (20-sim)	41	9
6 – SysML for DSE	46	46
7 – Editing and Running DSE in App	N/A	7
8 – SysML for Co-simulation	24	N/A
9 – Building Controllers in VDM	29	4
10 – Deploy the LFRController	28	N/A
11 – Building Controllers in PVSio-web	47	2
Total Steps	343	115

---

[9] See https://github.com/electron-userland/spectron.
[10] See https://mochajs.org/.
[11] See https://www.chaijs.com/.

4   Upgrading the Dependencies

The application is built on top of the Electron framework, which in turn uses *Node.js*[12]. We use the *npm*[13] package manager to download and manage the dependencies including *gulp*, which we use as a build system. There are hundreds of dependencies including the *electron-packager*, *spectron* and *spectron-fake-menu*. Each of these packages, in turn, depends on other packages and it is easy to find two npm packages requiring different versions of the same base package. Figure 2 illustrates a concise dependency scenario for one development snapshot of the application. While *spectron* is both a direct and nested dependency for the application, *yargs-parser* is completely a nested dependency. Despite the complexity the one finds, it is possible to list upgradable and security vulnerabilities posed by using outdated packages.

The *npm* command line interface provides a comprehensive summary of the security vulnerabilities arising out of the existing package versions. Figure 3 shows the audit report for *yargs-parser* package; the audit report suggests a range of packages to be used. We used this feature and the Angular update web tool to upgrade the INTO-CPS Application. Notably, an upgrade from Angular 2.0.0 to 7.2 and the Gulp build system to its version 4.

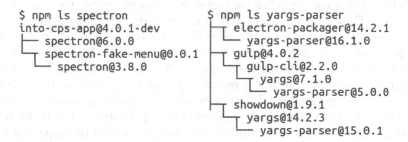

```
$ npm ls spectron                    $ npm ls yargs-parser
into-cps-app@4.0.1-dev               ┬ electron-packager@14.2.1
├── spectron@6.0.0                   │ └── yargs-parser@16.1.0
└─┬ spectron-fake-menu@0.0.1         ├─┬ gulp@4.0.2
  └── spectron@3.8.0                 │ └─┬ gulp-cli@2.2.0
                                     │   └─┬ yargs@7.1.0
                                     │     └── yargs-parser@5.0.0
                                     ├─┬ showdown@1.9.1
                                     │ └─┬ yargs@14.2.3
                                     │   └── yargs-parser@15.0.1
```

Fig. 2. A concise view of npm dependency problems encountered during upgrade of Electron application.

```
$ npm audit
                === npm audit security report ===
```

Low	Prototype Pollution
Package	yargs-parser
Patched in	>=13.1.2 <14.0.0 || >=15.0.1 <16.0.0 || >=18.1.2
Dependency of	gulp [dev]
Path	gulp > gulp-cli > yargs > yargs-parser
More info	https://npmjs.com/advisories/1500

Fig. 3. Security vulnerability report for yargs-parser package.

---

[12] See https://nodejs.org/.
[13] See https://www.npmjs.com/.

By upgrading npm packages to the latest possible versions and updating the code, we were able to reduce the number of security vulnerabilities in the application. After the software upgrade, the total security vulnerabilities have decreased from 73 major and 195 minor vulnerabilities to 3 major and 7 minor vulnerabilities. We have also mapped the deprecated features of npm packages used to features of the application. Based on this mapping, certain features have been either marked for deprecation or prioritized in the current development process.

In addition to the previous Jenkins based continuous integration (CI) server, we developed an additional *GitHub* release workflow. Each code update in the *git* repository triggers build process on the CI server and the software build status is visible to all the developers. The build status immediately highlights any problems due to the latest changes in the source code. The lead developer periodically updates the *master* branch of the git repository which automatically builds the latest version the application and release the same for Windows, Linux and Mac platforms.

5 Concluding Remarks and Future Work

The INTO-CPS Application is a user interface to an integrated toolchain used in the design of Cyber-Physical Systems. Updates to underlying tools and software library frameworks mandate continuous upgrades to the application. Our work made progress in regression testing and upgrading the application. The regression tests provide a check on the expected features of the application while the upgrade process significantly reduced the software vulnerabilities. Due to improved testing and successful upgrades, we are also making definitive progress in code reuse in desktop and web versions of the application. Shortly, we intend to unify the development of the desktop and web versions of the application. We also intend to track the test coverage based on the number of lines of code or functions, and deepen the testing procedures. Doing so will increase the usability and performance of the application and its associated CPS development paradigm, thus providing better user experiences.

Acknowledgements. We acknowledge the European Union for funding the INTO-CPS project (Grant Agreement 644047) which developed the open toolchain and the INTO-CPS Application, the Poul Due Jensen Foundation that has funded subsequent work on taking this forward towards the engineering of digital twins and the European Union for funding the HUBCAP (Grant Agreement 872698) project.

References

1. Amalio, N., Cavalcanti, A., Miyazawa, A., Payne, R., Woodcock, J.: Foundations of the SysML for CPS modelling. Technical report, INTO-CPS Deliverable, D2.2a, December 2016
2. Fitzgerald, J., Gamble, C., Larsen, P.G., Pierce, K., Woodcock, J.: Cyber-physical systems design: formal foundations, methods and integrated tool chains. In: FormaliSE: FME Workshop on Formal Methods in Software Engineering, ICSE 2015, Florence, Italy, May 2015
3. Gomes, C., Thule, C., Broman, D., Larsen, P.G., Vangheluwe, H.: Co-simulation: a survey. ACM Comput. Surv. **51**(3), 1–33 (2018)
4. Larsen, P.G., et al.: The integrated toolchain for cyber-physical systems (INTO-CPS): a guide. Technical report, INTO-CPS Association, October 2018. www.into-cps.org
5. Larsen, P.G., et al.: Frontiers in software engineering education. In: Bruel, J.M., Capozucca, A., Mazzara, M., Meyer, B., Naumchev, A., Sadovykh, A. (eds.) Frontiers in Software Engineering Education, pp. 196–213. Springer International Publishing, Cham (2020)
6. Macedo, H.D., Rasmussen, M.B., Thule, C., Larsen, P.G.: Migrating the INTO-CPS application to the cloud. In: Sekerinski, E., et al. (eds.) FM 2019. LNCS, vol. 12233, pp. 254–271. Springer, Cham (2020). https://doi.org/10.1007/978-3-030-54997-8_17
7. Pohlmann, U., Schäfer, W., Reddehase, H., Röckemann, J., Wagner, R.: Generating functional mockup units from software specifications. In: Modelica Conference (2012)
8. Rasmussen, M.B., Thule, C., Macedo, H.D., Larsen, P.G.: Migrating the INTO-CPS application to the cloud. In: Gamble, C., Couto, L.D. (eds.) Proceeding 17th Overture Workshop, pp. 47–61. Newcastle University Technical Report CS-TR-1530, October 2019
9. Thule, C., Lausdahl, K., Gomes, C., Meisl, G., Larsen, P.G.: Maestro: the INTO-CPS co-simulation framework. Simul. Model. Pract. Theor. **92**,45–61 (2019). http://www.sciencedirect.com/science/article/pii/S1569190X1830193X

Cosimulation-Based Control Synthesis

Adrien Le Coënt[(✉)][iD], Julien Alexandre dit Sandretto[iD],
and Alexandre Chapoutot[iD]

U2IS, Institut Polytechnique de Paris, ENSTA Paris, 828 Boulevard des maréchaux,
91762 Palaiseau Cedex, France
adrien.le-coent@ens-cachan.fr, {alexandre,chapoutot}@ensta.fr

Abstract. In this paper, we present a procedure for guaranteed control synthesis for nonlinear sampled switched systems which relies on an adaptive state-space tiling procedure. The computational complexity of the procedure being exponential in the dimension of the system, we explore the use of cosimulation for improving computation times and the scalabity of the method. We apply the procedure on a scalable case study of various dimensions, which is, to our knowledge, a significant step towards the scalability of formal control synthesis methods with respect to the state of the art.

Keywords: Switched systems · Guaranteed numerical integration · Interval analysis · Symbolic control synthesis · Cosimulation

1 Introduction

Model-based design [20,27] is an effective approach to tackle the increasing complexity of cyber-physical systems. In this approach, physical systems, *e.g.*, plant, are usually modelled by differential equations while computer parts are described by transition systems. Combining these models allows to simulate the behaviour of the whole model of the system in order to predict its behaviour to avoid faults or to synthesize control algorithms.

Safety critical cyber-physical systems require strong guarantees in their execution in order to assess the safety of the mission or the users. Formal methods can produce rigorous evidence for the safety of cyber-physical systems, *i.e.*, based on mathematical reasoning. For example, reachability analysis is an efficient technique to compute the set of reachable states of cyber-physical systems. Once, knowing the set of reachable states, the avoidance of bad states can be formally proved. The main feature of reachability analysis is its ability to propagate sets of values through dynamical systems instead of performing several numerical simulations.

This work was supported by the Chair Complex Systems Engineering - Ecole polytechnique, THALES, DGA, FX, Dassault Aviation, Naval Group Research, ENSTA Paris, Télécom Paris, and Fondation ParisTech.

© Springer Nature Switzerland AG 2021
L. Cleophas and M. Massink (Eds.): SEFM 2020 Workshops, LNCS 12524, pp. 318–333, 2021.
https://doi.org/10.1007/978-3-030-67220-1_24

One weakness of formal verification methods, in particular, reachability analysis, is the scalability with respect to the dimension (number of states) of cyber-physical systems. Applying a cosimulation approach to reachability analysis is attractive since it could broaden the class of problems which can be solved with this technique. A set-based approach of cosimulation to solve differential equations has been defined [21], we explore here its use in the context of formal control synthesis.

Contribution. We propose an extension of a controller synthesis algorithm for a particular class of cyber-physical systems, a.k.a. *nonlinear sampled switched systems*, it relies on a cosimulation approach for the required reachability analysis. A formal definition of the set-based cosimulation is given and then used in order to compute a safe controller for a model of an apartment with a controlled heating.

Related Work. Most of the recent work on set-valued integration of nonlinear ordinary differential equations is based on the upper bounding of the Lagrange remainders either in the framework of Taylor series or Runge-Kutta schemes [2,3,5,7,9,10,12,24]. Sets of states are generally represented as vectors of intervals, a.k.a. *boxes*, and are manipulated through interval arithmetic [25] or affine arithmetic [11]. Taylor expansions with Lagrange remainders are also used in the work of [3], which uses *polynomial zonotopes* for representing sets of states in addition to interval vectors.

The *guaranteed* or *validated* solution of ODEs using interval arithmetic is studied in the framework of Taylor series in [10,13,22,25,26], and Runge-Kutta schemes in [2,5,6,14,18]. The former is the oldest method used in interval analysis community because the expression of the remainder of Taylor series is simple to obtain. Nevertheless, the family of Runge-Kutta methods is very important in the field of numerical analysis. Indeed, Runge-Kutta methods have several interesting stability properties which make them suitable for an important class of problems. The recent work [1] implements Runge-Kutta based methods which prove their efficiency at low orders and for short simulations.

Cosimulation has been extensively studied in the past years [16,17], and has been reported in a number of industrial applications (see [16] for an extensive list domain applications and associated publications). However, most of the uses and tools developed rely on the FMI/FMU standard [4,8,28], which do not allow guaranteed simulation. To our knowledge, guaranteed cosimulation of systems has never been applied on controller synthesis method.

Organization of the Paper. Section 2 presents the mathematical model of sampled switched systems as well as an algorithm to synthesize a safe controller. Set-based simulation and its extension to cosimulation are presented in Sect. 3. Experimental results are presented in Sect. 4 before concluding in Sect. 5.

2   Control Synthesis of Switched Systems

A presentation of the mathematical model of sampled switched systems is given in Sect. 2.1. An algorithm to synthetize safe controllers is described in Sect. 2.2.

2.1   Switched Systems

Let us consider nonlinear switched systems such that

$$\dot{x}(t) = f_{\sigma(t)}(x(t), d(t)) \tag{1}$$

is defined for all $t \geq 0$, where $x(t) \in \mathbb{R}^n$ is the state of the system, $\sigma(\cdot) : \mathbb{R}^+ \longrightarrow U$ is the switching rule, and $d(t) \in \mathbb{R}^m$ is a bounded perturbation. The finite set $U = \{1, \ldots, N\}$ is the set of switching modes of the system. We focus on *sampled switched systems*, given a sampling period $\tau > 0$, switchings will periodically occur at times $\tau, 2\tau, \ldots$. Switchings depend only on time, and not on states, this is the main difference with hybrid systems.

The switching rule $\sigma(\cdot)$ is thus piecewise constant, we will consider that $\sigma(\cdot)$ is constant on the time interval $[(k-1)\tau, k\tau)$ for $k \geq 1$. We call *"pattern"* a finite sequence of modes $\pi = (i_1, i_2, \ldots, i_k) \in U^k$. With such a control pattern, and under a given perturbation d, we will denote by $\mathbf{x}(t; t_0, x_0, d, \pi)$ the solution at time $t \geq t_0$ of the system

$$
\begin{aligned}
\dot{x}(t) &= f_{\sigma(t)}(x(t), d(t)), \\
x(t_0) &= x_0, \\
\forall j \in \{1, \ldots, k\}, \ \sigma(t) &= i_j \in U \text{ for } t \in [t_0 + (j-1)\tau, t_0 + j\tau).
\end{aligned}
\tag{2}
$$

We address the problem of synthesizing a state-dependent switching rule $\tilde{\sigma}(\cdot)$ for Eq. (2) in order to verify some properties. This important problem is formalized as follows:

Problem 1 (Control Synthesis Problem). Let us consider a sampled switched system as defined in Eq. (2). Given three sets R, S, and B, with $R \cup B \subset S$ and $R \cap B = \emptyset$, find a rule $\tilde{\sigma}(\cdot)$ such that, for any $x(0) \in R$

- τ-stability[1]: $x(t)$ returns in R infinitely often, at some multiples of sampling time τ.
- *safety*: $x(t)$ always stays in $S \backslash B$.

In this problem, S is a *safety* set in which the state should always stay. The set R is a *recurrence* set, in which the state will return infinitely often, it is used to make the computation of a safety controller easier. The set B is an optional obstacle, or *avoid* set. Under the above-mentioned notation, we propose the main procedure of our approach which solves this problem by constructing a rule $\tilde{\sigma}(\cdot)$,

---

[1] This definition of stability is different from the stability in the Lyapunov sense.

such that for all $x_0 \in R$, and under the unknown bounded perturbation d, there exists $\pi = \tilde{\sigma}(\cdot) \in U^k$ for some k such that:

$$
\begin{cases}
& \mathbf{x}(t_0 + k\tau; t_0, x_0, d, \pi) \in R \\
\forall t \in [t_0, t_0 + k\tau], & \mathbf{x}(t; t_0, x_0, d, \pi) \in S \\
\forall t \in [t_0, t_0 + k\tau], & \mathbf{x}(t; t_0, x_0, d, \pi) \notin B.
\end{cases}
$$

Such a law permits to perform an infinite-time state-dependent control. The synthesis algorithm is described in Sect. 2.2 and involves guaranteed set-based integration presented in Sect. 3, the main underlying tool is interval analysis [25].

2.2 Controller Synthesis Algorithm

Before introducing the algorithms to synthetize controller of sampled switched systems, some preliminary definitions will be introduced.

Definition 1. *Let $X \subset \mathbb{R}^n$ be a box of the state space. Let $\pi = (i_1, i_2, \ldots, i_k) \in U^k$. The* successor set *of X via π, denoted by $Post_\pi(X)$, is the image of X induced by application of the pattern π, i.e., the solution at time $t = k\tau$ of*

$$
\begin{aligned}
\dot{x}(t) &= f_{\sigma(t)}(x(t), d(t)), \\
x(0) &= x_0 \in X, \\
\forall t \geq 0, \quad d(t) &\in [d], \\
\forall j \in \{1, \ldots, k\}, \quad \sigma(t) &= i_j \in U \text{ for } t \in [(j-1)\tau, j\tau).
\end{aligned}
\tag{3}
$$

Note that $Post_\pi(X)$ is usually hard to compute so an over-approximation will be computed instead in order to guarantee rigourous results.

Definition 2. *Let $X \subset \mathbb{R}^n$ be a box of the state space. Let $\pi = (i_1, i_2, \ldots, i_k) \in U^k$. We denote by $Tube_\pi(X)$ the union of boxes covering the trajectories of IVP (3), which construction is detailed in Sect. 3.*

Principle of the Algorithm. We describe the algorithm solving the control synthesis problem for nonlinear switched systems (see Problem 1, Sect. 2.1). Given the input boxes R, S, B, and given two positive integers P and D, the algorithm provides, when it succeeds, a decomposition Δ of R of the form $\{V_i, \pi_i\}_{i \in I}$ verifying the properties:

- $\bigcup_{i \in I} V_i = R$,
- $\forall i \in I, \ Post_{\pi_i}(V_i) \subseteq R$,
- $\forall i \in I, \ Tube_{\pi_i}(V_i) \subseteq S$,
- $\forall i \in I, \ Tube_{\pi_i}(V_i) \cap B = \emptyset$.

Decomposition $\Delta = \{V_i, \pi_i\}_{i \in I}$ is thus a set of boxes (V_i) covering R, each box being associated with a control pattern (π_i), and I is a set of indexes used for listing the covering boxes. The sub-boxes $\{V_i\}_{i \in I}$ are obtained by repeated bisection to produce a paving of R. At first, function *Decomposition* calls sub-function *Find_Pattern* which looks for a pattern π of length at most P such that $Post_\pi(R) \subseteq R$, $Tube_\pi(R) \subseteq S$ and $Tube_\pi(R) \cap B = \emptyset$. If such a pattern π is found, then a uniform control over R is found (see Fig. 1(a)). Otherwise, R is divided into two sub-boxes V_1, V_2, by bisecting R w.r.t. its longest dimension. Patterns are then searched to control these sub-boxes (see Fig. 1(b)). If for each V_i, function *Find_Pattern* manages to get a pattern π_i of length at most P verifying $Post_{\pi_i}(V_i) \subseteq R$, $Tube_{\pi_i}(V_i) \subseteq S$ and $Tube_{\pi_i}(V_i) \cap B = \emptyset$, then it is a success and algorithm stops. If, for some V_j, no such pattern is found, the procedure is recursively applied to V_j. It ends with success when every sub-box of R has a pattern verifying the latter conditions, or fails when the maximal depth of decomposition D is reached. The algorithmic form of functions *Decomposition* and *Find_Pattern* are given in Algorithm 1 and Algorithm 2 respectively.

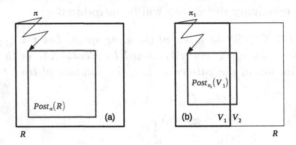

Fig. 1. Principle of the bisection method.

Algorithm 1. Algorithmic form of Function *Decomposition*.

Function: $Decomposition(W, R, S, B, D, P)$

Input: A box W, a box R, a box S, a box B, a degree D of bisection, a length P of input pattern

Output: $\langle \{(V_i, \pi_i)\}_i, True \rangle$ or $\langle \_, False \rangle$

$(\pi, b) := Find\_Pattern(W, R, S, B, P)$
if $b = True$ **then**
    return $\langle \{(W, \pi)\}, True \rangle$
else
    if $D = 0$ **then**
        return $\langle \_, False \rangle$
    else
        Divide equally W into (W_1, W_2)
        for $i = 1, 2$ **do**
            $(\Delta_i, b_i) := Decomposition(W_i, R, S, B, D - 1, P)$
        end for
        return $(\bigcup_{i=1,2} \Delta_i, \bigwedge_{i=1,2} b_i)$
    end if
end if

Our control synthesis method being well defined, we introduce the main algorithm of this paper, stated as follows:

Proposition 1. *Algorithm 1 with input (R, R, S, B, D, P) returns, when it successfully terminates, a decomposition $\{V_i, \pi_i\}_{i \in I}$ of R which solves Problem 1.*

Proof. Let $x_0 = x(t_0 = 0)$ be an initial condition belonging to R. If the decomposition has terminated successfully, we have $\bigcup_{i \in I} V_i = R$, and x_0 thus belongs to V_{i_0} for some $i_0 \in I$. We can thus apply the pattern π_{i_0} associated to V_{i_0}. Let us denote by k_0 the length of π_{i_0}. We have:

- $\mathbf{x}(k_0 \tau; 0, x_0, d, \pi_{i_0}) \in R$,
- $\forall t \in [0, k_0 \tau], \quad \mathbf{x}(t; 0, x_0, d, \pi_{i_0}) \in S$,
- $\forall t \in [0, k_0 \tau], \quad \mathbf{x}(t; 0, x_0, d, \pi_{i_0}) \notin B$.

Let $x_1 = \mathbf{x}(k_0 \tau; 0, x_0, d, \pi_{i_0}) \in R$ be the state reached after application of π_{i_0} and let $t_1 = k_0 \tau$. State x_1 belongs to R, it thus belongs to V_{i_1} for some $i_1 \in I$, and we can apply the associated pattern π_{i_1} of length k_1, leading to:

- $\mathbf{x}(t_1 + k_1 \tau; t_1, x_1, d, \pi_{i_1}) \in R$,
- $\forall t \in [t_1, t_1 + k_1 \tau], \quad \mathbf{x}(t; t_1, x_1, d, \pi_{i_1}) \in S$,
- $\forall t \in [t_1, t_1 + k_1 \tau], \quad \mathbf{x}(t; t_1, x_1, d, \pi_{i_1}) \notin B$.

We can then iterate this procedure from the new state

$$x_2 = \mathbf{x}(t_1 + k_1 \tau; t_1, x_1, d, \pi_{i_1}) \in R.$$

This can be repeated infinitely, yielding a sequence of points belonging to R $x_0, x_1, x_2, \ldots$ attained at times $t_0, t_1, t_2, \ldots$, when the patterns $\pi_{i_0}, \pi_{i_1}, \pi_{i_2}, \ldots$ are applied.

We furthermore have that all the trajectories stay in S and never cross B:

$$\forall t \in \mathbb{R}^+, \exists k \geq 0, \ t \in [t_k, t_{k+1}]$$

and

$$\forall t \in [t_k, t_{k+1}], \ \mathbf{x}(t; t_k, x_k, d, \pi_{i_k}) \in S, \ \mathbf{x}(t; t_k, x_k, d, \pi_{i_k}) \notin B.$$

The trajectories thus return infinitely often in R, while always staying in S and never crossing B. □

Remark 1. Note that it is possible to perform reachability from a set R_1 to another set R_2 by computing $Decomposition(R_1, R_2, S, B, D, P)$. The set R_1 is thus decomposed with the objective to send its sub-boxes into R_2, i.e., for a sub-box V of R_1, patterns π are searched with the objective $Post_\pi(V) \subseteq R_2$.

Remark 2. The search space of control patterns is the set of patterns of length at most P, i.e. $U \cup U^2 \cup \ldots U^P$. In a practical way, function $Find\_Pattern$ tests control patterns of length 1, then control patterns of length 2, iteratively up to length P. Patterns of length i are generated as combinatorial i-tuples. The set of

Algorithm 2. Algorithmic form of Function $Find\_Pattern$.

Function: $Find\_Pattern(W, R, S, B, P)$

Input: A box W, a box R, a box S, a box B, a length P of input pattern
Output: $\langle \pi, True \rangle$ or $\langle \_, False \rangle$

---

for $i = 1 \ldots P$ **do**
  $\Pi := U^i$ (the set of input patterns of length i)
  while Π is non empty **do**
    Select π in Π
    $\Pi := \Pi \setminus \{\pi\}$
    if $Post_\pi(W) \subseteq R$ **and** $Tube_\pi(W) \subseteq S$ **and** $Tube_\pi(W) \bigcap B = \emptyset$ **then**
      return $\langle \pi, True \rangle$
    end if
  end while
end for
return $\langle \_, False \rangle$

---

patterns of length i is U^i, its size is N^i. The complexity of function $Find\_Pattern$ is thus exponential with the length of control patterns P. The value of P leading to successful decompositions is unknown and depends on each system, but in most cases $P = 4$ leads to successful control synthesis. Longer sequences might be required if the dynamics is slow.

3   Set-Based Cosimulation

In this section, we explain how the $Post$ and $Tube$ operators can be computed in a distributed way through a cosimulation approach. We first explain the principle of interval analysis and standard guaranteed integration, we then suppose that the system can be written as the composition of components and explain our method for guaranteed cosimulation. In order to ease the reading of this section, we omit the notation of the switched modes σ and control sequences π associated to the $Post$ and $Tube$ operators.

Before presenting the details of interval analysis and cosimulation, let us introduce the following time periods:

- τ is the switching period,
- H is the communication period,
- h is the simulation period (or integration time-step).

We suppose that $h \leq H \leq \tau$, H is a multiple of h, and τ is a multiple of H. Consider $H = kh$ and $\tau = KH$ with $k, K \in \mathbb{N}_{>0}$, and an initial time t_0. On time intervals $[t_0, t_0 + \tau)$, the switching mode is constant. In case of cosimulation a model of cyber-physical systems is broken down into different *Simulation Units* (SU). Those SUs will exchange information at periodic rate, *i.e.*, at times $t_0, t_0 + H, \ldots, t_0 + KH$.

3.1   Interval Analysis

In this section, the main set-based tools that are required in this paper are presented.

Interval Artithmetic. The simplest and most common way to represent and manipulate sets of values is with *intervals*, see [25]. An interval $[x_i] = [\underline{x_i}, \overline{x_i}]$ defines the set of reals x_i such that $\underline{x_i} \leq x_i \leq \overline{x_i}$. $\mathbb{IR}$ denotes the set of all intervals over reals. The size or the width of $[x_i]$ is denoted by $w([x_i]) = \overline{x_i} - \underline{x_i}$.

Interval arithmetic extends to $\mathbb{IR}$ elementary functions over $\mathbb{R}$. For instance, the interval sum, *i.e.*, $[x_1]+[x_2] = [\underline{x_1}+\underline{x_2}, \overline{x_1}+\overline{x_2}]$, encloses the image of the sum function over its arguments. In general, an arithmetic operation $\diamond = \{+, -, \times, \div\}$ is associated to its interval extension such that:

$$[a] \diamond [b] \subset [\min\{\underline{a} \diamond \underline{b}, \overline{a} \diamond \underline{b}, \underline{a} \diamond \overline{b}, \overline{a} \diamond \overline{b}\}, \max\{\underline{a} \diamond \underline{b}, \overline{a} \diamond \underline{b}, \underline{a} \diamond \overline{b}, \overline{a} \diamond \overline{b}\}].$$

An interval vector or a *box* $[\mathbf{x}] \in \mathbb{IR}^n$, is a Cartesian product of n intervals. The enclosing property basically defines what is called an *interval extension* or an *inclusion function*.

Definition 3 (Inclusion function). *Consider a function $f : \mathbb{R}^n \to \mathbb{R}^m$, then $[f] : \mathbb{IR}^n \to \mathbb{IR}^m$ is said to be an extension of f to intervals if*

$$\forall [\mathbf{x}] \in \mathbb{IR}^n, \quad [f]([\mathbf{x}]) \supseteq \{f(\mathbf{x}), \mathbf{x} \in [\mathbf{x}]\}.$$

It is possible to define inclusion functions for all elementary functions such as $\times$, $\div$, sin, cos, exp, etc. The *natural inclusion function* is the simplest to obtain: all occurrences of the real variables are replaced by their interval counterpart and all arithmetic operations are evaluated using interval arithmetic. More sophisticated inclusion functions such as the centered form, or the Taylor inclusion function may also be used (see [19] for more details).

Combining the inclusion function and the rectangle rule, integral can be bounded following:

$$\int_a^b f(x)\, dx \in (b-a).[f]([a,b]).$$

Set-Based Simulation. Also named validated simulation or reachability, set-based simulation aims to compute the reachable tube of an Initial Value Problem with Ordinary Differential Equation (IVP-ODE) with a set-based approach and validated computations.

When dealing with validated computation, mathematical representation of an IVP-ODE is as follows:

$$\begin{cases} \dot{y}(t) = f(t, y(t), d(t)) \\ y(0) \in [y_0] \subseteq \mathbb{R}^n \\ d(t) \in [d] \subseteq \mathbb{R}^m. \end{cases} \tag{4}$$

We assume that $f : \mathbb{R} \times \mathbb{R}^n \to \mathbb{R}^n$ is continuous in t and globally Lipschitz in y, so Eq. (4) admits a unique solution for a given continuous perturbation trajectory. We furthermore suppose that d is bounded in the box $[d]$.

The set (expressed as a box) $[y_0]$ of initial conditions is usually used to model some (bounded) uncertainties. The set $[d]$ is used to model (bounded) perturbations. For a given initial condition $y_0 \in [y_0]$, and a given perturbation $d \in [d]$, the solution at time $t > 0$ when it exists is denoted $y(t; y_0, d)$. The goal, for *validated numerical integration* methods, is then to compute the set of solutions of Eq. (4), *i.e.*, the set of possible solutions at time t given the initial condition in the set of initial conditions $[y_0]$ and the perturbation lying in $[d]$:

$$y(t; [y_0], [d]) = \{ y(t; y_0) \mid y_0 \in [y_0], \ d(t) \in [d] \}. \tag{5}$$

Validated numerical integration schemes, exploiting set-membership framework, aims at producing the solution of the IVP-ODE that is the set defined in Eq. (5). It results in the computation of an over-approximation of $y(t; [y_0], [d])$.

The use of set-membership computation for the problem described above makes possible the design of an inclusion function for $[y](t; [y_0], [d])$, which is an over-approximation of $y(t; [y_0], [d])$ defined in Eq. (5). To do so, let us consider a sequence of time instants $t_1, \ldots, t_K$ with $t_{i+1} = t_i + h$ and a sequences of boxes $[y_1], \ldots, [y_K]$ such that $y(t_{i+1}; [y_i], [d]) \subseteq [y_{i+1}], \ \forall i \in [0, K-1]$ are computed. From $[y_i]$, computing the box $[y_{i+1}]$ is a classical 2-step method (see [23]):

- *Phase 1:* compute an a priori enclosure $\mathcal{P}^h([y_i], [d])$ of the set $\{ y(t_k; y_i, d) \mid t_k \in [t_i, t_i + h], y_i \in [y_i], d \in [d] \}$, such that $y(t_k; [y_i], [d])$ is guaranteed to exist;
- *Phase 2:* compute a tight enclosure of the solution $[y_{i+1}]$ at time t_{i+1}.

The a priori enclosure $\mathcal{P}^h([y_i], [d])$ computed in Phase 1 is referred to as a Picard box, since its computation relies on the Picard-Lindelöf operator and the Picard theorem (see [2,21] for more details). We omit the theoretical details, but a successful computation of this box ensures the existence and uniqueness of solutions over the time interval $[t_i, t_i + h]$ for the given box of initial conditions $[y_i]$ and perturbation box $[d]$. Two main approaches can be used to compute the tight enclosure in *Phase 2*. The first one, and the most used, is the Taylor method [25,26]. The second one, more recently studied, is the validated Runge-Kutta method [2]. *Guaranteed integration* or *reachability analysis* consists in computing a sequence of boxes that enclose the state of the system on a given time interval. For a given switched mode (the notation being omitted) and perturbation set $[d]$ on time interval $[t, t + \tau]$, given a time integration period h such that $\tau = Kh$, $(k = 1)$, the *Tube* operator is computed as the union of enclosures $\mathcal{P}^h([y_i], [d])$:

$$Tube([y_0]) = \bigcup_{i=1,\ldots,K} \mathcal{P}^h([y_i], [d]).$$

The post operator is the tight enclosure given at the final time:

$$Post([y_0]) = [y_K].$$

3.2 Cosimulation of Reachable Sets

The complexity of the computation of the Picard boxes, as well as the tightening of the solutions, is exponential in the dimension of the differential equation considered. As a result, reachability analysis lacks scalability with respect to the dimension of the system. In order to break the exponential complexity of those computations, a cosimulation approach can be used with the aim of computing these objects only on parts of the system.

Cosimulation aims at simulating components of a coupled system separately. In brief, the principle is to enable simulation of the coupled system through the composition of simulators, or simulation units (SUs) [17], each SU being dedicated to only a component of the system. SUs exchange information at some given communication times in order to ensure the simulation error does not grow uncontrollably.

Let us suppose that the dynamics can be decomposed as follows:

$$\dot{x}_1 \in f_1(t, x_1, u_1) \quad \text{with} \quad x_1(0) \in [x_1^0], \ u_1 \in [u_1],$$
$$\dot{x}_2 \in f_2(t, x_2, u_2) \quad \text{with} \quad x_2(0) \in [x_2^0], \ u_2 \in [u_2],$$
$$\ldots$$
$$\dot{x}_m \in f_m(t, x_m, u_m) \quad \text{with} \quad x_m(0) \in [x_m^0], \ u_m \in [u_m],$$
$$L(x_1, \ldots, x_m, u_1, \ldots, u_m) = 0,$$

where the state x is decomposed in m components $x = (x_1, \ldots, x_m)$, for all $j \in \{1, \ldots, m\}$, $x_j \in X_j$, $X_1 \times \cdots \times X_m = \mathbb{R}^d$, and L is a coupling function between the components. The coupling condition $L(x_1, \ldots, x_m, u_1, \ldots, u_m) = 0$ should hold at all time. From now on, we use index $j \in \{1, \ldots, m\}$ to denote subsystem j, and index i to denote a time interval starting at t_i. Note that, in order to increase the accuracy of the method, the decomposition should be made so as to minimize the number of shared variables between sub-systems.

In the most general case, coupling L is an algebraic condition. For our applications, the coupling is supposed to be given explicitly, *i.e*, u_j is given as function of the other state variables: $u_j = K_j(x_1, \ldots, x_m)$. Cosimulation then consists in computing *Post* operators for each sub-system separately, and doing a cross product to obtain the global state. To ensure a guaranteed computation, the inputs u_j can be considered as bounded perturbations. The difficulty lies in the determination of the size of the set in which the perturbations evolve, since it has to be determined before performing the simulation of the other sub-systems.

This can be done using the *cross-Picard operator*, introduced in [21]. The purpose of the cross-Picard operator is to over-approximate the solutions of all the sub-systems over a given time interval (the communication period, also called macro-step), using only local computations.

To compute these sets, we start by guessing a rough over-approximation $[p_j]$ of the solutions x_j over the next macro-step. This gives some rough over-approximations $[r_j] = K_j([p_1], \ldots, [p_m])$ of the perturbations u_j. We then compute local Picard boxes iteratively, until the proof of validity of the approximations is obtained for all sub-systems.

More precisely, let us denote by $\mathcal{P}_j^H([x_j], [u_j])$ the enclosure of the set of solutions of subsystem j over the time-interval $[t, t + H]$: $\{x_j(t_k; x_j, u_j) \mid t_k \in [t, t + H], x_j \in [x_j], u_j \in [u_j]\}$, where $[x_j]$ and $[u_j]$ are the boxes of initial conditions and perturbation for sub-system j. If we can prove that for all sub-systems $j \in \{1, \ldots, m\}$, $\mathcal{P}_j^H([x_j], K_j([p_1], \ldots, [p_m])) \subsetneq [p_j]$, then, by application of the Picard theorem, existence and uniqueness of global solutions is ensured for the time interval $[t, t + H]$. Fortunately, this condition is in practice easily met by application of a fixed point algorithm that tightens the rough initial guesses $[p_j]$ (see [21]). Once the Picard boxes are computed and proved safe, each sub-system j can, in parallel, compute its own solution safely on the time interval $[t, t + H]$ by considering u_j as a perturbation lying in $K_j([p_1], \ldots, [p_m])$. We denote the cross-Picard operator as the computation of the validated Picard boxes, the result being given as the cross-product of the Picard boxes.

Our approach for guaranteed cosimulation of the *Post* operator over the interval $[t, t + \tau]$ is thus summarized as follows:

1. Compute an over-approximation of the solutions on time interval $[t, t + H]$ (compute the cross-Picard operator),
2. Advance simulation of all subsystems in parallel (using a time step h) until time $t + H$, the inputs are considered as bounded perturbations in the sets returned in Step 1,
3. Update initial conditions and input values,
4. Repeat on interval $[t + H, t + 2H]$ until $[t + \tau - H, t + \tau]$.

3.3 Discussion on Meta-parameters

The different time periods involved in the synthesis and cosimulation procedures (h, H, and τ) play a crucial role in the accuracy of the reachability analysis, and thus in the success of the control synthesis. In mere words, a reachability analysis is performed each time a control sequence is tested. Improving the speed of the reachability analysis drastically improves control synthesis computation times, provided that the accuracy is high enough to allow control synthesis.

One of the key aspects is that the frequency at which we update the initial conditions and perturbation sets (the communication frequency $1/H$) should be as small as possible in order to increase the speed of the reachability analysis, but at the cost of the accuracy of cosimulation. The speed increase when using fewer communications is due to the fact that each communication time involves the application of a fixed-point algorithm to validate the perturbation sets, thus taking a non negligible amount of computation. However, using shorter communication periods means that the perturbation sets are smaller, and the cosimulation thus leads to tighter reachable tubes, making the synthesis easier. The largest communication period we can consider is actually the switching period $H = \tau$. If such a large communication period allows enough accuracy for control synthesis purposes, then this would lead to the best computation time gains. However, in practice, the switching period can be too large to avoid communication between switching times. If a communication is necessary between switchings, then, in order to maximize the use of the data exchange, communication frequency should be a multiple of the switching frequency.

We would like to point out that the integration time step h can actually be different for each simulation unit (for reachability analysis of separate components, once the perturbation sets are validated). The integration methods can be essentially different since we can even consider implicit and explicit methods in parallel. The only requirement is that the perturbation sets are proved safe (with the use of Picard operators). This means that complex systems involving stiff and nonstiff dynamics, or linear and nonlinear dynamics, can be divided in such a way that the computation power is dedicated to the more difficult parts to integrate. In our applications, we illustrate the scalability property of the proposed method, but industrial applications involving more complex dynamics could show even better improvements.

4   Experiments

4.1   Case Study

This case study is based on a simple model of a two-room apartment, heated by a heater in one of the rooms (adapted from [15]). Initially of dimension of the state space is 2, the case study is made scalable by concatenating two-room apartments in line, so that each room exchanges heat with its neighbouring rooms, and every other room is equipped with a heater.

In this example, the objective is to control the temperature of all rooms. There is heat exchange between neighbouring rooms and with the environment. The *continuous* dynamics of the system, *the temperature*, is given by

$$
\begin{pmatrix} \dot{T_1} \\ T_2 \\ \vdots \\ T_n \end{pmatrix} = A \begin{pmatrix} T_1 \\ T_2 \\ \vdots \\ T_n \end{pmatrix} + B_u.
$$

The dimension n is supposed to be even $n = 2m$. The non null coefficients $a_{i,j}$ of matrix A are:

$$a_{1,1} = -\alpha_r - \alpha_e - \alpha_f u_1$$
$$a_{2i+1,2i+1} = -2\alpha_r - \alpha_e \qquad\qquad i = 1, \ldots, m-1$$
$$a_{2i,2i} = -2\alpha_r - \alpha_e - \alpha_f u_i \qquad\qquad i = 1, \ldots, m-1$$
$$a_{2m,2m} = -\alpha_r - \alpha_e$$
$$a_{i,i+1} = a_{i+1,i} = \alpha_r \qquad\qquad i = 1, \ldots, 2m-1.$$

The non null coefficients $b_{i,j}$ of the input matrix B_u are:

$$b_{2i-1} = \alpha_e T_e + \alpha_f T_f u_{i+1} \qquad\qquad i = 1, \ldots, m$$
$$b_{2i} = \alpha_e T_e \qquad\qquad i = 1, \ldots, m.$$

Here T_i for $i = 1, \ldots, 2m$ is the temperature of room i, and the state of the system corresponds to $T = (T_1, \ldots, T_n)$. The control modes are given by variables u_j for $j = 1, \ldots, m$, each can take the values 0 or 1, depending on whether the heater in room $2j - 1$ (for $j = 1, \ldots, m$) is switched off or on. Hence, the number of switched modes is 2^m. Temperature T_e corresponds to the temperature of the environment, and T_f to the temperature of the heaters. The values of the different parameters are as follows: $\alpha_r = 5 \times 10^{-2}$, $\alpha_e = 5 \times 10^{-3}$, $\alpha_f = 8.3 \times 10^{-3}$, $T_e = 10$ and $T_f = 50$.

The control objective is to ensure τ-stability of the temperature in $R = [19, 21] \times \cdots \times [19, 21]$, while ensuring safety in $S = [18, 22] \times \cdots \times [18, 22]$, with a switching period $\tau = 10$. We don't consider any obstacle B in this example, the maximal length of patterns is set to $P = 4$, and the maximum depth of decomposition is $D = 2$.

4.2   Experimental Results

In order to validate our approach, we synthetize the control rule for the problem given in Sect. 4.1 for different number of rooms $n = 2, 4, 6, 8$. The results are gathered in Table 1. All the simulations are performed with the classical method RK4, an explicit Runge-Kutta method with four stages at the fourth order. Our choice for this method is based on its fame and on the fact that to find a control for the case study, an order greater than two is needed. Cosimulation consists in m simulations of systems of dimension two (three with an additional dimension for time). More precisely, if $m \geq 3$, system

$$\begin{pmatrix} \dot{T}_1 \\ T_2 \\ \vdots \\ T_n \end{pmatrix} = A \begin{pmatrix} T_1 \\ T_2 \\ \vdots \\ T_n \end{pmatrix} + B_u,$$

for $n = 2m$, is rewritten as m systems:

$$\begin{pmatrix} \dot{T_1} \\ \dot{T_2} \end{pmatrix} = A^1 \begin{pmatrix} T_1 \\ T_2 \end{pmatrix} + B_u^1 + D^1 (T_3),$$

$$\begin{pmatrix} \dot{T_3} \\ \dot{T_4} \end{pmatrix} = A^2 \begin{pmatrix} T_3 \\ T_4 \end{pmatrix} + B_u^2 + D^2 \begin{pmatrix} T_2 \\ T_5 \end{pmatrix},$$

$$\cdots$$

$$\begin{pmatrix} \dot{T_{2m-1}} \\ T_{2m} \end{pmatrix} = A^m \begin{pmatrix} T_{2m-1} \\ T_{2m} \end{pmatrix} + B_u^m + D^m (T_{2m-2}),$$

where D^i is a disturbance matrix composed of coefficients of A. One can see that each subsystem is perturbed by the adjacent rooms (there is one adjacent room for the first and last system, and two adjacent rooms for the others). There is one communication per switching period, meaning that $H = 5$ for $\tau = 10$.

These simulations are also performed in parallel using Open Multi-Processing API for Linux. Experiments are done on a bi-processor Intel(R) Xeon(R) CPU E5-2620 v3 @ 2.40 GHz with 12 cores each. A time out (T.O.) is fixed at three days, $i.e.$, 4320 min. Computation times seem important but the results are guaranteed and have to be computed only one time and offline.

Table 1. Results of synthesis for problem given in Sect. 4.1: computation times, in minutes, for centralized dynamics, with cosimulation and with parallelized cosimulation.

Number of rooms (n)	Centralized	Cosimulation	Cosimulation in parallel
2	0m43	–	–
4	2m28	2m30	1m58
6	185m	80m	42m
8	T.O	3606m	2072m

4.3 Discussion

Our method shows its efficiency, even with only 4 rooms if parallelization is used. A control rule can be synthetized for 8 rooms with cosimulation while no result can be obtained without our approach before time out. Cosimulation allows a very straightforward parallelization which reduces significantly computation time. Our experiments revealed the necessity of using one communication per switching period for this case study. Using none (communicating only at the beginning of a switching period) led to sets too wide for ensuring τ-stability.

5 Conclusion

In this paper, we presented a procedure for control synthesis that relies heavily on (guaranteed) reachability computations, its scalability being limited by the complexity of set-based integration. We proposed to use a guaranteed cosimulation to improve the control synthesis computation times. We illustrate the scalability of our method on a scalable case-study that shows the efficiency of our approach. As of now, some expertise is required for choosing a communication frequency that allows computation time gains as well as successful control synthesis. We would like to explore the possibility of automating the determination of a good communication frequency in the context of switching systems.

The current implementation of the procedure allows to simulate subsystems in parallel, but the cross-Picard computation (involving repeated applications of Picard operators) is still sequential due to memory management issues. Our future work will be devoted to the development of a parallel implementation of the cross-Picard computation. Such an implementation would hopefully mitigate the cost of communication times in the present procedure.

References

1. Alexandre dit Sandretto, J., Chapoutot, A.: DynIbex. https://perso.ensta-paris. fr/~chapoutot/dynibex/
2. Alexandre dit Sandretto, J., Chapoutot, A.: Validated explicit and implicit Runge-Kutta methods. Reliable Comput. **22**, 79 (2016)
3. Althoff, M.: Reachability analysis of nonlinear systems using conservative polynomialization and non-convex sets. In: Hybrid Systems: Computation and Control, pp. 173–182 (2013)
4. Arnold, M., Clauß, C., Schierz, T.: Error analysis and error estimates for co-simulation in FMI for model exchange and co-simulation v2.0. In: Schöps, S., Bartel, A., Günther, M., ter Maten, E.J.W., Müller, P.C. (eds.) Progress in Differential-Algebraic Equations. DEF, pp. 107–125. Springer, Heidelberg (2014). https://doi. org/10.1007/978-3-662-44926-4_6
5. Bouissou, O., Chapoutot, A., Djoudi, A.: Enclosing temporal evolution of dynamical systems using numerical methods. In: Brat, G., Rungta, N., Venet, A. (eds.) NFM 2013. LNCS, vol. 7871, pp. 108–123. Springer, Heidelberg (2013). https:// doi.org/10.1007/978-3-642-38088-4_8
6. Bouissou, O., Martel, M.: GRKLib: a guaranteed Runge Kutta library. In: Scientific Computing, Computer Arithmetic and Validated Numerics (2006)
7. Bouissou, O., Mimram, S., Chapoutot, A.: HySon: set-based simulation of hybrid systems. In: Rapid System Prototyping. IEEE (2012)
8. Broman, D., et al.: Determinate composition of FMUs for co-simulation. In: 2013 Proceedings of the International Conference on Embedded Software (EMSOFT), pp. 1–12. IEEE (2013)
9. Chen, X., Abraham, E., Sankaranarayanan, S.: Taylor model flowpipe construction for non-linear hybrid systems. In: IEEE 33rd Real-Time Systems Symposium, pp. 183–192. IEEE Computer Society (2012)
10. Chen, X., Ábrahám, E., Sankaranarayanan, S.: Flow*: an analyzer for non-linear hybrid systems. In: Sharygina, N., Veith, H. (eds.) CAV 2013. LNCS, vol. 8044, pp. 258–263. Springer, Heidelberg (2013). https://doi.org/10.1007/978-3-642-39799-8_18
11. de Figueiredo, L.H., Stolfi, J.: Self-validated numerical methods and applications. In: Brazilian Mathematics Colloquium Monographs, IMPA/CNPq (1997)

12. Alexandre dit Sandretto, J., Chapoutot, A.: Validated simulation of differential algebraic equations with Runge-Kutta methods. Reliable Comput. **22**, 57 (2016)
13. Dzetkulič, T.: Rigorous integration of non-linear ordinary differential equations in Chebyshev basis. Numer. Algorithms **69**(1), 183–205 (2015). https://doi.org/10.1007/s11075-014-9889-x
14. Gajda, K., Jankowska, M., Marciniak, A., Szyszka, B.: A survey of interval Runge–Kutta and multistep methods for solving the initial value problem. In: Wyrzykowski, R., Dongarra, J., Karczewski, K., Wasniewski, J. (eds.) PPAM 2007. LNCS, vol. 4967, pp. 1361–1371. Springer, Heidelberg (2008). https://doi.org/10.1007/978-3-540-68111-3_144
15. Girard, A.: Low-complexity switching controllers for safety using symbolic models. IFAC Proc. Vol. **45**(9), 82–87 (2012)
16. Gomes, C., Thule, C., Broman, D., Larsen, P.G., Vangheluwe, H.: Co-simulation: state of the art. arXiv preprint arXiv:1702.00686 (2017)
17. Gomes, C., Thule, C., Broman, D., Larsen, P.G., Vangheluwe, H.: Co-simulation: a survey. ACM Comput. Surv. (CSUR) **51**(3), 1–33 (2018)
18. Immler, F.: Verified reachability analysis of continuous systems. In: Baier, C., Tinelli, C. (eds.) TACAS 2015. LNCS, vol. 9035, pp. 37–51. Springer, Heidelberg (2015). https://doi.org/10.1007/978-3-662-46681-0_3
19. Jaulin, L., Kieffer, M., Didrit, O., Walter, E.: Applied Interval Analysis. Springer, London (2001). https://doi.org/10.1007/978-1-4471-0249-6
20. Jensen, J.C., Chang, D.H., Lee, E.A.: A model-based design methodology for cyber-physical systems. In: 2011 7th International Wireless Communications and Mobile Computing Conference, pp. 1666–1671. IEEE (2011)
21. Le Coënt, A., Alexandre dit Sandretto, J., Chapoutot, A.: Guaranteed cosimulation of cyber-physical systems. hal-02505237 https://hal.archives-ouvertes.fr/hal-02505237 (2020)
22. Lin, Y., Stadtherr, M.A.: Validated solutions of initial value problems for parametric odes. Appl. Numer. Math. **57**(10), 1145–1162 (2007)
23. Lohner, R.J.: Enclosing the solutions of ordinary initial and boundary value problems. In: Computer Arithmetic, pp. 255–286 (1987)
24. Makino, K., Berz, M.: Rigorous integration of flows and ODEs using Taylor models. In: Proceedings of the 2009 Conference on Symbolic Numeric Computation, SNC 2009, pp. 79–84. ACM, New York (2009)
25. Moore, R.E.: Interval Analysis. Series in Automatic Computation. Prentice Hall, Englewood Cliffs (1966)
26. Nedialkov, N.S., Jackson, K.R., Corliss, G.F.: Validated solutions of initial value problems for ordinary differential equations. Appl. Math. Comput. **105**(1), 21–68 (1999)
27. Nielsen, C.B., Larsen, P.G., Fitzgerald, J., Woodcock, J., Peleska, J.: Systems of systems engineering: basic concepts, model-based techniques, and research directions. ACM Comput. Surv. (CSUR) **48**(2), 1–41 (2015)
28. Schierz, T., Arnold, M., Clauß, C.: Co-simulation with communication step size control in an FMI compatible master algorithm. In: Proceedings of the 9th International MODELICA Conference, Munich, Germany, 3–5 September 2012, no. 076, pp. 205–214. Linköping University Electronic Press (2012)

Author Index

Printed in the United States
By Bookmasters